Mitigating Climate Change

The power of

We the People

Adeniyi A. Afonja

Sineli Books

Mansfield, Texas, U.S.A.; Luton, U.K.

ISBN: 978-0-9985843-1-7

Sineli Books

Mansfield, Texas, U.S.A.; Luton, U.K.
info@sinelibooks.com
Printed in the United States of America
First Impression 2020

Table of contents

Preface/Summary

The Human-Environment Systems comprise four major spheres: *lithosphere*, *hydrosphere*, *atmosphere*, and *biosphere*. The complex interactions between these spheres to a large extent control the variables that sustain life on earth. The functions are controlled by natural regulatory processes and the boundaries between them may be clear or ill defined. Events in the first two layers of the atmosphere are critical to the sustenance of life on earth because they host the natural processes which regulate events that sustain life: temperature, rainfall, human and plant respiration, the natural carbon cycle, weather, climate, protection from damaging radiation from the Sun, etc. The lower atmospheric layer (*troposphere*) comprises gases that extend from the Earth's surface to about 15-20 km into space, in particular, oxygen and carbon dioxide which support human and plant life respectively. The layer also hosts most of the natural activities that determine the local weather - winds, clouds, humidities, precipitation, rainfall, heat waves, etc. The next layer (*stratosphere*) which extends to about 55 kilometers into space hosts a mixture of gases *greenhouse gases* - which control and regulate the Earth's average temperature, keeping it at an average of around 13-15°C. Without this regulation, the Earth's average temperature would be around minus 18°C, too cold to sustain life on earth. Also nature sustains a thin *ozone layer* in the lower part of the stratosphere through a continuous chain of chemical reactions that use up most of the dangerous rays of the Sun thereby preventing them from reaching the Earth and causing severe health issues. The environment that is controlled by natural forces which have the capacity to dilute, absorb or dissipate natural pollutants is known as *natural environment* and it is imperfect in many ways: for example, extreme weather, flooding, volcanoes are caused by known or unknown natural phenomena. Slight changes in the Earth's orbital parameters can cause significant changes in the Earth's climate system, and this explains why land areas that were fertile thousands of years ago are now deserts and interchanges between land and ocean areas have occurred many times in the Earth's history. However, human activities, particularly in the last hundred years or so have been releasing *anthropogenic emissions* that compromise the natural regulatory processes of the Human-Environment Systems. Emissions of greenhouse gases mainly from energy production and use and agriculture enhance the heat-trapping capabilities of the natural stratospheric greenhouse gases, causing global warming; halocarbons which destroy the protective ozone layer come mainly from refrigeration and air-conditioning; particulate/gaseous aerosols emanating from energy production and use, transportation, agriculture, domestic use of traditional energy, impact directly on human and ecosystems health. Many natural processes release toxic emissions which end up in the Earth's atmosphere and can cause problems including climate change and human health issues: desert dust, volcanoes, carbon dioxide from human respiratory systems, methane from vegetation decay and animal respiratory systems, etc. Severe weather, heat waves, flooding, draught, weather and climate change have always been part of human existence but they have always occurred in different places at different times. However, many human activities that release anthropogenic emissions (energy production and use, agriculture and land use) have intensified since the early part of the last century to the extent that they are now making these consequences more frequent, more widespread, and more devastating. Literature on environmental pollution and climate change is extensive and the science is complex but the main issues are summarized as follows:

- The world is warming, the average temperature has increased by around 1°C compared with pre-industrial levels, a primary indicator of climate change. Extensive scientific data indicate (with a high degree of confidence) that human-related emissions have been the dominant cause of the observed global warming since the mid-20th century. The negative impacts are gradual, global, could take decades or hundreds of years to manifest; energized events are largely widespread and independent of the source of pollution. The small average temperature rise is enough to enhance a wide range of natural phenomena and projections predict a rise of 2-3°C by the end of the 21st century under all possible emission scenarios if pollution continues unabated and climate mitigation continues at the currently slow pace. The negative impacts of global warming are already evident and will become more severe: oceans which occupy around three-quarters of the Earth's surface are warming; warm waters will energize tropical winds to become storms and transform into hurricanes that cause severe damage when they touch down thousands of kilometers from origin; severe weather and heat waves are becoming more frequent, more severe, more enduring, and more widespread; heat waves are becoming more common and will intensify; warmer and acidified oceans will have devastating consequences for aquatic life; ocean water levels will rise possibly by up to two meters by the end of the century due to heating and melting of ice caps and glaciers, causing severe flooding which could threaten coastal cities worldwide; acid rain will damage infrastructure and crops. Human-related anthropogenic emissionss are not the cause of any of these phenomena which are part of the natural Earth-Environment System but there is ample scientific evidence that they are potent enhancers and will continue to exacerbate the consequences. Various recent studies have shown that emissions need to decline by approximately 25% and 55% from current levels over the next decade or so in order to put the world on a least-cost pathway to limiting global warming to 2°C and 1.5°C respectively and avoid the potentially devastating consequences of climate change.

- Most human-related pollutions from energy production and use, transportation, agriculture and land use first enter the troposphere before eventually rising to the stratosphere where they enhance global warming and destroy the protective ozone layer. However, in the few weeks that they reside in the troposphere, they can cause severe damage because this part of the atmosphere is in direct contact with life on earth and events in the layer largely control local weather and the quality of the local environment. Some of the negative consequences of tropospheric (ambient) pollution are the smog that blankets most urban areas around the world, and the toxic household environment in many countries the combined effects of which have been associated with around 7.5 million deaths annually. Other ecosystems (plant, animal, aquatic life) are also negatively impacted. In effect, the problem of ambient pollution is much more urgent than climate change.

Energy-related emissions mirror the trends in energy consumption which in turn reflect global economic, population and prosperity trends, and both will continue to rise in response to these primary drivers. Most of the human-related emissions come from energy use and agriculture, both of which are critical to human survival and development. Both also lie at the core of the global economy which is projected to double over the next two decades; population will rise by 25-30%; prosperity and urbanization will grow rapidly in most regions of the world, and the global vehicle population will also double. Commensurate rises in the demand for energy and agricultural products and associated anthropogenic emissions are inevitable. The United Nations Systems estimate that around four-fifths of the global population still lack access to modern energy, 80% are below the poverty threshold, nearly one billion of people are currently hungry, and every third person is malnourished. The situation will likely get worse with the projected growth in population, increasingly scarce agricultural resources (arable land and other vital resources), and increasingly frequent, widespread and intensified disasters and pandemics. Climate change, associated weather variability and desertification will make the negative impacts of these stressors more severe and lead to wider deployment of high-carbon energy and agricultural technologies, particularly in emerging economies where options are often limited.

Fossil energy has supplied over 80% of global primary energy for five decades and all recent projections show no significant change over the next two decades or so. Although renewable energy will play a much more prominent role in power generation, fossil fuels will continue to dominate the global economy in the foreseeable future largely because no viable substitute has been found to replace fossil energy in most products that are critical to human development: energy and raw materials for industrial manufacture, transportation, building; production of primary metals, cement, plastics, fertilizers; and the chemical building blocks for pharmaceuticals and many industrial/consumer products. While the future demand for transportation oil is projected peak in about a decade due to increasing market penetration of electric vehicles and the forces of supply and demand activated by increasing efficiencies and lower energy intensities across the entire spectrum of the global economy, the chemicals sector (that uses oil, gas and coal as raw materials as well as energy sources) has been the main driver of global primary energy demand growth in the last decade or so and will account for more than half of the future growth in demand. The global demand for natural gas remains very strong as a lower-carbon alternative to coal in heat and power generation; and coal remains a primary fuel across most regions of the world, powering nearly 40% of global power generation (much higher in many countries) and more than three quarters of global output of steel, cement, and vital industrial chemicals.

The fossil energy sector (upstream and downstream) is one of the world's largest employers of industrial labour and the economic lifeline of many countries across all regions of the world. Crude oil and refined products are the world's two leading export products and investments in exploration and infrastructure are very strong all over the world: oil, gas and converted products remain the most traded commodities worldwide; new reserves are being developed and exploited; natural gas pipeline networks are being extended across country borders in all regions of the world; and there has been an exponential growth in international trade of liquefied natural gas (LNG) over the last decade or so, driven by huge investments from the United States, Russia, Australia, Qatar, Malaysia, Nigeria, and many other countries. Asian countries, Australia, Russian Federation, United States and South Africa accounted for over 90% of the global coal production and trade in 2019. It is difficult to imagine that any country could contemplate a reduction in fossil fuel output and trade because of potential damage to the environment. Improvement of efficiencies across the entire spectrum of energy production and use and an exponential rise in the deployment of renewable energy are key options for mitigating human-related environmental pollutions and climate change.

Numerous studies have shown that the world can achieve carbon neutrality or zero net carbon in two to three decades without shutting down the fossil industry. This would position the world on a feasible path to the Paris-2015 sustainable environment goal of limiting the global average temperature rise to around 2-1.5 C by the end of the century. Carbon neutrality describes a sustainable environment in which human-related emissions are sucked out of the atmosphere continuously by natural processes such as plant photosynthesis and other carbon sinks as well as technological innovations, to the extent that there is no net growth in the concentrations of the natural greenhouse gases. Main options and key drivers of anthropogenic emissions abatement include optimization of energy production and use; technology innovations, particularly those that improve efficiencies across all sectors of the global economy; decarbonization of energy through much more widespread use of lower carbon energy sources, in particular, renewables; provision of adequate and modern low-carbon energy for around 80% of the global population that currently lack adequate access; and reduction of people's carbon footprint through moderation of conspicuous consumption driven currently by 'want' and 'impulse' rather than 'need'. Many new technologies are incubating and will emerge which could radically change the dynamics of global pollution: for example, technologies for production of plastics from biomaterials rather than fossil fuels (bioplastics which degrade within months of disposal); mass tree-planting and cultivation of algae that usefully convert carbon dioxide to carbohydrates and proteins; sucking carbon dioxide out of the atmosphere and converting to useful or benign products; decarbonization of primary metals and cement production, etc. The major responsibility for decarbonizing global energy lies with two regions which contributed three-quarters of the total energy-related emissions in 2017: Asia (53%) and the Americas (22%) and accounted for 85% of the net increase in emissions in 2018. Just two countries: China and the United States accounted for 44% of the total global energy-related emissions in 2017, contributing 29% and 15% respectively, with coal contributing the largest share of 81% in China compared with 28% in the United States where oil accounted for the largest share of 42%. Five countries: China, United States, India, Russia and Japan accounted for nearly 60% of total global emissions while Africa contributed only 3.6%.

Climate change is a major global issue today, and one of the most controversial. The debate is full of well-mixed ideals, myths and realities. The pro-climate change world has an aggressive global campaign machinery against fossil energy use which is believed to be the cause of global warming and climate change. The sceptics are just as strong: either they do not believe that the climate is changing (climate denial), or they think human activities have nothing to do with it. Infusion of politics inspired by intra-country exigencies and empowered by the virile social media has further complicated the issues and it has become difficult to distinguish between realities, facts, and fiction. Ideals and myths are strong weapons for sensitizing and motivating global mitigation action on climate change, but it is also useful to know what is real, feasible or possible. Global debate is also strengthening on who should be held responsible for degrading the environment and what should be the price to be paid, and the focus is on the global fossil energy companies, with lawsuits beginning to emerge. The first issue is that no science has proved conclusively that human activities are entirely responsible for climate change since there are also many potential natural causes. This explains why virtually all conclusions reached by climate experts are expressed in varying degrees of probability. However, energy companies deserve to be held accountable for avoidable pollutions such as oil spills, gas flaring and landscape degradation, and society is also accountable for its increasingly insatiable appetite for energy and consumer products.

Reliable primary and end-use energy supplies are indispensable to modern societies and human development; they drive the global economy just as economic growth drives the demand for even more energy. However, what is also indispensable is a concerted global effort towards putting associated emissions into consistent decline in order to place the world on an achievable path towards a sustainable environment. Fossil energy currently dominates the global primary energy mix, accounting for around 80% over the last four decades and projected to remain dominant, probably at around the same level for decades to come. Renewable energy, in particular for power and heat generation will play a critical role in decarbonizing energy but production of many vital inputs that will drive development still depend critically on fossil fuels: solar PVI and concentration systems, electric storage batteries, fertilizers for bioenergy, materials of construction for renewable energy infrastructure, etc. Furthermore, the persistently low and fluctuating global oil prices are potent depressants of investments in renewable energy. The current Covid-19 global pandemic has resulted in a significant decline of human-related emissions by about 15% within weeks largely because economic activities especially road and air transportation and manufacturing have been severely disrupted, but it has also led to near-total collapse of the global economy. The situation is unsustainable and the consequences will be multi-dimensional, severe, far-reaching and enduring. Already, there are unprecedented job losses across the world and across all sectors of the economy. This is a clear preview of what would happen if there were any severe interruptions in global primary energy supply that cause demand to outstrip supply, reminiscent of the early 1970s when a severe oil supply shortage brought energy security to the forefront and provided the primary stimulant for the proliferation of nuclear power generation and massive investments in local fossil fuel resources. Emissions are currently reduced significantly, urban environments are unusually clean, but a sharp rebound of 10% has already occurred since the gradual resumption of economic activities. However, this monumental shock is an invaluable opportunity for the world to rethink strategies in the process of re-opening the economy, in particular, the energy sector which needs to be done in a way that puts the world on a stronger footing towards an environmentally sustainable future. The International Energy Agency (IEA, 2020b) has recently released a Sustainable Recovery strategic plan which details options for the guidance of countries as they consider recovery plans which will shape investment, infrastructure and industry development strategies for decades to come. One key conclusion of the study is that fossil fuel use will rebound although the demand for oil, natural gas and coal could drop by 4-8% in 2020 but it is unclear whether the decline can be sustained. New investments in the fossil energy sector are currently on hold but they are vital to the sustenance of existing facilities. Furthermore, the fossil energy sector employed 50% of the around 40 million people around the world who worked directly in production, transport and distribution of fossil fuels, and crude and refined oil led global commodity trade in 2019. However, investments in renewable energy have remained strong, the sector has high employment potential and will likely grow at a much faster rate than previously projected because of the expected growth in demand for electricity to meet the new normals such as significantly increased home-based work, virtual meetings, e-commerce, and other digital services. The report presents comprehensive options that could rapidly transform the global energy system and provide a feasible pathway towards a carbon-neutral world. The world needs to face the reality of continued dominance of fossil fuels in the global primary energy mix in the foreseeable future and focus on a pragmatic approach which involves the management of energy production, conversion and use in an environmentally sustainable manner.

The very complex problems of energy-related pollution and climate change and various pathways to a carbon-neutral world have been discussed in depth in a companion book: *Fossil fuels and the environment*. This volume is a less technical presentation of the main issues and focuses on the close interdependency of energy, global economy and human development. Numerous studies and projections by reputable independent international organizations which show that, in spite of continued dominance of fossil energy, the world can still attain carbon neutrality within two decades and present several potential pathways are reviewed in some depth. The most promising options focus on four key mitigation variables: greatly enhanced efficiency improvements across the entire spectrum of energy production and use in all major sectors of the global economy; much more aggressive decarbonization of energy, in particular power and heat generation; significant reduction of societal/personal carbon footprint through moderation of energy use and consumer product consumption that is guided by need rather than want; and activation of potentially powerful societal instruments of influence on the enactment of strong and resilient climate-friendly policies that survive political cycles.

The global distribution of anthropogenic emissions is grossly inequitable and there is wide variability between social stratifications: although the world average carbon footprint per capita per year is about 5 metric tons, there is a very wide disparity between the rich and the poor (carbon inequality). The poorest half of the global population (around 3.5 billion people) account for around 10% of total global emissions attributed to individual consumption; about 30% of emissions are attributable to the richest 10% of people around the world whose average footprint is about 11 times as high as the poorest half of the population, and 60 times as high as the poorest 10%; also, the average footprint of the richest 1% of people globally could be 175 times that of the poorest 10% (Gore, 2015). Around 80% of the global population which will also account for most of the future growth in population currently lack access to adequate modern energy and there is little doubt that energy equity will drive the future growth of energy use, mostly in emerging nations. If the current nearly 8 billion people could reduce personal footprint by just 20% on average, emissions would be reduced by 8 Gigatonnes, around half of the decline required to move the world to carbon neutrality by 2040. However, the data presented above show that lifestyle changes would be most effective only if the richest across the world (but concentrated in the developed economies) decide to moderate their excesses. North America is the world's largest consumer of primary energy: 240 Gigajoules/capita compared with the European Union (140), OECD (73), Africa (15), and the world average (76). The power to reduce global emissions through lifestyle changes and promote behavioural plasticity that spreads public support lies critically with two sectors of the global population: the wealthy and the youth. The wealthy are high-carbon individuals and can cut their carbon footprints drastically without any significant loss of comfort. Adolescents are poised to establish lifelong lifestyle patterns; they still have the freedom to make large behavioural choices that will structure the rest of their lives; their behavioural shifts have the potential to be more rapid and widespread; they are likely to suffer the greatest negative impacts of climate change; hence they are an important target group for promoting high-impact climate mitigation actions through lifestyle changes. Young people from all over the world are raising their voices on issues which range from climate change to social issues such as inequality, corruption, freedom; they are mobilizing to tear down the toxic social structures built by past generations. Millennials (Generation-Y) are the largest living generation and the largest age group in the work force in many countries (and the Zoomers (Generation-Z) are poised to join), they are mobilizing across all regions and weaponizing their extraordinary appetite for the social media. The youth now form the progressive wings of political parties in many countries of the world and are becoming a force to reckon with; they have succeeded in installing many young country leaders and politicians; they are bringing social issues to the forefront and forcing society to confront the perils of inaction; the most effective global climate change movement to date was initiated by and is led by a teenager; and the youth are making a significant difference on all these issues, evident in the words of Emmanuel Macron, the French President: *"When you are a leader and every week you have young people demonstrate with such a message, you cannot remain neutral. They helped me change."* There is little doubt that the People's Movement, strongly energized by the 'youthquake' is one of the most powerful instruments that can change and reposition the world on the pathway to a sustainable environment.

Adeniyi A. Afonja
Emeritus Professor of Materials Science & Engineering
& Energy Consultant
April, 2020

New Delhi, India

Shanghai, China

Los Angeles. U.S.A.

Moscow, Russia

Urban smog

Chapter 1

The Earth and its environment

1.1.   INTRODUCTION

Climate change is a major global issue today, and one of the most controversial. The debate is full of well-mixed ideals, myths and realities. The pro-climate change world has an aggressive global campaign machinery against fossil energy use which is believed to be the cause of global warming and climate change. The sceptics are just as strong: either they do not believe that the climate is changing, or they think human activities have nothing to do with it. Infusion of politics inspired by intra-country exigencies has further complicated the issues and it has become difficult to distinguish between realities, facts, and fiction. Perhaps the best way to start a discussion on the environment is to examine the dynamics of the natural environment which has sustained life on earth for millions of years in order to place in proper context the net effect of current human anthropogenic degradation.

1.2.   THE EARTH'S NATURAL ENVIRONMENT

The Earth is central to a global system commonly defined in terms of four major spheres: *lithosphere*, *hydrosphere*, *atmosphere*, and *biosphere*. The hydrosphere may be sub-divided into fresh and frozen water, hence the common fifth member: *cryosphere*. The lithosphere comprises the Earth's crust (from the surface to a depth of about 100 km), and the upper part of the solid mantle that extends to a depth of about 3000 km. The hydrosphere comprises water (liquid or frozen) that exists under or over the surface of planet Earth - oceans, lakes, streams, glaciers, ground waters. The lower atmosphere (*troposphere*) comprises layers of gases that extend from the Earth's surface to about 15-20 km into space, and hosts most of the natural activities that determine the local weather - winds, clouds, precipitation, etc. The next atmospheric layer which extends to about 50 km is the *stratosphere* which plays a critical role in the regulation of temperature on earth and screening off much of the Sun's potentially dangerous rays. Although these are the two most important atmospheric layers that make life on earth possible, there are several other layers above the stratosphere which have different characteristics and different influences on the Earth's natural environment. The biosphere refers to parts of the land, sea, and atmosphere that host all living organisms, microorganisms and plants.

The atmosphere (in particular, the first two layers) sustains life on earth through the supply of oxygen needed by most organisms for respiration; carbon dioxide required by vegetation (plants, algae etc.) for photosynthesis; regulation of the Earth's temperature, weather, climate; and protection of the Earth from potentially damaging effects of the Sun's ultraviolet radiation. The four major spheres that make up the ecological system are characterized by intricate intra and interactions, all interdependent and interconnected. The interaction between these spheres to a large extent determines the weather and climate; the functions are controlled by natural regulatory processes; and the boundaries between them may be clear or ill defined. For example, while the boundaries between organisms and vegetation are fairly well defined there are no clear boundaries between air, water, climate, energy, radiation, etc. The complex interactions between the spheres are major drivers of changes in a wide range of natural and human systems: weather, climate system, ecosystems, natural resources, human well-being, and the dynamics of the interactions are influenced by both natural and human factors. The environment that is controlled by natural forces which have the capacity to dilute, absorb or dissipate pollutants is known as *natural environment*. The natural environment is imperfect in many ways, for example, extreme weather, flooding, volcanoes, earthquakes are all natural occurrences which have always been part of the ecosystem, caused by known or unknown natural phenomena.

The Earth-Environment systems have co-existed for millions of years and the way they interact determines the sustainability of life on Earth. The Sun generates enormous energy through sustained natural chemical processes and radiates it to the Earth and other planets; physico-chemical activities in the atmosphere moderate the Sun's energy and rays that reach the Earth's surface without which life would not be sustainable; the Earth's rotation around the Sun determines climate; the winds move global surface energy around, picking up moisture from the oceans and dumping rain on land, largely controlling the weather. Many natural processes like slight changes in the Earth's orbital characteristics, gravity, internal structure, and other unknown phenomena can alter both the weather and climate on earth significantly. Extensive archeo-geological evidence shows that this has occurred frequently and periodically for millions of years and caused profound changes on earth: some desert areas of the world today, including the Sahara desert were at one time fertile lands; many current land areas were once occupied by oceans; and some of the oceans today were dry lands. The Earth-Environment Systems have natural processes of regulating events on earth that sustain life - temperature, rainfall, human and plant respiration, the natural carbon cycle, protection from damaging radiation from the Sun, etc.

1.2.1.   The local (tropospheric, ambient) environment

The lowest level of the atmosphere that is in contact with life on earth is known as the troposphere and extends upwards for 15-20 kilometers. Most natural activities which determine and control the local weather take place in this layer: winds, clouds, humidity, rainfall, heat waves, etc. Activities in the layer are dynamic and are influenced by many factors notably wind dynamics and terrestrial topography. This explains why the weather can change dramatically within short periods of time, even within hours. Also, events in this layer have the most direct and instant impact on life on earth. Most air-borne emissions from human activities are discharged into this layer of the atmosphere, some drop off within days, some remain as smudge for weeks, while some gaseous and ultrafine particulate components may ascend to the middle atmospheric layer where they alter the natural dynamics of nature's climate regulatory mechanisms. In effect, the impacts of events in the troposphere are largely local but the effects of emissions which end up in the upper atmospheric layer will resonate across all regions of the world.

1.2.2.   The global (stratospheric) environment

The stratosphere is the layer of the atmosphere immediately above the troposphere and extends upwards to about 55 kilometers. The layer contains some natural gases in minor quantities (carbon dioxide, methane, nitrogen oxides, ozone, etc) which play profound roles in regulating events that make life on earth possible and sustainable. The natural gases in the layer are fairly well-mixed and human-related emissions (*anthropogenic emissions*) that eventually reach the layer mix fairly evenly with the natural gases, thereby creating similar impacts across all regions of the world. Events in this layer largely determine the average global temperature, global climate pattern which is regional, comparatively much less variable than weather and may remain fairly constant for decades. It is the layer that makes it possible for life to survive on earth, through two key regulation systems: the *greenhouse effect* and *ozone layer protection*.

1.2.2.1.   The Greenhouse Effect

About two-thirds of the solar short wave radiation reaching the Earth daily is absorbed and the balance is reflected into space as infrared long-wave radiation. Part of the reflected energy passes through the atmosphere back into space while the balance is trapped in the stratosphere by a mixture of minor natural gases known as *greenhouse gases (GHGs)*, and reflected back to earth as appropriate, to control global surface temperatures. This natural process ensures that a significant part of the infrared energy reflected from the Earth's surface in day time is absorbed and retained in the stratosphere. When the Earth's surface temperature drops at night, heat is radiated from the gas layer back to Earth. These processes help regulate the Earth's temperature, keeping it at an average of around 13-15°C. Without this regulation, temperatures on the Earth would be around minus 18°C, much too cold to sustain life. Natural greenhouse gases comprise mainly carbon dioxide, methane and nitrous oxide, all coming from natural processes such as animal and plant respiration and animal and vegetation decay. Moisture is also a major greenhouse gas but the natural processes of regulation make it non-polluting. The concentrations of GHGs in the stratosphere had remained fairly constant for millennia, thereby keeping the Earth's average temperature fairly constant.

1.2.2.2.   Ozone Layer protection

Natural ozone is produced in the lower part of the stratosphere through a series of chemical processes: natural oxygen dissociates into unstable atoms, energized by the harmful portions of the Sun's energy (ultraviolet UV-A and UV-B); a molecule and an atom of oxygen recombine to form a molecule of ozone (another unstable gas), releasing less harmful energy; ozone dissociates into one molecule and an atom of oxygen, using more of the Sun's harmful rays. In effect, there is a continuous chain of chemical reactions involving formation and dissociation of ozone which sustains a thin layer of ozone mostly in the lower part of the stratosphere (the atmospheric layer immediately above the troposphere that is in direct contact with the Earth). Chemical reactions in this layer play a key role in screening out potentially dangerous constituents of the Sun's energy. Without this natural protection, the negative impact on life on earth would be severe. UV (A & B) radiation which energizes the above chemical reactions is the very high ionizing component of the Sun's energy and has been shown to be harmful to humans, animals and other living organisms, damaging DNA, proteins, lipids and membranes and causing many human ailments including skin cancer (Melanoma).

1.3.   THE EARTH'S ANTHROPOGENIC ENVIRONMENT

Environmental pollution is the introduction into the atmosphere of contaminants which can alter significantly the natural balance of the environment systems, with potentially serious consequences. Pollution can be in many forms, including gases, particulates, chemical compounds, or in energy form - heat, noise, light, radioactivity, etc. There are many naturally occurring contaminants - emissions coming from volcano eruptions, sand storms, vegetation decays, etc., but the environment has in place a system of processing and balancing, which is crucial to the sustenance of life on Earth. For example, carbon dioxide exhaled by humans and animals is utilized by plants that produce oxygen, carbohydrates and proteins which humans and animals need to survive. However, human activities which produce greenhouse gases have accentuated emissions in the last two hundred years or so, coming from energy use, agriculture, livestock production, etc., thereby disrupting the natural balances (Figures 1.1 to 1.3). The environment contaminated by human-related pollution is known as the *anthropogenic environment*.

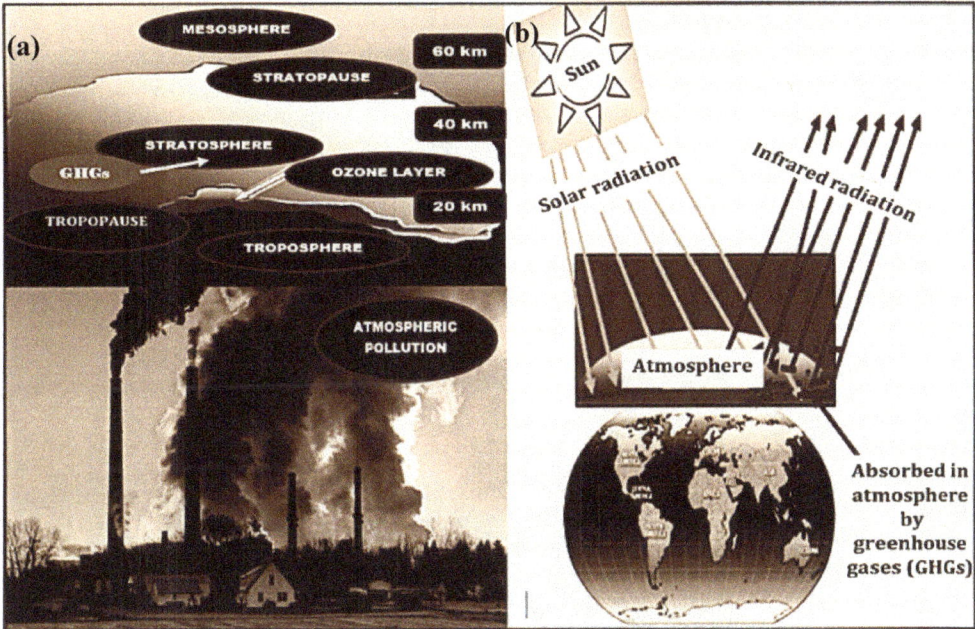

Figure 1.1 (a) Schematics of the Earth and its atmospheres (b) the Greenhouse effect *(Afonja, 2017).*

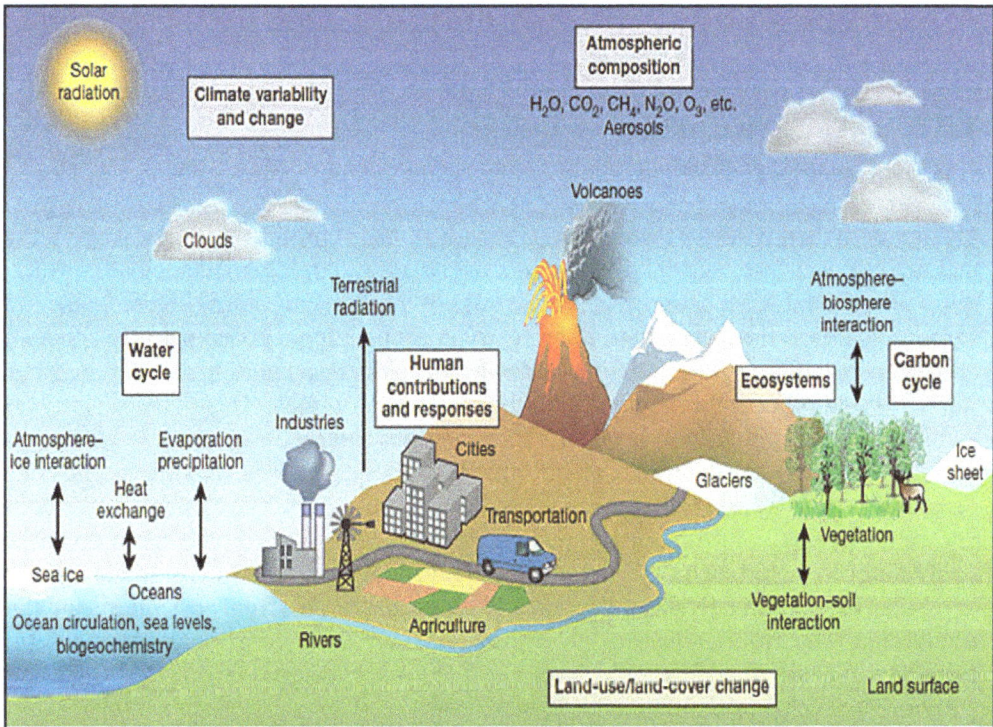

Figure 1.2 Major natural and anthropogenic processes and influences on the climate system addressed in scenarios. *(gfdl.noaa.gov).*

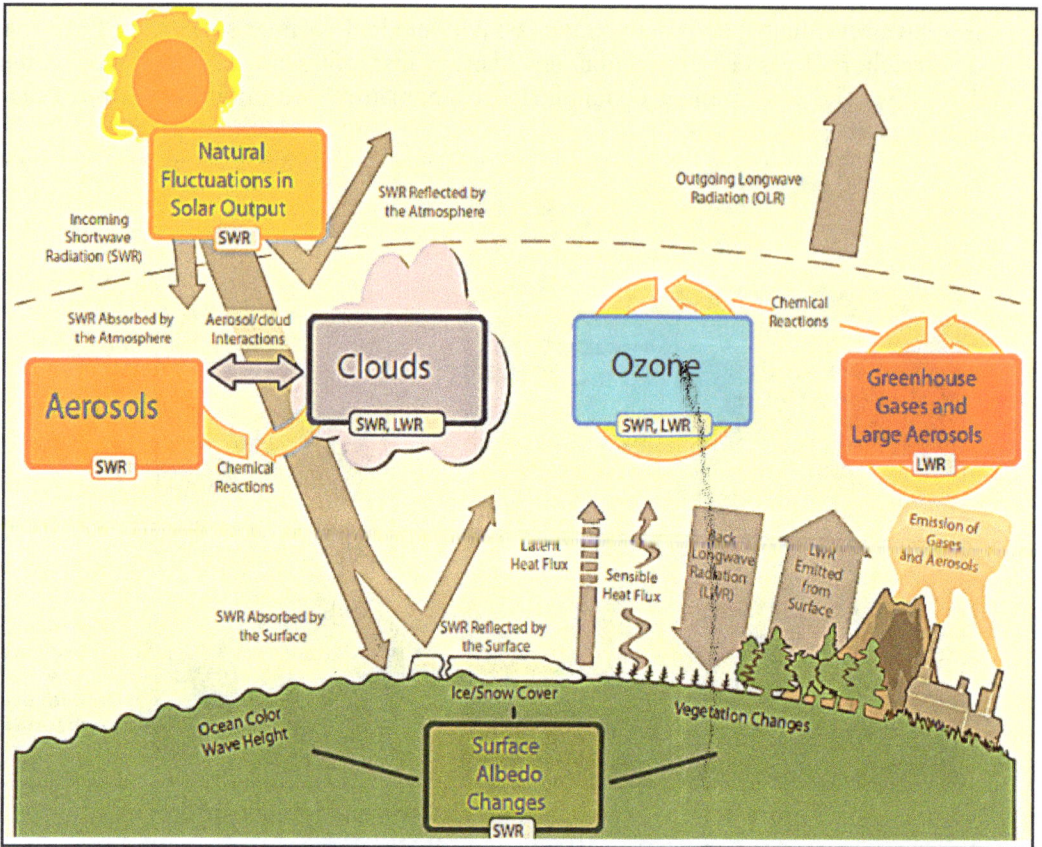

Figure 1.3 The radiative balance between incoming solar short-wave radiation and reflected long-wave radiation by the Earth's surface *(IPCC-WG1, 2013).*

1.3.1. Anthropogenic Greenhouse Gases (GHGs)

Human activities produce three main greenhouse gases primarily from energy production and use but also from agriculture and land use: carbon dioxide, methane and nitrous oxide, with carbon dioxide accounting for around 90% of the total emissions (Figure 1.4a). Although carbon dioxide is the least potent in terms of environmental damage per unit concentration, its dominance makes it the reference gas and benchmark against which the environmental impacts of all other anthropogenic greenhouse gases are rated. This system has established a combined measure known as *carbon dioxide equivalent* (CO_2eq or CO_2e) that represents all anthropogenic emissions. In effect, any reference to carbon dioxide in environmental discourse implies a basket of all the major anthropogenic gases. The carbon dioxide equivalent allows the potency of different greenhouse gases to be compared on a common basis, rated with reference to carbon dioxide which is assigned unity. Around 74% of total anthropogenic emissions came from energy production and use in 2015, 13% from agriculture and livestock production, 8% from industrial processes, and 5% from large-scale biomass burning, post-burn decay, landfills, and indirect emissions coming from reactions between anthropogenic chemical compounds (Figure 1.4b).

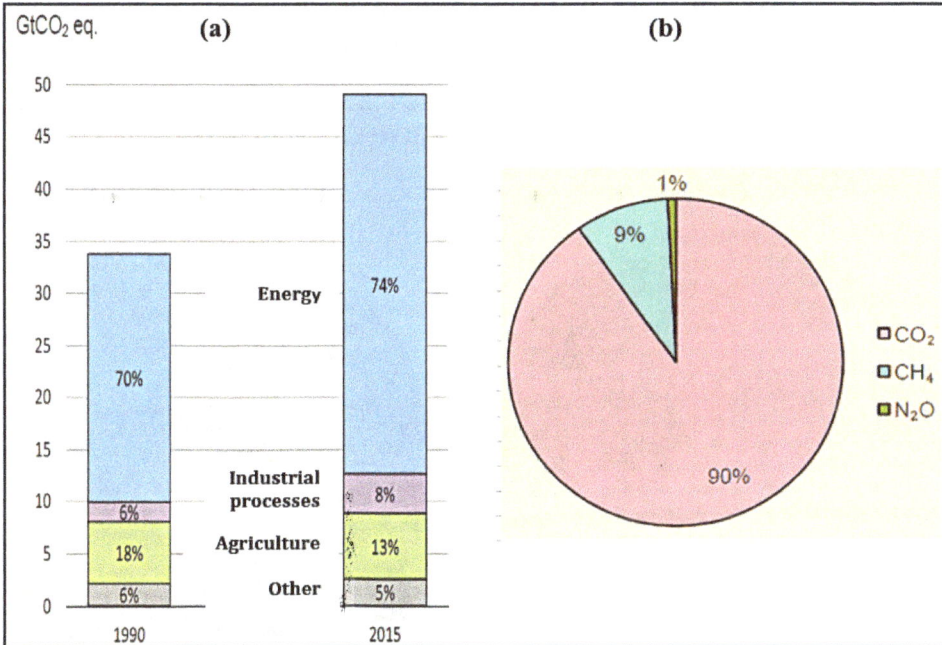

Figure 1.4 (a) Energy emissions by source (b) Global anthropogenic GHG emissions *(IEA, 2019d).*

Advances in atmospheric sciences have identified many of the variables which control the Earth's natural phenomena, and concluded that many human activities, particularly in the last two centuries or so are interfering with the natural processes by which equilibrium is maintained, with undesirable consequences, and environmental pollution coming mainly from energy production and use is the primary focus, although agriculture is also coming under intense scrutiny. Also, there is ample scientific evidence confirming that human activities are producing emissions known collectively as F-gases (mainly fluorocarbons used in refrigeration, air conditioning and some other industrial processes). Apart from the fact that they are potent GHGs, they react with ozone in the lower part of the stratosphere and reduce the concentration, thereby creating 'ozone holes' through which dangerous components of the Sun's rays can penetrate and reach the Earth.

Records of global anthropogenic emissions from 1850 and numerous scientific reports show that the atmospheric concentrations of carbon dioxide (CO_2eq) have been increasing gradually over the past century compared with pre-industrial levels but there has been a significant rise in the rate of increase from the 1950s when industrial activities intensified. Significant increases have also occurred in the levels of the other main greenhouse gases - methane (CH_4) and nitrous oxide (N_2O). Pre-industrial levels of CO_2eq which ranged between 180 and 280 ppm had increased to 407.4 ppm in 2018. The annual levels of anthropogenic emissions from energy production and use have increased steadily over the last fifty years or so, from 14 $GtCO_2$eq in 1970 to 33 $GtCO_2$eq in 2019 and there are no signs of peaking (Figure 1.5). Global CO_2 emissions from energy and industry (including from land-use change) reached a record high of 53.5 $GtCO_2$eq in 2017 and projections predict a rise of about 10% by 2040. Clearly the world is not on a feasible path to sustainable environment which requires that global GHG emissions in 2030 are 25% lower than in 2017 in order to limit global warming to 2°C or below by the end of the century. A reduction of around 55% would be required by 2040 to achieve a goal of 1.5°C (IPCC, 2018). Just a few industrialized countries produce more than half of the total global emissions. China alone accounts for around a quarter, more than twice emissions from the United States and three times the level of the 28 countries in the European Union.

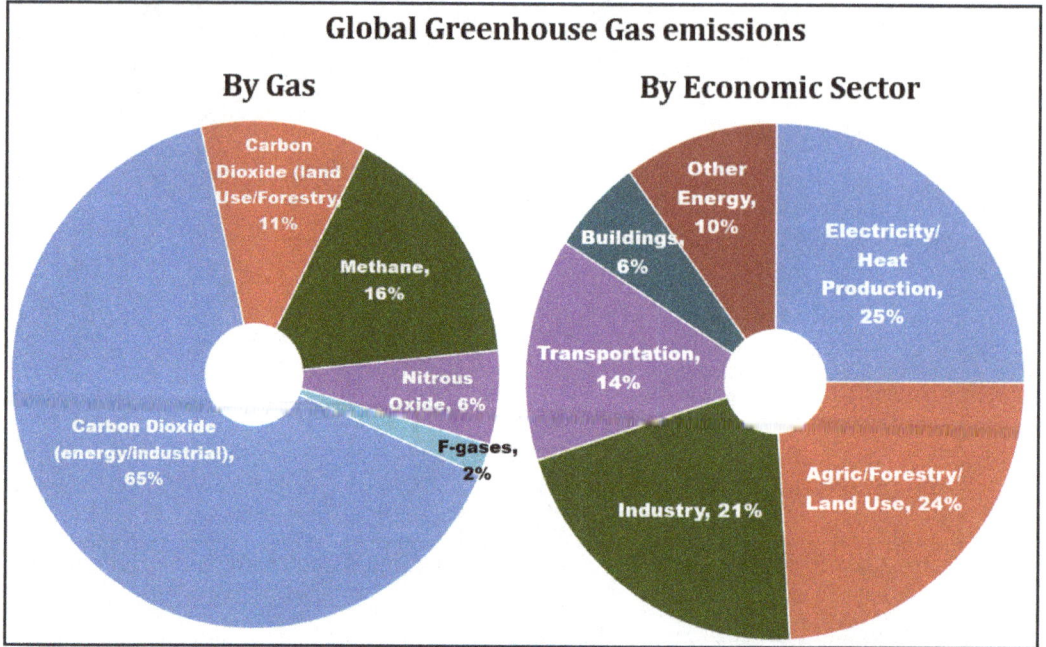

Global Greenhouse Gas emissions

By Gas

By Economic Sector

Carbon Dioxide (land Use/Forestry, 11%

Methane, 16%

Nitrous Oxide, 6%

Carbon Dioxide (energy/industrial), 65%

F-gases, 2%

Other Energy, 10%

Buildings, 6%

Electricity/ Heat Production, 25%

Transportation, 14%

Industry, 21%

Agric/Forestry/ Land Use, 24%

Figure 1.5a Global greenhouse gas emissions by gas and economic sector *(IPCC, 2014).*

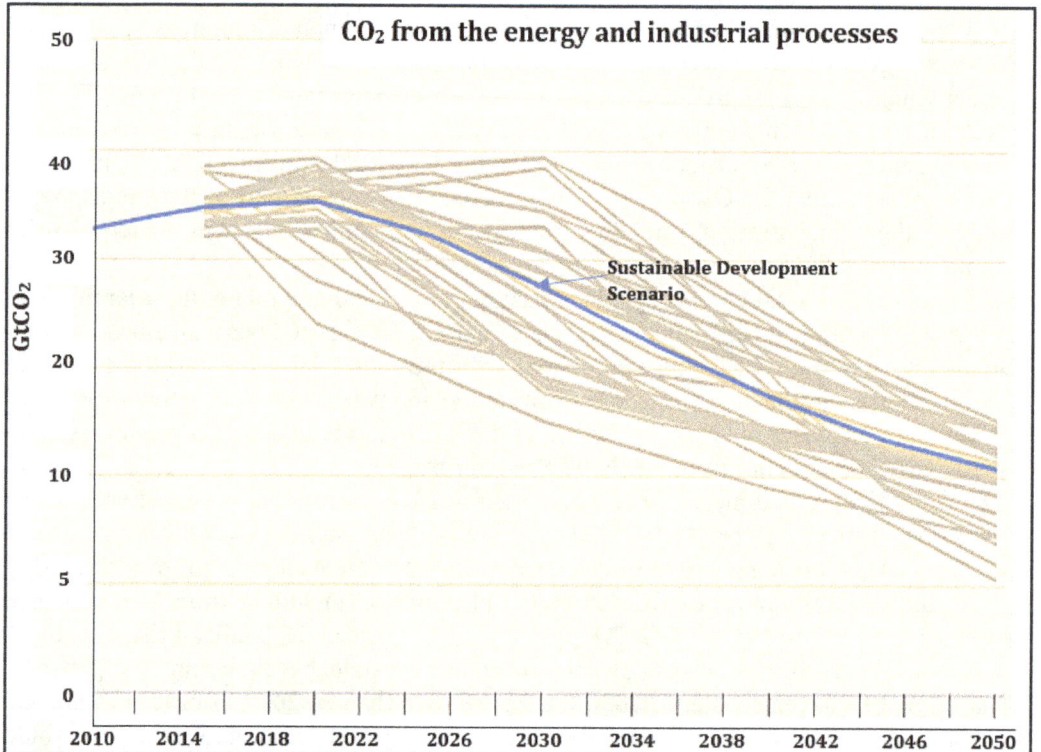

CO$_2$ from the energy and industrial processes

GtCO$_2$

Sustainable Development Scenario

2010   2014   2018   2022   2026   2030   2034   2038   2042   2046   2050

Figure 1.5b Sustainable Development Scenario (SDS) of Energy-related carbon dioxide emissions *(IEA, 2019d).*

1.3.2. Other anthropogenic pollutions

Although anthropogenic greenhouse gases have been the main focus of global environmental mitigation action, human activities produce other pollutants some of which could have even more severe consequences because of the direct impact on human health: for example, aerosol pollution, solid waste/e-waste pollution.

1.3.2.1. *Aerosol pollution*

Pollutants (natural or human-related) comprising gases, particulates and chemical compounds which emanate from human activities end up in the troposphere (up to 10-15 kilometers above the Earth's surface), the layer of the atmosphere that is in direct contact with life on earth. Most of these pollutants last only a few days to a few weeks in the atmosphere before they drop out, hence the name: short-lived climate pollutants (SLCP). However, they contribute a substantial proportion of the additional positive radiative forcing that is believed to be causing anthropogenic climate change. Also, the effects of pollution at this level on life on earth can be serious. Apart from causing a wide range of human health issues, the smog formed over towns and cities interferes with the natural cloud systems and local weather, obscures vision and affects agricultural crop yields. Tropospheric pollution, often referred to as *ambient air pollution* or *atmospheric aerosols*, is the contamination of the indoor or outdoor environment by any chemical, physical or biological agent that modifies the natural characteristics of the atmosphere. It originates from both natural and anthropogenic sources: natural sources include dust storms, volcanoes and forest and grassland fires, but the contribution is small compared to human activities. Heat and power generation, transportation, industrial activities (manufacturing, mining, oil production and refining, etc.), agriculture, municipal and agricultural waste incineration, landfills, and household biofuel combustion for cooking and lighting, are common sources of anthropogenic ambient pollution.

Atmospheric aerosols are a complex colloidal mixture of minute solid particulate matter suspended in liquid droplets, air or another gas. Aerosols reside in the lower atmosphere (up to around 15 km) and may be visible or otherwise depending on the density and composition. There are many sources of aerosols, some are natural (*background aerosols*), while others are anthropogenic (*ambient aerosols*). Atmospheric aerosols, whether natural or anthropogenic, originate from two different pathways: emissions of primary particulate matter, and formation of secondary particulate matter from gaseous precursors. The main constituents of the atmospheric aerosol are inorganic species (such as sulphate, nitrate, ammonium, sea salt); organic species (also termed organic aerosol or OA); black carbon (BC), a distinct type of carbonaceous material formed from the incomplete combustion of fossil and biomass-based fuels under certain conditions; mineral species (mostly desert dust); and primary biological aerosol particles (PBAPs). Aerosols are always present in the atmosphere but in extremely variable concentrations, due to the very large heterogeneity in aerosol sources and their relatively short residence in the atmosphere (a few hours to a few weeks). The vast majority of aerosols are not visible to the naked eye because of their microscopic size, but they become visible as haze or smog when the concentration is sufficiently high (Boucher, 2015). The major component of atmospheric aerosol (sulphate aerosols) emanates from the burning of fossil fuels, (in particular, coal and oil), biofuels, and biomass.

Mineral dust, sea salt, BC and PBAPs are introduced into the atmosphere as primary particles, whereas non-sea-salt sulphate, nitrate and ammonium are predominantly from secondary aerosol formation processes. Organic aerosols have both primary and secondary sources, influenced by both natural and anthropogenic events. The majority of BC, sulphate, nitrate and ammonium come from anthropogenic sources, whereas sea salt, most mineral dust and PBAPs are predominantly of natural origin. Atmospheric aerosols are largely generated locally, but can be transported a long way from their source if the winds are strong. Once they are airborne, particles can change in size and composition as a result of condensation, evaporation, chemical reaction or coagulation with other particles. Also, the effect on the environment is largely local and the most noticeable impact is the formation of haze/smog which scatters and absorbs sunlight, and obscures visibility. The health impacts are even more serious, accounting for millions of human deaths annually and interfering with the weather and natural balance of the ecosystems.

Particulate matter (PM) refers to inhalable particles, composed of sulphates, nitrates, ammonia, sodium chloride, organic carbon, black carbon, mineral dust or water. The health risks associated with particulate matter are substantial and well documented. Particulate matter is also the most widely used indicator to assess the health effects from exposure to ambient air pollution. Fine particulate carbon also known as soot originates from incomplete combustion of carbonaceous fuels - fossil fuels, wood, biomass etc. The finest component, (2.5 microns and below), known as $PM_{2.5}$ is the major component of soot and the most hazardous, prompting the World Health Organization (WHO) to set a safe limit of 10 $\mu g/m^3$ for $PM_{2.5}$ and 10 $\mu g/m^3$ for PM_{10} in ambient environment. Black carbon (BC) is the most strongly light-absorbing component of $PM_{2.5}$ and indeed the most light-absorbing airborne particle in the atmosphere. It is also the most potent health hazard because it is fine enough to be inhaled and can bypass most of the human body defense systems. BC represents about 10% of PM mass globally but concentrations can vary widely depending on the sources, location, time of day, and atmospheric transport and loss mechanisms. For example, concentrations in urban areas could be over a hundred times of values for remote, rural areas and emissions from prolific sources such as diesel engines could be up to 80% BC (Bond *et al.*, 2007; 2013). Other components of outdoor pollution are brown carbon, volatile organic carbon, sulphur dioxide, nitrous oxide, methane, ozone and carbon dioxide.

Aerosols are products of natural phenomena as well as human activities, in particular, fossil fuel use. Those that are the results of direct emission (black carbon, organic carbon, sea salt, dust) are often grouped as primary aerosols. However, aerosols also get into the atmosphere as products of chemical reactions (secondary aerosols). Secondary aerosols are in two main groups: secondary inorganic aerosols (SIAs) - sulphate, nitrate, ammonium - and secondary organic aerosols (SOAs). Secondary inorganic aerosols are products of reactions involving sulphur dioxide, ammonic and nitric oxide emissions while SOAs are the results of chemical reactions of non-methane hydrocarbons and their products with the hydroxyl radical (OH), ozone (O_3), or nitrate (NO_3). Although many hydrocarbons in the atmosphere are of biogenic origin, anthropogenic pollutants are believed to act as catalysts that promote their conversion into SOAs. Aerosols containing black carbon, organic carbon and sulfates as well as methane (CH_4) and ozone (O_3) do not last long in the atmosphere, often only a few days, hence they are commonly identified as short-lived climate forcers (SLCF) or short-lived climate pollutants (SLCP). However, in spite of the short atmospheric lifetime, black carbon is one of the largest contributors to global warming, surpassed only by carbon dioxide. Human activities enhance the greenhouse effect directly by emitting GHGs such as CO_2, CH_4, N_2O and chlorofluorocarbons (CFCs). In addition, pollutants such as carbon monoxide (CO), volatile organic compounds (VOC), nitrogen oxides (NO_x) and sulphur dioxide (SO_2), which themselves are negligible GHGs, have an indirect impact on the greenhouse effect by altering, through atmospheric chemical reactions, the abundance of important gases and the amount of reactive gases such as methane (CH_4) and ozone (O_3), and/or by acting as precursors of secondary aerosols. Clouds affect the climate system in a variety of ways: they produce precipitation (rain and snow) that is necessary for most life on land; they warm the atmosphere as water vapor condenses. Clouds strongly affect the flows of both sunlight (warming the planet) and infrared light (cooling the planet as it is radiated to space) through the atmosphere. Clouds also contain powerful updraughts that can rapidly carry air from near the surface to great heights, carrying energy, moisture, momentum, trace gases, and aerosol particles. Ambient pollution can alter the natural cloud system significantly, thereby contributing to weather variability, and, eventually to climate change.

1.3.2.2. *Solid waste pollution*

The global economy is rising and access to modern appliances and electronic gadgets is growing exponentially. As a result of rising prosperity and rapid technological advances that are driving innovation and rapid obsolescence, people can afford to upgrade frequently and this is leading to the dumping of home equipment (solid/e-waste). Appliances are designed to give ten to fifteen years of useful service but are now being replaced with newer models within a few years. Electronic waste is produced in staggering quantities, estimated at about 54 million metric tons globally in 2019, an average of 7.3 kg per capita and rising by an average of 2.5 million metric tons a year (Global E-waste Monitor 2020). Asia accounted for the largest share (25 Mt), followed by the Americas (13.1 Mt), Europe (12 Mt), Africa (2.9 Mt), and Oceania (0.7 MT). Electronic waste contains many valuable metals such as gold, copper, silicon, but also contains potentially hazardous chemicals and materials - lead, cadmium, chromium, plastic, brominated flame retardants, polychlorinated biphenyls (PCBs), etc. Only about 17% of the e-waste generated was formally collected for recycling and the fate of the balance is unknown. Europe accounted for the highest recycling rate of 43%, followed by Asia (12%), the Americas (9.4%), Oceania (8.8%) and Africa (0.9%). Few countries have formal collection and recycling policies in place but enforcement policies especially on local recycling are lax: some of the e-waste collected in developed countries end up in landfills while the bulk is shipped to developing countries for second-life use or recycling by an unregulated informal sector using primitive recycling techniques such as burning cables in poorly ventilated homes to recover copper. This results in a significant risk of release of toxic aerosols into the atmosphere, and exposure of recyclers who are often women and children to hazardous environment (WHO, 2018; Grant *et al*. 2013; Noel-Brune, 2013). Also, e-waste used as landfill may eventually contaminate soil, agro-products and water.

Polymers are products of petrochemicals derived from fossil fuels and, since invention in the 1950s, have taken over virtually every aspect of human life, from packaging to clothing. One of the most attractive properties is that they are unreactive, which makes them suitable for food and chemicals storage. Unfortunately, this property also makes it difficult to dispose of polymer waste. Most polymers are not biodegradable and can remain intact for several hundred years in landfills. When incinerated, polymers release a lot of energy, black smoke, carbon dioxide, and black carbon aerosols into the atmosphere. Furthermore, the bulk of plastic waste from the developed countries is being exported to emerging nations where they are sorted manually and recycled or incinerated largely by the informal sector usually under unhealthy conditions. China which had been the main destination has closed its doors and plastic waste is now being diverted to poor Asian countries.

1.4.   HUMAN CONTRIBUTION TO ENVIRONMENTAL POLLUTION

It is clear from the discussion above that environmental pollution is a natural process but there are also many complex ways in which the negative impacts on earth are mitigated, to the extent that the Earth's average temperature which largely determines climate has remained under control for millions of years. However, while many changes in weather and climatic patterns observed today may have been caused by natural phenomena, there is ample scientific evidence that human interference can exacerbate the spread, frequency, intensity and consequences. Urban smudge can destabilize local weather significantly; just one degree centigrade rise in ocean water temperatures which could result from enhanced greenhouse gas emissions can impact severely on the frequency and intensity of heat waves, cause flooding due to a rise in ocean levels, and energize storms and tornadoes. Recent research has shown that the physical processes in the ocean and atmosphere that produce strong El Ninos are being supercharged by warmer ocean waters caused by human-induced climate change and will become more frequent, stronger, causing more extreme weather events (severe storms, heat waves, floods) particularly in the United States and other North American countries (Wang *et al*., 2019).

1.5.   ENVIRONMENTAL POLLUTION INDICATORS

A number of environmental pollution indicators have emerged over the years which provide useful methods of quantifying human contribution to environmental pollution.

1.5.1.   Carbon intensity

Carbon Intensity (also called carbon intensity index or emission intensity, or specific emissions index) is the amount of metric tons of carbon dioxide or equivalent produced per unit of output or an activity, for example per Megajoule of energy produced, per capita or per GDP, or even per human activity like driving a car over a distance of a hundred kilometers. Population and GDP are the major determinants of primary energy consumption, and therefore a country's energy-related emissions, but two factors may mitigate carbon intensity, namely *energy intensity* (amount of energy input per unit output, determined by process efficiency) and *fuel mix* (which determines carbon emissions per unit of energy). Energy intensity/GDP gives some indication of a country's level of economic development and the extent of adoption of energy-efficient technologies, but simply moving away from high-energy intensity production, like is happening in many developed countries can also lower energy intensity. On the other hand, emerging countries which are increasingly hosting primary production of goods that fill most of the world's demand, such as primary metals, polymers, cement, chemicals, petrochemicals will have high energy intensities largely because of the inherent energy-intensive nature of these production processes, but also because these countries often cannot afford to adopt the most efficient technologies. Emissions intensity also depends on fuel mix (carbon content of energy consumed). Coal has the highest carbon content of fossil fuels, followed by oil; natural gas has the lowest carbon content, while most renewables and nuclear energy have relatively low carbon contents. The developed countries are decarbonizing energy by moving away from coal and substituting less carbon-intensive natural gas and renewables, mainly for power generation. On the contrary, coal is the only readily available and affordable fuel (for power generation, steam raising, home heating, etc.) in many emerging countries, hence carbon intensity will be relatively high. Carbon intensity is also useful in comparing the negative impact of anthropogenic greenhouse gases from various sources on the environment. Values of carbon intensity can be calculated for any primary energy, electric power plants per unit of energy produced using different fuels, for heat generating units per joule of heat, and for various production processes. For example, the emissions intensity of wood in power generation is 0.39 $kgCO_2eq/kWh$, corresponding values for other primary fuels are coal 0.36-0.34 depending on the maturity, crude oil 0.26, natural gas 0.22, diesel oil 0.27, petrol/gasoline 0.25. Carbon intensity calculations also make it possible to compare the environmental impact of different power generation systems: the carbon intensity of a coal-fueled power plant is 888-1054 metric tons of CO_2eq/GWh (on a lifetime basis) depending on coal grade, and comparative average values for other fuels and energy sources are oil 733, natural gas 499, solar PV 85, biomass 45, nuclear 29, wind 26, hydro 26 (Figure 1.6b).

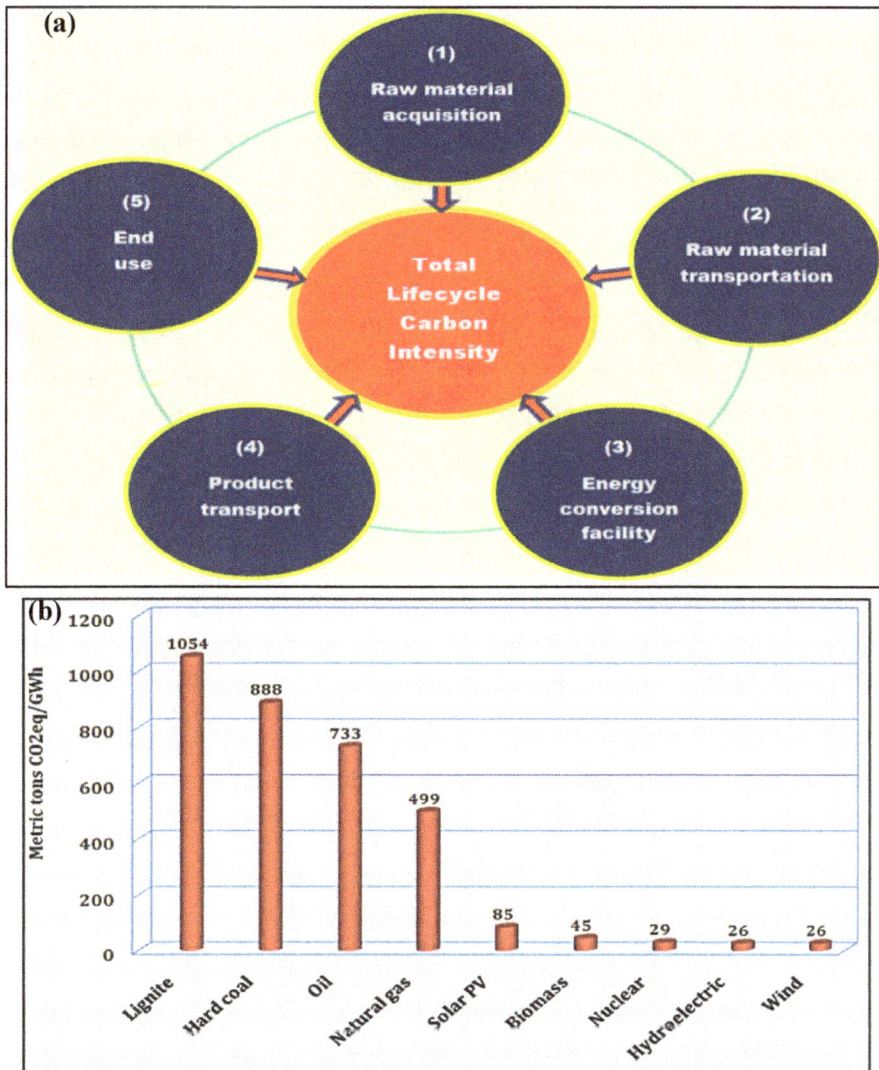

Figure 1.6 (a) Cradle-to-Grave Life Cycle Analysis of fossil fuels. (b) Lifecycle analysis of GHG emissions by fuel in electric power generation. *(Data from WNA, 2011).*

1.5.2.   Carbon footprint

Carbon footprint literally means an estimate of the amount of direct or indirect contribution of greenhouse gas emissions from any source over a time frame (usually one year), for example, events, products, organizations or persons. A comprehensive, quantitative assessment based on this definition would not be possible because of inadequate knowledge about the complex interactions between contributing natural and anthropogenic processes. For example, every person inhales oxygen and exhales carbon dioxide; plants use carbon dioxide from the atmosphere for photosynthesis, producing plant/animal carbohydrates and proteins as well as oxygen. When plants die and are buried under ideal conditions of temperature and pressure, they are transformed into coal; dead sea animals, fish et cetera are transformed by microorganisms into oil and gas. In effect, fossil fuels are part of the natural carbon cycle. Virtually every daily activity contributes to a person's anthropogenic carbon footprint: use of all household goods and materials which require energy for manufacture or as raw material (for example, plastics, synthetic clothing, carpets, furniture, bedding, personal care products, pharmaceuticals), energy use in home heating and cooling, transportation, etc. A carbon footprint is measured in metric tons of carbon dioxide equivalent (tCO_2eq)/year, and carbon footprint calculations consider all six of the Kyoto Protocol greenhouse gases: carbon dioxide (CO_2), methane (CH_4), nitrous oxide (N_2O), hydrofluorocarbons (HFCs), perfluorocarbons (PFCs) and sulphur hexafluoride (SF_6). Carbon footprint makes it possible to compare the relative negative impact of fuels, other energy sources, manufacturing processes, indeed all human anthropogenic activities on the environment.

Although accurate calculation is problematic, many approximate methods have been developed and carbon footprint calculation has become a powerful tool for assessing the impact of human (including personal) behavior on global warming. It should be noted that carbon represents all greenhouse gases which are now quantified as carbon dioxide equivalents. The foods and goods that humans buy and use everyday, polymer shopping bags, travels, sports/recreation equipment all have carbon footprints which must be accounted for in calculating personal contribution. For example, ten liters of petrol or diesel burnt in a car, or of oil used in home heating contribute 23-27 kg of carbon dioxide to the atmosphere; the manufacture of three empty one-liter plastic bottles of water/soft drink or twenty plastic shopping bags releases about 1 kg of CO_2eq into the atmosphere; a cow releases around 100 kilograms of methane from its digestive system into the atmosphere annually (equivalent to 2 metric tons of CO_2eq because methane is 20 times more potent than carbon dioxide). In effect, four metric tons of carbon footprint need to be shared among people who consume the meat from a two-year old slaughtered cow, or milk produced by the cow. On the other hand, supporting recycling or abstaining from single-use plastic products can reduce personal carbon footprint significantly. Carbon footprint (positive or negative) can be calculated for virtually any human activity: industrial and commercial activities; production and use of consumer goods; commercial and personal transportation; home energy use; use of computers, cell phones, the Internet; tree planting and gardening; support for waste recycling, etc. Clearly, it is near impossible to completely eliminate personal carbon footprint but it can be minimized in many ways: rationalizing energy use, cutting out unnecessary driving, opting for energy-efficient vehicles, opting for energy-efficient homes, fittings and appliances, or supporting district recycling efforts, are effective ways of reducing personal carbon footprint, and promoting personal contribution to global efforts at mitigating energy-related emissions. The average personal carbon footprint varies widely by region and country: the average for Africa was 1.0 in 2018 compared with 9.0 for OECD (IEA, 2019; Wikipedia, 2019) (Figure 1.7).

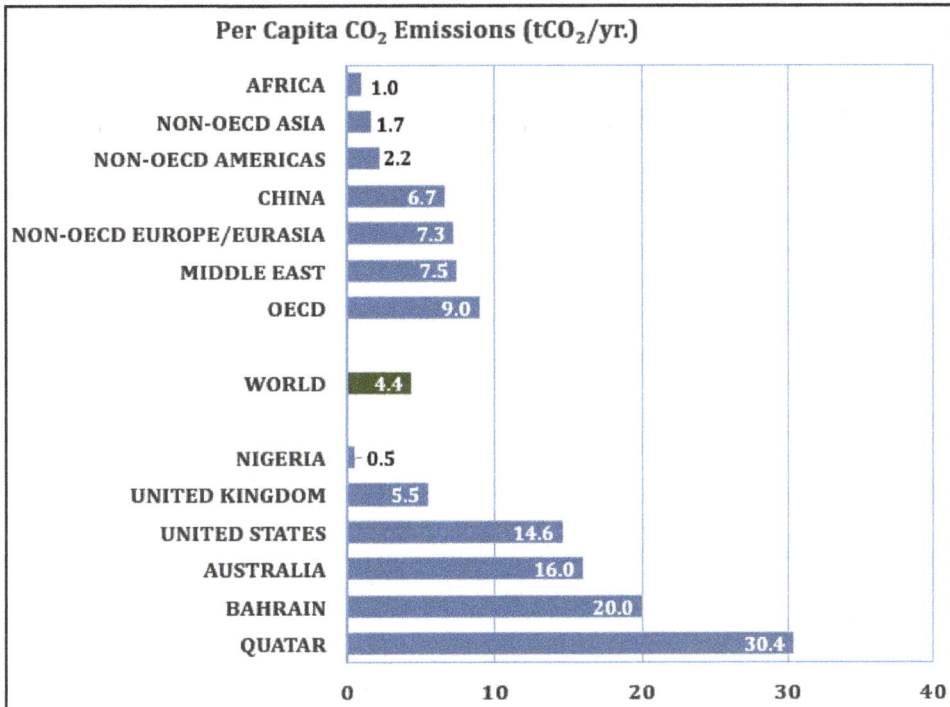

Figure 1.7 Per Capita carbon dioxide emissions by country and region, 2017. *(Data from IEA, 2019d).*

1.5.3. Carbon neutrality

Carbon is a critical element that sustains life on earth. It is the primary element in the human-animal-plant lifecycle: plants process carbon dioxide from the atmosphere into carbohydrates (around 50% carbon), animals feed on plants to produce protein (75-85% carbon), and humans feed on both, exhaling carbon dioxide into the atmosphere; humans, animals and plants die, returning carbon to the Earth; the carbon is the precursor to fossil fuels. As discussed earlier, carbon dioxide is a major greenhouse gas which helps regulate temperature on earth, but there is an optimal concentration beyond which it becomes anthropogenic. Carbon neutrality is an environmental mitigation effort which seeks to balance atmospheric input and output of carbon dioxide in the natural carbon cycle in order to preserve the natural concentration, by actions that seek to offset anthropogenic carbon dioxide. Examples include massive planting of trees which can utilize much of the emissions from energy generation or installation of solar power units that can meet fully the power needs of homes, utilities, small communities, and possibly feed excess power to public grids. An industrial enterprise can become carbon neutral by capturing and sequestrating all the carbon dioxide produced, and other more complex possibilities include offsetting carbon dioxide produced in one area of activity by actions such as efficiency improvement and lower carbon energy production (for example, those that involve material and heat recycling) that reduce emissions from other areas; or by purchasing carbon credits from other enterprises which have lower or negative carbon footprint. People can also strive to be carbon neutral by optimizing energy and use in homes and transportation, minimizing meat and dairy product consumption, practicing organic gardening, and actively supporting district *refuse, reduce, reuse, recycle* policies on consumer products.

1.5.4.   Lifecycle carbon footprint

Total Life-Cycle Analysis (LCA) is an accounting process that considers the environmental impacts of all stages of a particular process or product chain. For example, LCA for an energy resource involves consideration of emissions at every stage from production to utilization, from material and fuel mining, manufacturing of components, construction, installation, de-commissioning, to waste management. A typical cradle-to-grave LCA of fossil fuels begins with extraction of oil, natural gas or coal and ends with electricity or fuel delivered to the consumer (Figure 1.6a). Lifecycle analysis of carbon intensity is a very useful method of comparing anthropogenic emissions from generating plants using different fuels. Calculations of the total life cycle emissions of a fuel take into account emissions at the mining stage, transportation, processing, ultimate energy utilization, and recycling of solid or heat by-products. Emissions from the production of materials of construction such as steel and cement used in dam construction, energy required to produce silicon and other inputs of solar energy production, materials used in the transportation and distribution of natural gas and fuel oil, fugitive emission during production and transportation, are also taken into account. For example, solar energy is considered a zero emission energy resource in the traditional evaluation process, however, life-cycle-analysis presents a different picture: the overall carbon footprint is significant because enormous energy (usually supplied by coal-fired plants) is required for the production of silicon used in the manufacture of photovoltaic (PV) solar cells, materials for collectors, storage battery, associated equipment, building and infrastructure, the copper used extensively in solar power generation etc. Also, large land areas that need to be cleared for solar farms have negative carbon footprint. In effect, there is nothing like near-total generation of electricity from zero carbon sources (solar, hydroelectric, nuclear, wind) as claimed by some developed countries and environmentalists - all energy sources have significant carbon footprints.

It is also a common assumption that natural gas-fired power generating plants emit around 50% lower environmental pollutants compared with coal. However, total life-cycle analysis shows that, although the carbon footprint of natural gas at point-of-use is significantly lower than that of coal, the ultimate carbon footprint depends on losses in production and transportation. Excessive release of gas (mainly highly potent methane) into the atmosphere during production, flaring and transportation can narrow or even eliminate the relative advantage of natural gas compared with coal especially when the coal plant is of comparable efficiency. In fact, it is estimated that, unless emissions of methane (the main component of natural gas) at the gas production stage can be kept below 2-3%, there is no relative advantage, since methane has twenty times more heat-trapping capability than carbon dioxide which is the predominant pollutant in coal-fired plants. Life cycle analysis also involves the quantification of all emissions in the chain, including emissions saved (displacement emission) where for example the by-products of one process are used to replace products of another process, thereby avoiding the emissions associated with those replaced products: for example, using gas released during oil production for fueling captive power generation plant instead of flaring. Steel, aluminium, copper, lead, tin and indeed most metals are 100% recyclable and require less energy. For example, the energy required in processing recycled steel is about 30% less than required for processing virgin steel raw materials. Recycled aluminium only consumes about 10% of the energy required to produce the material from bauxite ore. In effect recycling of material and waste heat can significantly lower the ultimate carbon footprint of many products. In summary, the true assessment of the environmental impact of a product requires a cradle-to-grave analysis; it should not only take account of direct emissions from the manufacturing and transportation to retailers but must also consider a host of indirect emissions such as those caused by the production of the raw materials used in the production of the good.

1.5.5.   Radiative forcing

The Earth's climate is largely determined by the radiant energy (sunlight) received from the Sun. The Earth may absorb all the energy received, or some of it in which case the balance is reflected back into space. The balance between the absorbed and reflected solar energy determines the average global temperature which in turn largely determines the Earth's climate. Radiative forcing (RF) is a quantitative measure of the difference between energy received by the Earth from the Sun and the amount radiated back into space (net change in energy balance in response to some external perturbation). Positive RF means that the Earth receives more energy than it radiates and negative RF implies that the Earth loses more energy into space than it receives. Net energy gain (positive RF) will cause global warming while negative RF means that the Earth cools. In effect, zero RF promotes a sustainable environment. The RF concept is valuable for comparing the influence on global mean surface temperature (GMST) of most agents affecting the Earth's radiation balance. Human activities have changed and continue to change the Earth's surface topography and atmospheric composition. Some of these changes have direct or indirect effects on the energy balance of the Earth and are thus drivers of climate change.

Greenhouse gases are present in the atmosphere in very minor quantities compared with oxygen and nitrogen which make up over 99% of the gases in the atmosphere. However their regulatory role is critical, and there are many natural events that can move RF either way. However, any human activities that increase the concentrations will also cause (mostly positive) radiative forcing. The concentrations of these gases in the atmosphere have increased exponentially over the last hundred years or so due to human activities (anthropogenic emissions), in particular, use of fossil energy, and is believed to be the main cause of positive radiative forcing and gradual increase in the average global temperature over the last two hundred years or so, since the exploitation of fossil fuel resources started to intensify.

1.5.6.   Global warming potentials and atmospheric lifetimes

Not all of greenhouse gases make an equal contribution to the greenhouse effect. For example, one molecule of methane (CH_4) has 20 times the impact of a molecule of carbon dioxide for a 100-year time scale; nitrous oxide (N_2O) 300 times; ground-level ozone 2,000 times; and a chlorofluorocarbon molecule has from 13,000 to 20,000 times the impact of a molecule of carbon dioxide. Greenhouse gases may remain in the atmosphere for very short periods or for as long as 150 years. Global Warming Potentials are quantified measures of the globally averaged relative, cumulative, radiative forcing impacts (both direct and indirect integrated over a period of time) of the emission of a unit mass of a particular greenhouse gas with reference to carbon dioxide (IPCC 1996, 2007). Carbon dioxide has a variable atmospheric lifetime, and cannot be specified precisely. Being by far the largest anthropogenic gas emission in terms of quantity (around 90%), the gas is assigned a unit global warming potential (GWP) over all time, and serves as the baseline. In effect, GWP of any other greenhouse gas is a measure of how well the gas absorbs reflected solar energy from the Earth, preventing it from immediately escaping into the atmosphere, compared with carbon dioxide. The higher the GWP, the more positive the radiative forcing (RF), and the more warming the gas causes.

As discussed earlier, carbon dioxide equivalent (CO$_2$eq) is a method of placing emissions of various radiative forcing agents on a common footing in accounting for their effects on the environment. It describes, for a given mixture and amount of greenhouse gases, the amount of CO$_2$eq that would have the same global warming ability, when measured over a specified time period, usually 100 years. Most quoted values represent a basket of greenhouse gases listed in the Annex A to the Kyoto Protocol which was the first major global mitigation action on anthropogenic greenhouse gas emissions. Each of the greenhouse gases can remain in the atmosphere for different amounts of time, ranging from a few years to thousands of years. Methane remains in the atmosphere for 10-12 years before it degrades, nitrous oxide has a lifespan of 100-120 years, and carbon tetrafluoride (tetrafluoromethane, (CF$_4$) may remain in the atmosphere for up to 50,000 years. All these gases remain in the atmosphere long enough to become well mixed, in effect, the concentration in any atmospheric location across regions remains approximately constant regardless of the source of the emissions. A gas which has a high radiative forcing but a short life compared with carbon dioxide could have a high GWP on a 20-year time scale or a low value on a 100-year time scale. On the other hand, a gas that has longer atmospheric life than the reference gas will have higher GWP with the time scale. (IPCC, 2014) (Table 1.1).

Direct effects known as *direct radiative forcing* occur when the gas itself is a greenhouse gas and *indirect radiative forcing* refers to situations when chemical transformations involving the original gas produce a gas or gases that are greenhouse gases, or when a gas influences other radiatively important processes such as the atmospheric lifetimes of other gases. For example, methane has both direct and indirect effects, an atmospheric lifetime of 12 ± 3 years and a GWP of 56 over 20 years, 21 over 100 years and 7.6 over 500 years. The decrease in GWP at longer times is because methane is degraded to water and carbon dioxide through chemical reactions in the atmosphere. The effect of radiative forcing varies substantially in space and time, and therefore the influence on climate response may also vary. For example, carbon dioxide has the largest forcing in the subtropics, decreasing towards the poles, with the largest forcing in warm and dry regions and smaller values in moist regions and in high-altitude regions (Taylor *et al.*, 2011; IPCC, 2013).

Table 1.1 Global warming potentials (GWP) and atmospheric lifetimes (yrs.) of select greenhouse gases *(Extracted from IPCC, 2007; 2014).*

Gas	Chemical formula	Global warming potential (GWP) for given time scale		
		20-year	100-year	500-year
Carbon dioxide	CO$_2$	1	1	1
Methane	CH$_4$	72	25	7.6
Nitrous oxide	N$_2$O	289	298	153
*CFC-12	CCl$_2$F$_2$	11,000	10,900	5,200
*HCFC-22 Freon-22)	CHClF$_2$	5,160	1,810	549
Tetrafluoromethane	CF$_4$	5,210	7,390	11,200
Hexafluoroethane	C$_2$F$_6$	8,630	12,200	18,200
Sulphur hexafluoride	SF$_6$	16,300	22,800	32,600
Nitrogen trifluoride	NF$_3$	12,300	17,200	20,700
*Gases used in refrigeration and air conditioning				

Chapter 2

Energy and human development

2.1.   INTRODUCTION

Human development has always been tied to the availability of appropriate energy. Industrial and economic development which began with the Industrial Revolution of the late 1700s stimulated the transition from hand tool production and animal power to new mass manufacturing processes, a development which would not have been possible without the emergence of the coal industry. The availability of coal energy led to a revolution in agriculture, textile and metal production, and transportation. Advances in agricultural techniques and industrial production led to major improvements in the quality of life, which in turn stimulated growth in energy demand and the development of mass production technologies. Economic and social development tend to go hand-in-hand with energy sector transformation, and urbanization, a major consequence of economic growth, rapidly transforms the traditional agrarian economy into industrialization and a knowledge-based economy. This shift requires modern energy to thrive, while also providing the incentive for investment in modern energy infrastructures. Not many nations have succeeded in growing their economies and human development without the dual strategy of modernizing the economy and developing a sustainable modern energy infrastructure.

The global economy has been transforming from rural/agrarian to urbanized modern production, earning power and quality of life have been rising, stimulating expansion of energy infrastructure, and giving prominence to the interdependence of energy and human development. The gross domestic product (GDP) could double within the next two decades, with increased rural-urban migration and high levels of economic growth which means rising living standards and upward class mobility. The global fleet of vehicles will likely double to around 2 billion over the next twenty years; global population will grow from 7 billion today to around 9 billion; and 1.2 billion people mostly in the developing world will still lack access to adequate modern energy. A dynamic growth of secure and sustainable energy is critical to fueling these projected changes in the global dynamics and promoting universal access to modern energy. The World Energy Council (WEC, 2019) defines three core dimensions that determine energy sustainability, a triple challenge of ensuring that the world has access to affordable and reliable modern energy supplies while reducing energy-related anthropogenic emissions to moderate the risk of environmental pollution:

* Energy security

* Energy equity

* Environmental sustainability

Effective management of these variables promotes transition to a more sustainable, environmentally-sensitive energy future and the extent to which an individual country can reconcile these often conflicting key dimensions known as 'Energy Trilemma' determines its prosperity, economic growth and human development. The World Energy Council (WEC) identifies three key interconnected policy areas needed to balance the Energy Trilemma: coherent, predictable and resilient energy policies that can be sustained across the political cycles; stable regulatory and legal frameworks for long-term investment; and public and private partnerships that stimulate research, development and innovation. WEC ranks 125 countries annually, overall and per dimension in terms of their ability to provide a secure, affordable and environmentally sustainable energy system.

2.2.   ENERGY SECURITY

Access to adequate energy is a crucial determinant of a country's economic growth and prosperity, and every country has access to several of the primary sources – oil, natural gas, coal, biomass, hydro, wind, solar, geothermal, nuclear energy. However, global oil and gas resources which constitute around 60% of the total world's primary energy supply are located in a few countries, but coal deposits are much more widely spread. Around 70% of global reserves of oil and 50% of gas resources are located in the Middle East and North-Central America. Most of the oil and gas reserves are located away from countries with the highest demand, and in politically unstable regions. This explains why global trades in energy, in particular, oil and gas have become prominent pawns in geo-politics and major indicators of the health of the world's economy. The global oil crisis of the early 1970s accentuated the need for domestic energy security and most countries have developed strategic plans for the management of primary energy supply from domestic and external sources, notably the development of domestic energy infrastructure to meet current and future local energy demand, and strategic stockpiling, thereby ensuring substantial insulation from external supply source disruptions. The need for energy security stimulated the development and proliferation of nuclear power which is largely independent of developments in the global oil market. Nuclear plants only need small quantities of fuel to run for years and can be externally sourced while most other inputs are locally available.

The world economy has been growing rapidly, driven largely by growth in productivity (GDP per person), and the rapid growth of the economies of many emerging countries over the last decade or so, projected to grow even faster over the next two decades. Energy production and use are becoming more efficient, thereby driving down energy demand per unit of GDP (energy intensity) across all sectors of the economy and all regions of the world. Global population, an important driver of energy demand, has doubled from 3.7 billion in 1970 to 7.4 billion in 2017, with the increase concentrated in Africa, India, Southeast Asia and the Middle East where most of the projected economic and energy demand growth will also occur (Figure 2.1). Also, fossil energy production, export and associated job creation have become prime variables in the economies of many emerging and developed countries across all regions of the world. Recent projections indicate that the world economy will double within the next twenty years but there will be significant disparities between economic regions: while growth in OECD countries will be around 1.6%, the growth rate in non-OECD countries will more than double, about 4%. Global population will reach nearly ten billion by 2050 (UN, 2019) and developing countries will account for most of the growth. Continued rapid economic and population growth and the implied rapid growth in modern energy demand present very formidable challenges for sustainable development, particularly in the developing world, and holds important implications for economic and social development and environmental sustainability. Growth of energy-intensive industries, rising urbanization and growth in social status in these regions will likely fuel a rapid growth in energy demand by around a third over the same period. Also the developed world is moving rapidly from production of energy-intensive primary materials (metals, cement, chemicals, petrochemicals, fertilizers, etc.) to light-engineering and services, thus becoming increasingly dependent on the emerging world for supply of these critical materials. Inevitably these regions will account for most of the future rise in energy demand and associated pollutions.

The inequity in the global distribution of fossil primary energy reserves, coupled with the strategic role of energy in human development have made the security of primary energy sources a leading priority for all nations of the world, and an important factor in global energy market dynamics. Emerging countries that hold most of the resources seek to maximize income from sale of crude oil and natural gas in the international market, but spend most of the earnings importing finished products, including refined oil products from the developed countries at up to tenfold the price they received from crude oil sales. The developed countries also use strategic reserves to control global oil pricing by holding large stocks when oil is cheap and releasing to the international market as appropriate to force down high prices. Furthermore, rapid deployments of advanced oil prospecting and production technologies particularly in North America are changing the dynamics of global oil and gas production.

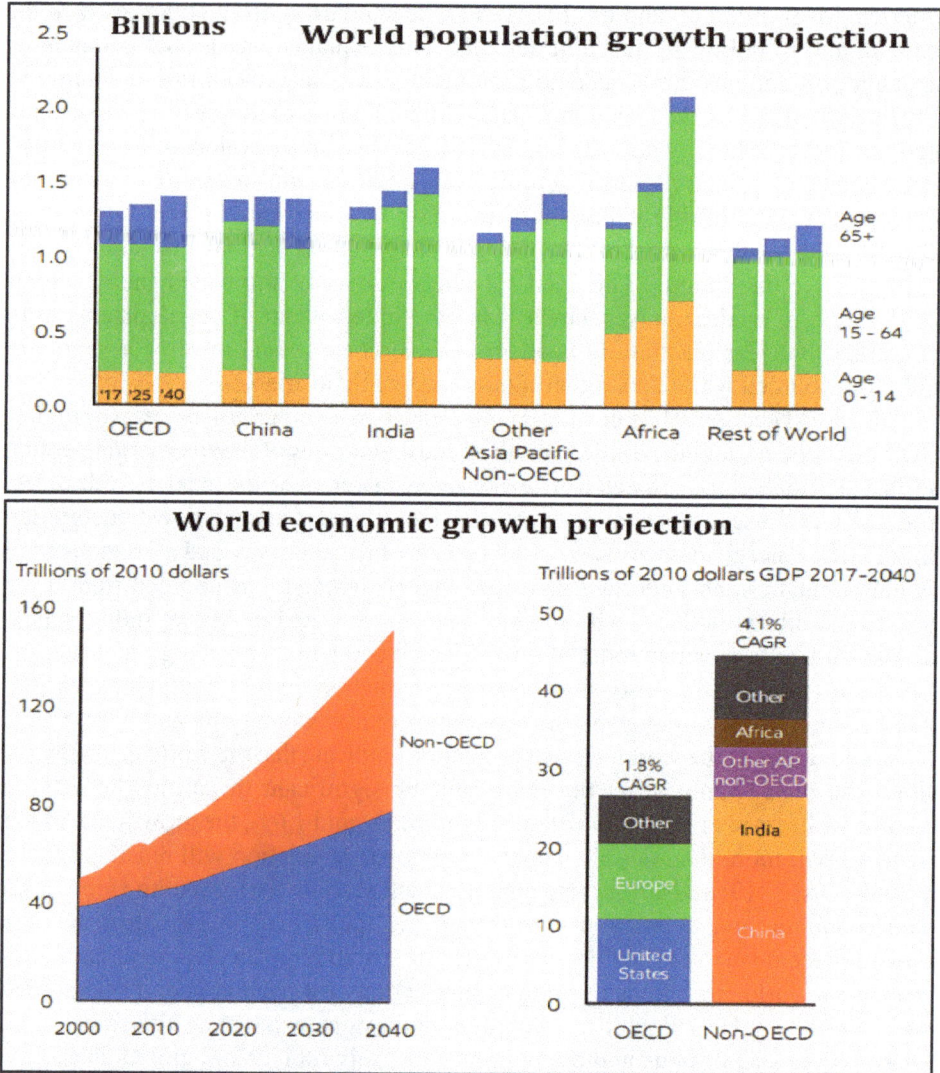

Figure 2.1 Projection of world population and economic growth *(ExxonMobil, 2019; EIA, 2019).*

While only a few countries have oil and gas resources, over a hundred countries have substantial coal reserves and many more resources remain unexplored. For these countries, coal has become the main resource for modernizing and proliferating internal modern energy supply, and ensuring energy security. For example, Asia has been leading the world's economic growth for two decades and the rate is expected to be faster over the next two decades, yet the region has very limited oil and gas resources, but holds a substantial proportion of the global coal resources. It is inevitable therefore that, for strategic and security reasons, coal will remain the primary fuel of the region in the foreseeable future, although solar and nuclear power deployments are also becoming increasingly significant. It should be noted also that the economies of many countries across all regions of the world depend critically on the production, processing and export of fossil fuels in terms of export earnings and employment: the United States, Russia, Australia, China, Indonesia, Norway, Poland, the United Kingdom, and many countries in the Middle East, South America and Africa. It is difficult to imagine that any of these countries will ever shut down or scale down production because of the environment. In fact, most of the countries are making massive investments and expanding their fossil fuel infrastructures internally and across country borders.

2.3.   ENERGY EQUITY/ACCESS

Global demand for energy is driven by two key factors: population growth and economic growth. All recent outlooks and projections over the next two to three decades indicate that both will be driven by the developing world due to rapid population growth and increasing prosperity in the regions. Increasing urbanization and emergence of a large and growing middle class in the regions will largely shape the global economic and energy demand trends. The global population is projected to increase by around 22% by 2040, the most profound growth being in Africa. The region's population is among the fastest growing and youngest in the world, increasing at more than twice the global average over the last two decades, and accounting for around 17% of the total global population. One-in-two people added to the world population between today and 2040 will be African, more than half a billion people will be added to the urban population, and the continent becomes the world's most populous region by 2023, overtaking China and India (IEA, 2019b). The pace of urbanization on the continent is unprecedented: in the last two decades the number of people living in cities has increased by 90% and the trend is projected to continue over the next two decades. China's GDP per capita is projected to triple, reaching about 75% of the OECD; India's GDP should grow by about 55% while Africa's is projected to increase by around 50%. The expanding middle class in these regions means that billions of people will aspire to improve their living conditions, and access to modern energy will be a critical enabler. Access to modern energy is a prerequisite for economic, technological and human development, and is essential for humanity to thrive because it stimulates the development of modern infrastructure that in turn grows large-scale economic enterprises, promotes urbanization, access to jobs, and raises the earning power and overall development of the people. This development transforms 'access' to 'consumption' and further stimulates demand for energy and development of additional modern energy infrastructure. Economic and social development thus tend to go hand-in-hand with energy sector transformation. Economic activities promote urbanization which in turn helps to move labor from rural agrarian subsistence economies to high-income wage economies that promote social mobility and allow people to afford modern energy and devices that use energy. Without this development, human development remains low. Many countries are energy poor, with around 80% of the world's population living in countries where average modern energy consumption is less than the United Nations' minimum of 1.9 metric tons of oil equivalent (or 100 Gigajoules of energy) per person per year. Most of the countries below this minimum are also in the lowest quarter of the United Nations' human development ranking. Furthermore, the fuel mix in these regions is dominated by traditional biofuels and waste (Figures 2.2 and 2.3).

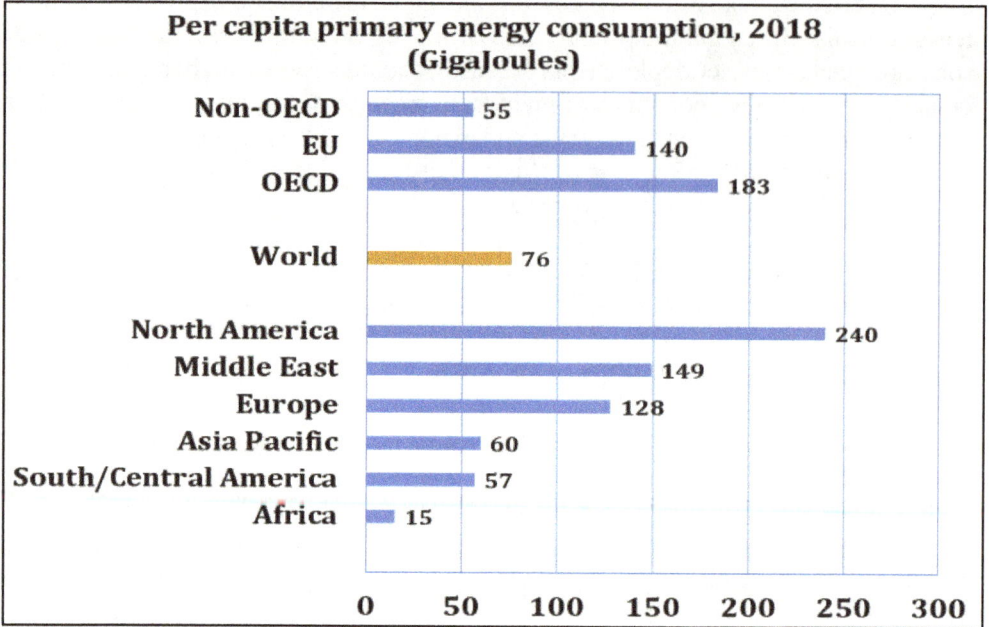

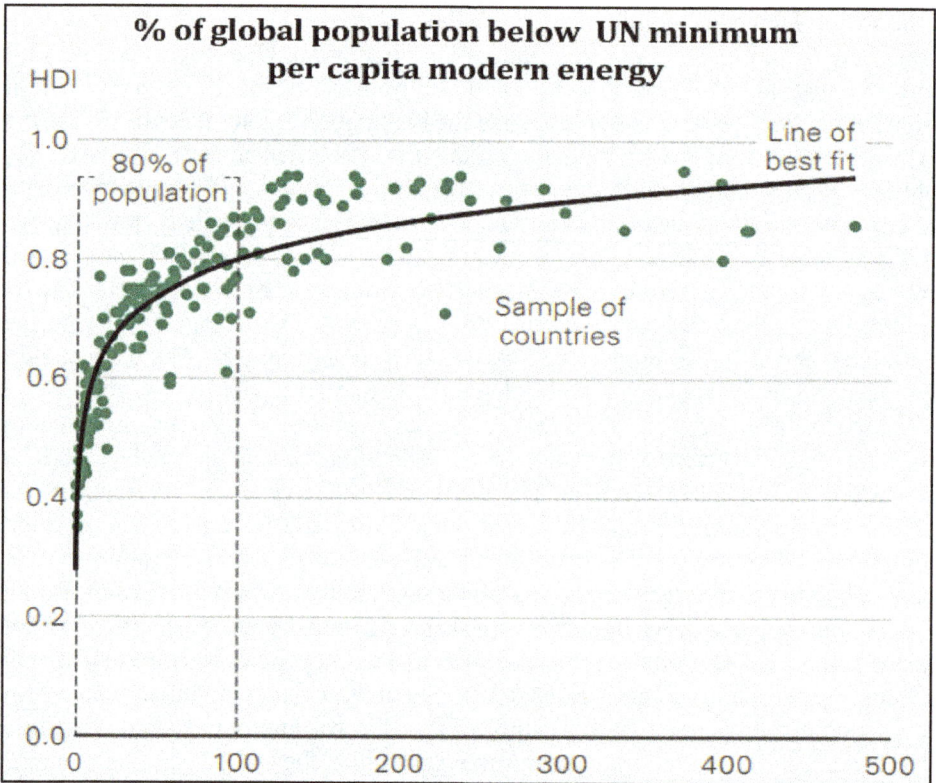

Figure 2.2 (a) World per capita primary energy consumption, 2018; (b) Human development index and energy consumption per head, 2017. *(BP 2019; UN, 2019).*

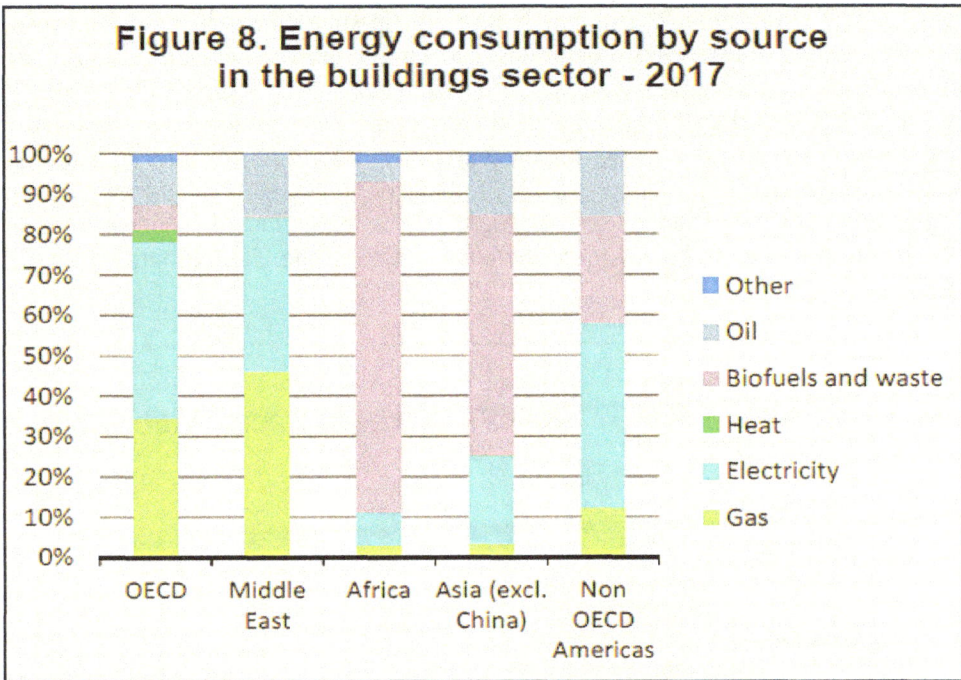

Figure 2.3 Energy consumption in the building sector by source and region *(IEA, 2019d).*

Access to modern energy, in particular, electricity and modern domestic energy, is a prime requirement for the achievement of economic growth, human development, and environmental sustainability. Modern energy is an important determinant of human health, as it plays a critical role in healthcare delivery capabilities and in the development of clean and safe household environment. In recognition of this, the United Nations set a target of 'sustainable modern energy for all by 2030' as part of the sustainable development goals adopted by 193 nations in 2015. Even before then, most nations had already developed strategies for improving their modern energy infrastructures, focusing on access to electricity. The number of people without access to electricity fell to 1.1 billion (about 15% of the global population) in 2016 from 1.7 billion in 2000. The electricity access deficit is overwhelmingly concentrated in sub-Saharan Africa where 62.5% (609 million people) are still without electricity.

Although access to electricity is the most prominent as an explicit objective of most national development strategies, access to modern household energy for cooking and heating is even more urgent because of its direct impact on household health. However, compared with access to electricity, little has changed about access to modern cooking energy over the last two decades: around 2.8 billion people (38% of the world's population) still lack access to clean household energy, same as in 2000. Developing Asia and sub-Saharan Africa dominate the global totals, and, although the numbers for developing Asia are larger, their share of the population is significantly lower than in sub-Saharan Africa (43% compared with around 80%). Nearly 3 billion people worldwide still rely on the use of traditional solid biomass while another 300 million cook with kerosene and coal, with serious environmental consequences. Around 68% live in developing Asia while another 30% live in sub-Saharan Africa. Over 80% of the expansion in world economic output in the next twenty years will occur in the developing regions, with China and India accounting for around half of the growth. The prospects for Africa are less bright (in spite of enormous wealth of potential energy resources) due to weak productivity and a highly-stressed modernization effort caused by the rapidly expanding population. Africa accounts for a low share of the world's energy demand and a high share of the population without access to modern energy services (Figure 2.4). Most of the countries in the region survive on export of raw minerals and then spend all the returns and more in importing the products of processed minerals, thereby remaining perpetually indebted to the developed nations. While the region will account for almost half of the increase in global population, the contribution to the GDP growth will be less than 10%.

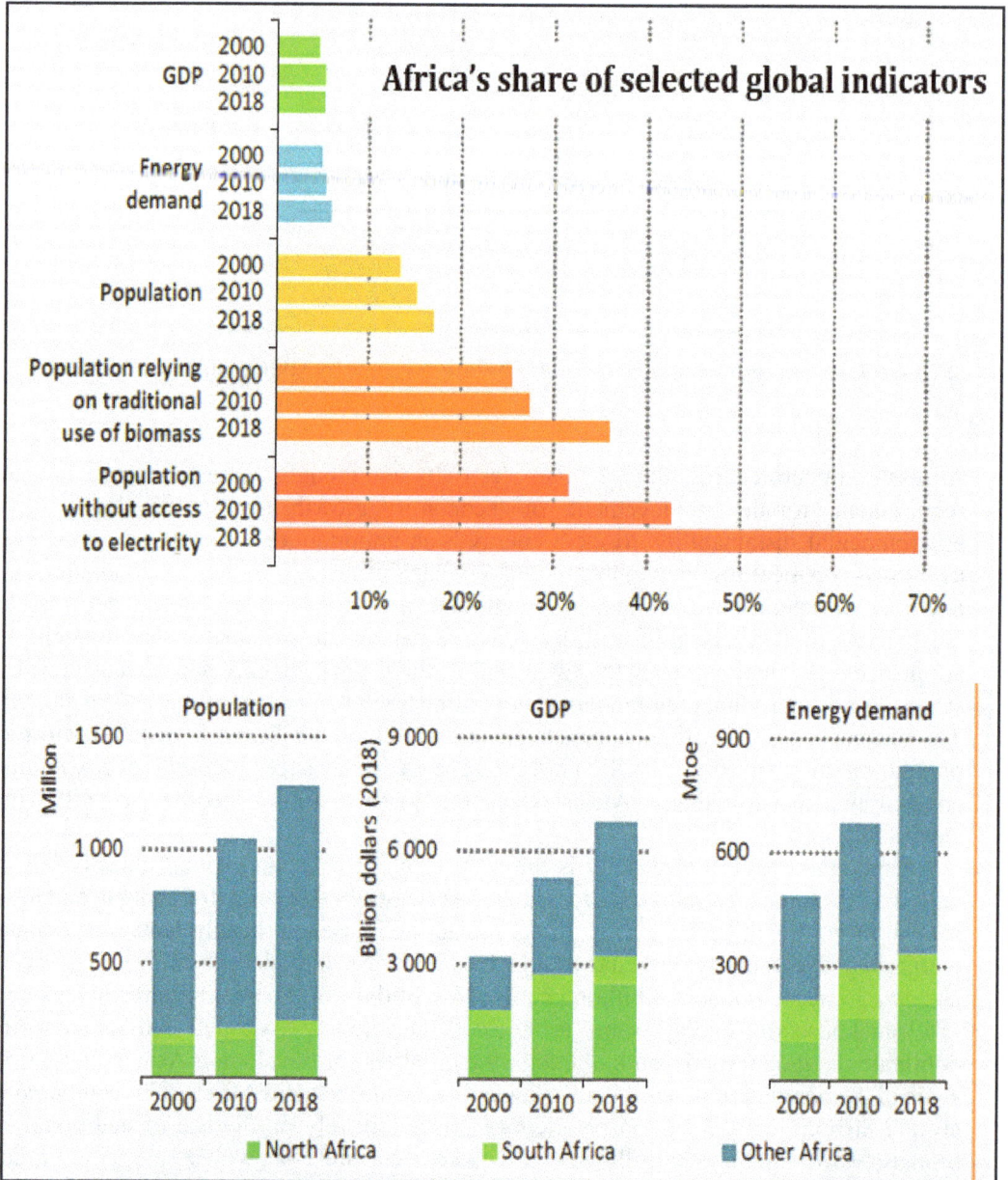

Figure 2.4   Selected human development indicators for Africa *(IEA, 2019b).*

The United Nations projects that the population of Africa will double to 2.4 billion by 2050 compared with its population in 2016, and about 50% of all newborns worldwide will be born in Africa by the end of the century. These profound demographic changes will drive economic growth, increased urbanization, infrastructure development and energy demand. Economic development that raises the level of education and economic prosperity in the region will be crucial because it promotes rural-urban migration and smaller families. Even with the projected rapid increase in access to modern energy over the text two or three decades, around two-thirds of the global population mostly in the developing world, led by sub-Saharan Africa will still remain below the minimum per capita modern energy consumption, a critical propellant for human development. All recent outlooks and projections for the global economic/GDP growth and energy demand over the next two to three decades indicate that both will be driven by the developing world due to rapid population growth and increasing prosperity in the regions. Increasing urbanization and emergence of a large and growing middle class in the regions will largely shape the future global economic and energy demand trends. Over 80% of the expansion in world economic output will occur in the developing regions, with China and India accounting for around half of the growth. The prospects for Africa are less predictable: progress in human development will largely depend on political stability of the region and strong, resilient economic policies that create a conducive environment for international investments. In spite of the fact that the region will account for almost half of the increase in global population over the next two decades, the contribution to the GDP growth will remain low.

Energy equity (access to modern energy by all) becomes sustainable, only if it is affordable, powered by rapid economic growth, urbanization, and funding capacity. Energy consumption in non-OECD countries is expected to increase by about 41% between 2015 and 2040 in contrast to a 9% increase in OECD countries (EIA, 2017). Around one billion vehicles including 229 million commercial vehicles will be added to the global vehicle population, mostly in emerging nations, and growth in industrialization will be driven largely by energy-intensive primary industries in these regions. For many countries in the group, providing adequate modern energy and infrastructure for these projected developments are very formidable and challenging goals. Many countries with low gross domestic product (GDP) and low rankings on the energy equity dimension are in a dilemma on how to promote productive access, affordability and at the same time create a conducive environment for investment in energy infrastructure. Some countries try to do this by introducing energy subsidies but this has not been very successful. While they may be beneficial in the short-term, subsidies can be costly to deploy, difficult to manage, contentious to remove, and most often do not benefit the poor for whom it is intended. Furthermore, they do not stimulate private sector investment in the long-term, and infrastructure provided mainly by the public sector tends to be unprofitable and poorly maintained. Possible solutions include the development of renewable energy-fueled distributed generation, stand-alone units for households and businesses, mini-grid/off-grid systems to serve rural communities that cannot be cost-effectively connected to the grid, and introduction of innovative policies and business models that can attract private sector investment. The policy of pay-as-you-go adopted by cell phone companies across Africa has been a phenomenal success and has stimulated strong foreign investments in the sector. The same model could work well for energy provision, especially as many are spending much more than they should on self-power generation because they use imported generators and fuels. Deployment of stand-alone solar energy systems for homes, communities, enterprises, institutions has great potential since many of the countries are located in areas of high and sustained solar energy intensity. Development has been slow because there are no coherent and sustainable policy instruments that could attract investments. Africa's energy prospects will depend on the way government and regional policies shape investment flows and regional cooperation, the continent's inclusive and sustainable vision for accelerated economic and industrial development, and the availability and affordability of modern energy sources (electricity and clean cooking) (IEA, 2019b).

2.4.   ENVIRONMENTAL SUSTAINABILITY

The environment becomes sustainable when the world is able to minimize human-related emissions to the extent that the Earth's natural control systems are largely unperturbed. This is achievable only if the world becomes carbon neutral by succeeding in sucking out of the atmosphere as much carbon as is emitted by human activities. However, the projected dynamic growth of the economies and population largely in the emerging nations over the next two decades will make the achievement of this goal difficult because they stimulate growth in urbanization, rising living standards across all regions, with more people being able to afford personal transport and appliances which require energy. Global energy demand will rise by around 25-35% and fossil fuel use will remain dominant, accounting for nearly 80% in 2040, roughly the same as in 1970 (IEA, 2019a). Most of the industrial growth in the emerging nations will be in the energy-intensive primary industries - iron and steel, non-ferrous metals, cement, chemicals, petrochemicals all of which afford little choice in terms of energy mix. In effect, delivering on the projected increase in energy demand in ways that mitigate the risk of environmental pollution will be a major challenge for many countries over the next two decades or so, especially when choice of potential primary energy sources is limited as is the case in many developing countries. Feasible options include continuing shift to less carbon intensive power generation whenever possible, and increased energy efficiency in every sector of the economy. Effective proliferation of these options across regions could stop the rise in CO_2 emissions by the 2030s despite increasing energy use. The major challenge is the fact that most of the countries in the emerging economies that will account for most of the future growth of the global population, economy, and energy demand, do not have much choice in terms of energy mix, neither can they afford to pay for high-efficiency technologies. Furthermore, these countries are increasingly becoming major suppliers of critical primary materials, semi-finished industrial components and finished consumer products to the global economy and most of the technologies involved are energy and carbon intensive, offering relatively few efficiency improvement opportunities. Over the last decade or two, the developed world has been moving away from energy-intensive manufacturing: services and light manufacturing from imported components now account for 80% or more of GDP output of many countries in the regions. In effect, the emerging world now accounts for the supply of most primary products to the global economy (iron and steel, non-ferrous metals, cement, fertilizers, chemicals, petrochemicals, pharmaceuticals, electronic components, minerals, etc.). Also, around 80% or more of consumer products in some developed nations come from the emerging nations. While this new dynamic will help the developed world to lower pollution, particularly in urban areas, there will be no benefit for climate change mitigation because the effect of human-related pollution is largely independent of the source and the negative consequences of future emissions which will come mainly from the emerging nations will impact the entire world. Furthermore, lack of diversity in GDP mix is also making the developed world increasingly vulnerable since any interruption in the flow of raw materials, components and sub-assemblies from the emerging world could quickly collapse their economies.

Climate change is a global issue and effective mitigation requires global cooperation, especially because the negative impacts of pollutions are independent of the source. In fact, many hurricanes originate from tropical waters as storms and the number formed is influenced by local weather, wind and cloud dynamics; they travel over very long distances, energized by warming oceans, transforming into hurricanes, causing havoc along the way, but the most devastating impact on touchdown occurs thousands of kilometers from source. Although most countries presented action plans at the Paris-2015 conference, recent progress assessment shows limited progress: events in most countries are driven largely by local economic, social and political exigencies and it is unlikely that this will change in the foreseeable future. Progress towards a sustainable global environment will depend largely on technology innovations which improve efficiencies across all sectors of the global economy, decarbonize energy, capture and utilize or safely store anthropogenic emissions. Emerging nations will also need help in many ways, in particular, adoption of lower carbon power generation technologies and improved efficiencies in industrial production and transportation.

2.5. GLOBAL ENERGY TRILEMMA RANKING

The world in undergoing an unprecedented energy transition, from centralized systems based on carbon-intensive fossil fuels to more decentralized and flexible low-carbon energy, driven by the twin imperatives of providing adequate energy for the world's growing economy and prosperity while mitigating climate change. Priorities and resources vary widely across nations and regions of the world and, to a large extent determine the success of each country in reconciling the conflicting trilemma of energy security, energy equity, and environmental sustainability, a prerequisite for a viable pathway to a carbon-neutral global environment. The World Energy Council (WEC) reviews country challenges and progress annually and rates them on the basis of performance, based on the three core dimensions of the energy trilemma. Ratings (A-C) are awarded on performance in each of the three cores, as well as overall scores (between 1 and 10). About 125 countries are then ranked to reflect the extent to which each country has succeeded in reconciling the triple challenge of energy security, equity and sustainability, a critical goal that promotes transition to a more sustainable, environmentally-sensitive energy future (Figure 2.5). The European Union led the world in 2019 as it had done in the previous five years, claiming nine of the top ten positions in the overall rating and five of the top ten with triple-A scores. On the other hand, only five countries in the bottom twenty are not in Africa. The success of the European Union is attributable to the unique structure of the Union, central control of environmental issues and the institution of enforceable policy instruments. The key challenge faced by sub-Saharan Africa is modern energy access, despite significant energy resources and renewables' potential. Around 65% of the total population still lack access to electricity and most still depend on traditional bio-energy for domestic needs. Most of the countries in the region are unable create a conducive business environment that could attract investment, or build the institutional capacity needed to unlock their resource potentials. Also, the fast rate of population increase severely limits the impact of investments in development infrastructure. Furthermore, the region needs to evolve unique development models that adequately address their peculiar circumstances. The region needs to adopt cost-effective development of its abundant energy and human resources along with expanded adoption of decentralized grids, distributed generation, and intra-regional cooperation.

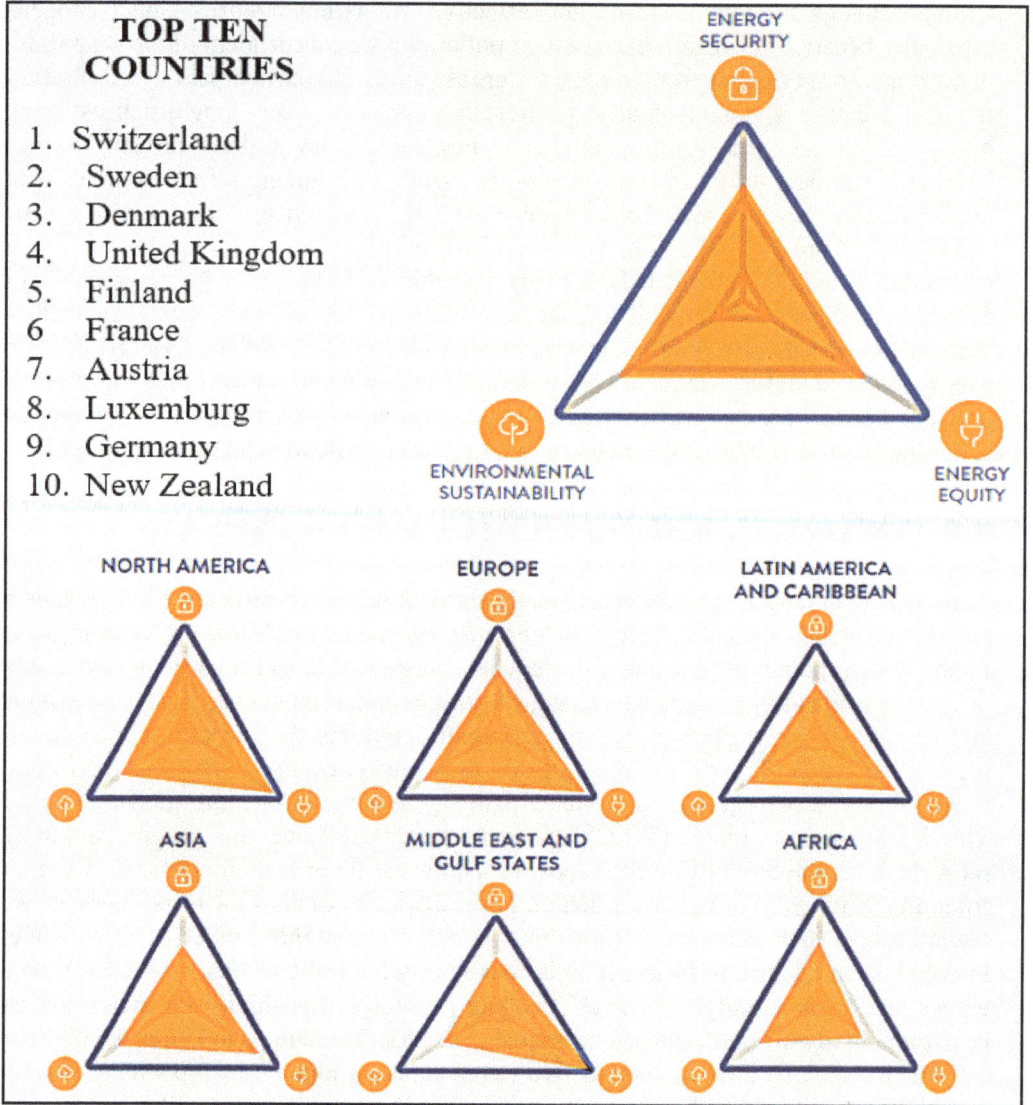

Figure 2.5 Global Energy Trilemma: sustainable environment ratings, 2019 *(WEC, 2019).*

Chapter 3

World primary energy resources and supply

3.1. INTRODUCTION

Energy is the prime mover of economic development and availability in sustainable and acceptable forms will to a large extent control the pace of world economic and human development in the foreseeable future. The world is richly endowed with many natural primary energy resources - renewables (solar, wind, hydro, biomass, geothermal, tidal, gravitational), and non-renewables (fossil coal, oil, natural gas, and nuclear) (Figure 3.1). The major challenge is how to manage these resources to provide a balanced mix to satisfy the strongly growing world energy requirements in a sustainable manner that does not continue to degrade the environment. Only four of the global primary energy resources are available everywhere, abundant, free and inexhaustible: solar, wind, gravitational and biomass energy. Hydro and geothermal energy sources are also free, but not every country has access to these resources. Fossil fuel and nuclear energy resources are non-renewable, and global distribution is inequitable, with relatively few countries holding most of the resources.

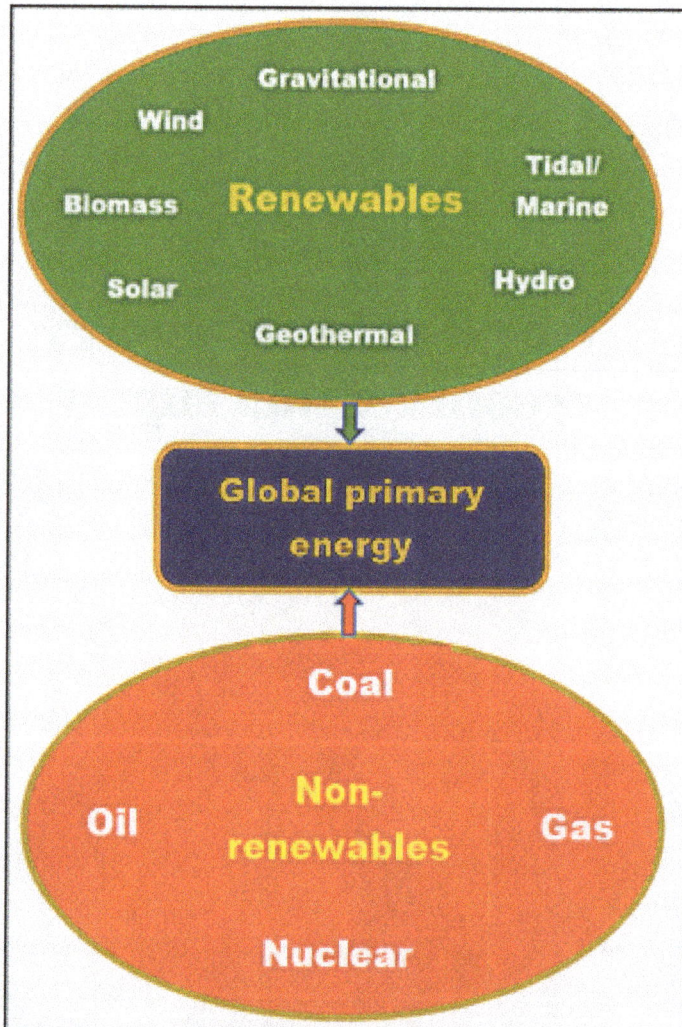

Figure 3.1 Global primary energy resources

All natural energy sources are classified as *primary energy* and, while some can be used directly - solar, wind, coal, natural gas, biomass, - most end-use energy sources are products of technological processes that convert primary energy into useable forms required for powering machinery, automobiles, etc. These are known as secondary, converted or end-use energy sources, and include petrol/gasoline, diesel, jet fuel, synthetic liquid and gaseous fuels, lubricants, heat and electricity. Without these energy sources, the rate of global economic and human development over the last 300 years or so would not have been achievable.

3.2. GLOBAL ENERGY RESOURCES AND RESERVES

The world is richly endowed with a wide range of primary energy resources. The Sun's energy is available everywhere although in varying intensities and availability; fossil fuels are abundant but inequitably distributed across regions; most regions have suitable hydroenergy sites; but relatively few have exploitable geothermal sites. Fossil fuels have dominated the global primary energy mix since the Industrial Revolution of the late 17th century and still account for over 80% currently. The global resources of non-renewable energy have not been adequately quantified because of severe limitations in current geological and geophysical exploration technologies as well as exploitation technologies, hence proven resources of non-renewable energy also known as *reserves* are believed to be only a small fraction of the total global resources which are yet to be discovered and evaluated (Figure 3.2-3.5). This explains why reserves are constantly being revised upwards regularly in spite of the increasing rates of exploitation.

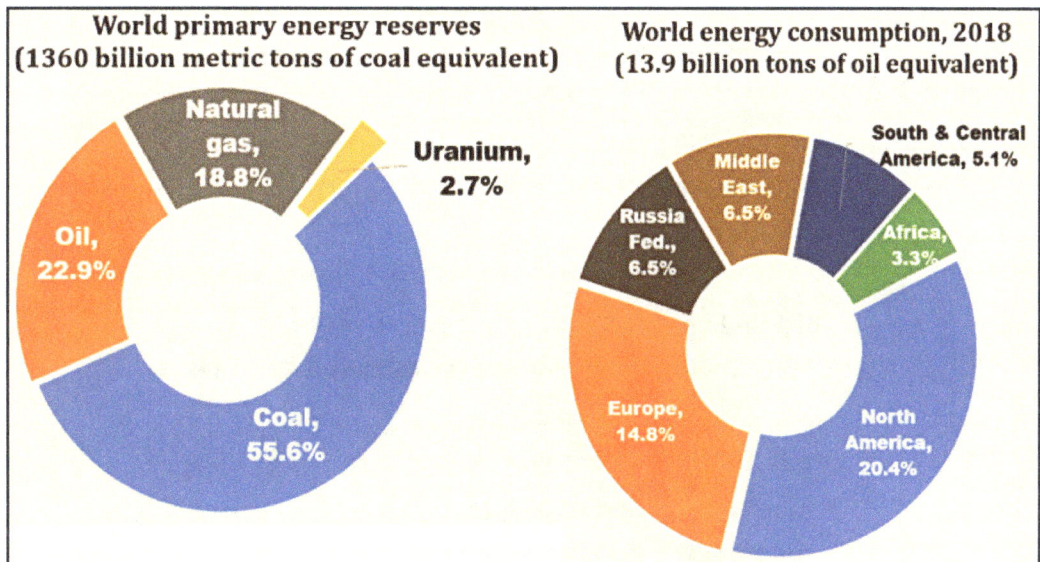

Figure 3.2 World proven primary energy reserves and consumption, 2018 *(WNA, 2018; BP, 2019b).*

Global oil reserves by region and country

Region	%
Europe	0.8%
Africa	7.2%
CIS	8.4%
North America	13.7%
South & Central America	18.8%
Middle East	48.3%
Rest of the world	61.9%
Iraq	8.5%
Iran	9.0%
Canada	9.7%
Saudi Arabia	17.2%
Venezuela	17.5%

0%   10%   20%   30%   40%   50%   60%   70%

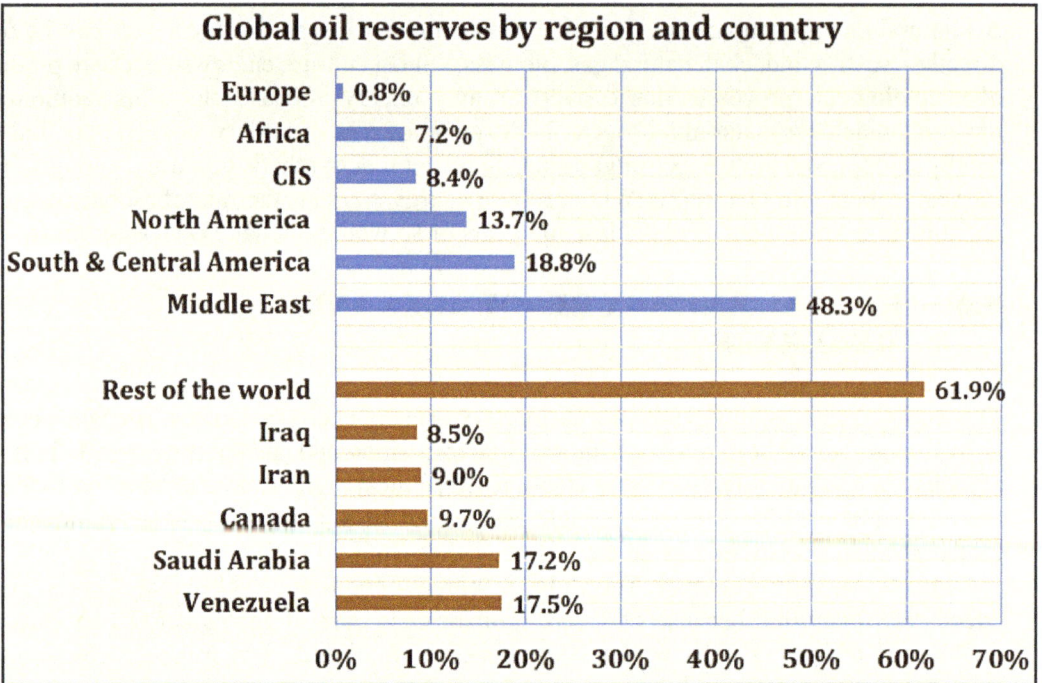

Figure 3.3   World proven oil reserves by region and country, 2018
(WNA, 2018; BP, 2019b).

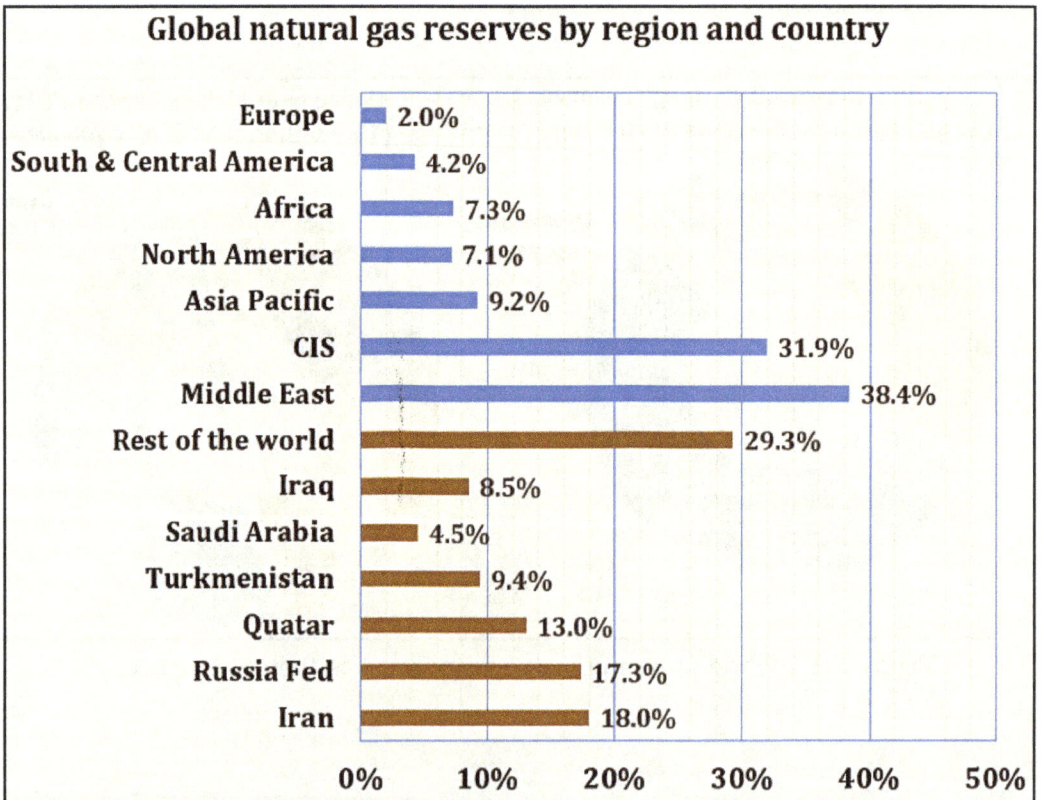

Global natural gas reserves by region and country

Region	%
Europe	2.0%
South & Central America	4.2%
Africa	7.3%
North America	7.1%
Asia Pacific	9.2%
CIS	31.9%
Middle East	38.4%
Rest of the world	29.3%
Iraq	8.5%
Saudi Arabia	4.5%
Turkmenistan	9.4%
Quatar	13.0%
Russia Fed	17.3%
Iran	18.0%

0%   10%   20%   30%   40%   50%

Figure 3.4   World proven natural gas reserves by region and country, 2018.
(WNA, 2018; BP, 2019b).

Global coal reserves by region and country

Region/Country	Percentage
South & Central America	1.3%
Middle East/Africa	1.4%
Europe	12.8%
CIS	17.9%
North America	24.5%
Asia Pacific	42.0%
Rest of the world	20.8%
Indonesia	3.5%
India	9.6%
China	13.2%
Australia	14.0%
Russian Fed	15.2%
United States	23.7%

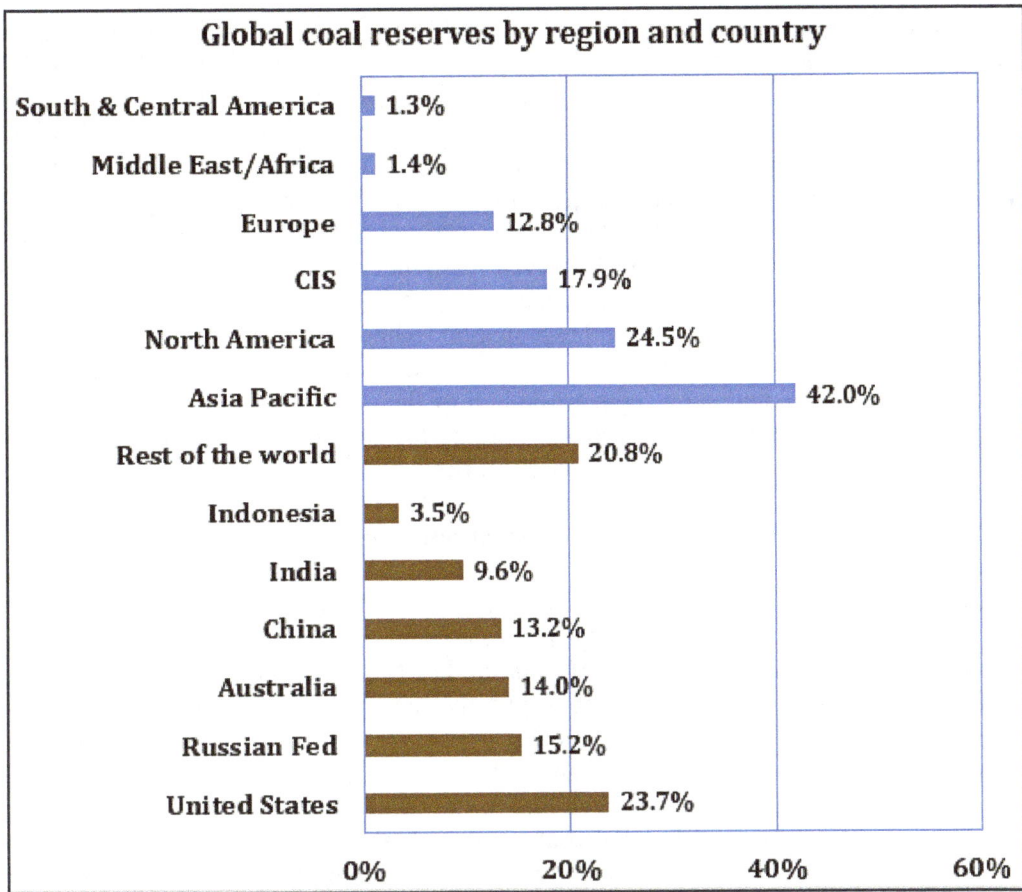

Figure 3.5 World proven coal reserves by region and country, 2018.
(WNA, 2018; BP, 2019b).

3.2.1 Fossil fuels

Fossil fuels (coal, oil and gas) were formed from pre-historic plants and sea animals that lived millions of years ago. When they died, they got buried under layers of swamp mud, ocean beds, rock and sand. With the aid of microorganisms, they decomposed slowly into organic materials and sank to deeper depths, often hundreds of meters and sometimes over a thousand meters deep, subjected to high temperatures and pressures. The formation of fossil fuels (coal, oil and gas) is a continuous though exceptionally slow process and only mature deposits which were initiated around 300 million years ago are being exploited currently. However, with the exponentially increasing rate of exploitation, mature deposits may become increasingly rare. Furthermore, the potential effects of human activities such as deforestation and ocean pollution on the natural processes of formation, especially on the initiation of potentially suitable sites (mires) and the microorganisms that break down the carbonaceous deposits are unclear. However, in view of the fact that deposits have been formed continuously for hundreds of millions of years and the potentially disruptive of human activities intensified only in the last 200-300 years or so, it is possible that many currently immature deposits will become available to future generations.

In reality, all primary energy sources are renewable, the difference being in the time scale: while it would take hundreds of millions of years to replace fossil fuels being exploited currently, solar, wind, hydro, geothermal are continuously available and are therefore classified as renewables. Fossil fuels constitute around 97% of the proved global non-renewable primary energy reserves, with nuclear energy making up the balance. On average, oil and gas reserves are adequate to meet global demand for around fifty years (up to 120 years in some countries). Coal will last longer, over a hundred and fifty years on the average (up to 350 years in some countries). However, fossil fuel resources that have been discovered and proved are believed to be only a fraction of the total global resources, due to limitations in currently available exploration and exploitation technologies. Most of the world's oil and gas reserves are off-shore and current exploration and drilling technologies are capable of no more than about 3 kilometers depth while half of the oceans which represent about three quarters of the planet are much deeper. Emerging technologies are locating new resources and providing access to reserves that were considered unrecoverable previously. For example, the recoverable coal reserves using currently available advanced mining technologies are believed to be less than 5% of the total global resources. Furthermore, deposits that were previously regarded as inaccessible or uneconomical but which meet other required criteria are now being gasified *in-situ* to generate heat and electricity.

Fossil fuels have supplied virtually all the global primary energy requirements for around four centuries, accounting for over 80% in the last fifty years or so. However, the global distribution of resources is inequitable: just five countries hold over 60% of global oil reserves (Venezuela, Saudi Arabia, Canada, Iran, Iraq); five countries (Russian Federation, Iran, Qatar, Turkmenistan, Saudi Arabia) also hold over 60% of the world's natural gas reserves; while five countries (United States, Russian Federation, China, Australia, India) hold nearly 80% of the total global coal reserves. It should be noted however that global proven reserves of these primary energy resources are believed to be only a small proportion of resources which are yet to be located, proven and moved to reserves. For example, Africa which had remained largely unexplored is increasingly coming to focus and 30% of new global oil and gas discoveries in the last few years have been in the region.

3.2.2. Nuclear energy

The inequitable distribution of non-renewable energy, the need for energy equity and security, and pressures of environmental sustainability (which implies more energy for less carbon emissions) constitute formidable challenges for most countries. Most of the growth in global primary and end-use energy demand will be in emerging countries which will also account for most of the growth in global GDP expected to double over the next two to three decades and population which is projected to increase by around 25%. For many of these countries, *energy equity* is the prime challenge and local primary energy resources, (in particular coal, the most polluting fossil fuel) will remain the first and probably the only choice. Developed nations face a different challenge: *energy security* which has become prominent since the global oil market crisis of the 1970s and energized the development of nuclear energy for peaceful applications. Nuclear power generation became well-established in the 1960s and grew at a fast pace over the next two decades, fast-tracked by the global energy crisis. For three decades, the OECD dominated nuclear power generation, accounting for nearly 93% of the total global nuclear electricity generation capacity in 2015. However, two serious accidents (Three Mile Island in 1979, Chernobyl in 1986) marked the beginning of a steady decline which reached its lowest after another accident in Fukushima, Japan in 2011. Nuclear share of global power generation declined from peak 18% in 1996 to under 11% in 2011 and, in spite of the fact that Japan has reactivated 5 of its 42 nuclear power reactors and some other countries have commissioned new plants, nuclear share of global power generation was still below 11% in 2018.

Sixteen countries depend on nuclear energy for at least one quarter of their electricity, and the United States leads the world in terms of nuclear power output, accounting for nearly a third, but the proportion in the domestic power generation fuel mix was less than 20% in 2018. In contrast, over 70% of the total power output in France in the same year was from nuclear plants. Also, many states that do not have nuclear power plants depend in part on nuclear-generated power supplied through regional transmission grids. However, the future of nuclear power generation is unclear. Japan is one of the leading countries in the deployment of nuclear generation capacity, the third largest in the world. Prior to the Fukushima accident in 2011, the country filled around a quarter of its electric power demand with nuclear electricity, but this dropped to less than 2% in 2013 and, although some plants are already back in operation and many more are being evaluated, it is unlikely that share of nuclear power in the country's energy mix will ever rise to the pre-accident level again, especially with the country's increased interest in renewable energy and investments in liquefied natural gas projects in other countries.

The United States, the world's largest nuclear power generator and producer of a third of the world's nuclear electricity from its around a hundred reactors, has shut down about a third in the last two years and initiated construction of only two new plants; United Kingdom shut down 30, Germany 29, Japan 18, although most of the plants shut down permanently still had many years of useful life. Projections for the United States indicate that the share of nuclear energy in power generation will decline from the current 19% to 12% by 2050 due largely to rises in the use of renewables and natural gas. Also, many other countries that have historically accounted for the majority of nuclear power development are de-commissioning ageing nuclear power plants and slowing down on new construction, Russia being the only exception, with 24 commissioned or planned new nuclear power plants as of 2019. However the urgent need to meet growing demand for power and global pressure to decarbonize electricity have been fueling interest in nuclear power generation in the developing world, in particular Asia over the last two decades, and many plants are either under construction or planned. Forty three plants were either commissioned, under construction or planned in China in 2019, 14 in India, 4 in Egypt. Many countries in the Middle East, Asia and Africa are also actively exploring the potential of nuclear power generation.

The fact that a nuclear plant can operate continuously for years anywhere in the world, fueled by relatively small quantities of readily available imported uranium is a major attraction as a means of rapidly increasing power supply to meet the growing local demand especially in the developing world. Also, although uranium (the primary fuel for nuclear energy) constitutes less than 3% of the total world energy reserves (in metric tons of coal equivalent), only small quantities are required to generate enormous energy: one metric ton of uranium energy is equivalent to about 20,000 metric tons of coal energy. Uranium resources are also distributed inequitably, with over half of the proven global reserves located in Australasia but this does not appear to be a problem since relatively small quantities needed can be imported and stockpiled. However, there is growing concern on the ability of emerging nations to manage extremely dangerous spent fuel rods, nuclear waste, and accidents, considering the experience of the developed nations most of which are simply stockpiling extremely dangerous waste dating back to the nuclear weapon projects of the

1940s: there is still no environmentally sustainable method of nuclear waste disposal and serious environmental issues of the nuclear accidents dating back forty years are still being dealt with today. Plans to bury nuclear waste at suitable sites have been consistently resisted by environmentalists and communities. Substantial nuclear waste has been buried or dumped in the oceans and, although ocean dumping was prohibited in 1994, the damage is already done because dumped waste which may remain toxic and potent for centuries was encased in concrete or copper canisters. Experience on the resistance of these materials to radiation in complex ocean and earth environments is very limited, around a few decades or so. It is unclear therefore how long canisters used to store radioactive waste will last before they start to disintegrate and release toxic waste into soils, aquifers, oceans. Furthermore buried nuclear waste could be disturbed and exposed by earth movements and earthquakes, posing serious environmental problems for future generations. However, China and India which will host most of the proposed new plants are nuclear powers with significant experience in nuclear technology and can handle some of the potential emergencies that are prominent features of nuclear power generation.

3.2.3.   Renewable energy

Renewable energy is the group of energy sources that derive from resources which are naturally replenished and can be used directly or converted into other useful forms of energy: solar, biomass, hydro, wind, tidal, geothermal. Technically, as discussed earlier, fossil fuels which are classified as non-renewable also fall in this category, the difference being the timescale. While it would take 350 to 400 million years to replenish the fossil fuel resources currently being exploited, energy is available on a continuous basis from the Sun, the wind, the tides, and around three quarters of the Earth's surface is covered by water which could be harnessed for hydropower or tidal energy. Also, while fossil fuel resources are concentrated in relatively few geographical locations, renewable energy resources are available worldwide, they are abundant, widely available, free and, if fully exploited could supply all the global primary energy requirements. Energy from the Sun (solar energy) is probably the most dominant of the renewable energy resources. It is available everywhere in the world, though at variable times and intensities; it can be used directly as a source of heat for drying; it can also be converted into heat or electricity through solar panels and inverters. The Sun's energy that falls on the Earth's surface daily is enormous and it is estimated that capturing less than 0.02% will provide enough energy to meet the world's total annual primary energy requirements (Tester *et al.*, 2005).

Bioenergy is a group of fuels of biological origin. The group includes wood which is perhaps the oldest source of energy used by mankind apart from solar energy. Other fuels in the group are charcoal, organic waste from forestry and agriculture, municipal solid waste, animal manure, human waste, and biofuels derived from crops including soybean, corn, sugarcane and vegetables. Bioenergy is abundant, can be sourced from organic crops and waste and used for raising steam for industrial and home heating, or for generating electric power. Biomass can also be converted to premium fuels (ethanol, propanol, butanol, etc) and chemicals. Most countries have rivers that can be impounded to produce electricity and many have sea coasts where the winds or tidal waves are sufficiently strong and consistent to drive turbines for power generation. Geothermal energy comes from the hot core of the Earth and is being tapped in many countries to heat homes, pools, and generate electricity. The Earth's gravity is not a form of primary energy but is a source of potential energy that can be applied usefully. If for example a mass is lifted up it acquires potential energy due to gravity, commensurate with its height. When it is dropped, it releases the potential energy in addition to kinetic energy from acceleration due to gravity. This energy system is used often in crushing and demolition processes. Biomass contains stored energy from the Sun in plants in form of carbon produced through photosynthesis. Animals and humans that feed on plants also excrete waste that is rich in carbon. The carbon can be converted into combustible gas in digesters and used for heating or cooking, or processed biochemically to obtain biofuels. Biomass resources can be widely regenerated on an annual timescale and some countries, notably Brazil and the United States are doing it on a large scale. Geological activities in the Earth's crust which generate enormous geothermal energy occur on a continuous basis and are becoming a significant source of heat and electricity in some countries which have suitable sites.

3.2.3.1. *Bioenergy*

Traditionally, bioenergy has been produced by direct combustion of biomass (organic matter). Using such fuels remains the primary energy source for many people in the developing world but, in view of the inefficient combustion leading to harmful emissions with serious health implications, this traditional biomass use is unsustainable. However, there are many pathways to the conversion of organic biomass to solid, liquid and gaseous forms that can more conveniently provide useful, environment-friendly energy, and possibly replace fossil fuels in many applications. Many technologies have now become available for converting biomass to biofuels which are more efficient, more environmentally sustainable, and can be utilized or transported more conveniently. Also, a wide range of biomass feedstocks can be combusted under controlled and environment-friendly conditions to generate heat and electricity. These include organic waste, municipal waste, agro-waste, crops grown specifically for energy, such as corn, soybean, wheat, sugar and vegetable oils, and nonfood crops such as ligno-cellulosic plants and grasses.

Most biofuels are made by fermentation of sugar derived from wheat, corn, sugar cane, sugar beets, molasses, etc., using microorganisms and enzymes. Starch (from corn, potato, fruit waste), vegetable oils and animal fats are also fermented to produce alcohols, mostly ethanol but propanol, butanol and biodiesel are also produced. While butanol can be used in gasoline engines directly as a replacement for gasoline, ethanol is blended with gasoline, and fossil diesel and biodiesel are also blended. Traditional use remains the largest end use of biomass and biowaste, but modern use is growing, particularly in the heat sector. Biogas (mainly methane, also known as landfill gas) that forms when garbage, agricultural waste and human waste decompose is becoming important as a source of household energy, and small digesters are in use especially in developing countries. Proliferation of biofuel use holds great promise as an effective option for decarbonizing the heat and transportation sectors of the global economy which account for 80% of the total final energy consumption: gasoline blended with bio-ethanol and diesel containing up to 20% biodiesel are already available in fuel stations in some countries; aviation fuel is being blended with biofuels; Brazil has developed internal combustion engines that can run entirely on biofuels; solid biofuels are used increasingly in generation of heat for the housing and industrial sectors.

3.2.3.2. *Solar energy*

The Sun's radiation about 60% of which reaches the Earth is the most common primary energy source both in abundance and geographical availability. The annual potential of solar energy is several times larger than the total world primary energy consumption (UNDP, 2000). Solar energy use has been boosted by significant developments in solar PV technology, notably, sharp cost reductions and strong policy support. Power generation has been the main beneficiary of solar energy, and China leads the world, with India following closely. In 2018, solar PV power generation capacity grew by 50%, with China alone accounting for almost half of this expansion. The record low generation cost per kilowatt-hour is becoming increasingly comparable or lower than generation cost of new, modern gas and coal power plants. Solar energy deployment has the greatest potential for decarbonizing power generation, because of the declining costs of solar PV, but also because of the very wide variety of deployment options, from small units that power cell phones, utility-scale stand-alone household units, to large power plants. The global PV generating capacity has grown five-fold in the last five years, due largely to dramatic cost reductions of PVs, around 50% over the period. Variability is the main challenge and capacity utilization could be as low as 15-20%, but rapid developments in battery storage technologies and advanced control systems that help integrate solar PV power with power grids will help to a large extent in mitigating this problem. Currently, even the most advanced battery banks have maximum storage capacities of around 1200 megawatt-hours, enough to power around 250,000 homes for 2-4 hours, a major constraint to the full exploitation of solar energy. For this reason, large solar power plants are usually configured in parallel with fossil fuel power plants (mostly gas-fired).

3.2.3.3 *Wind energy*

Wind energy has been used for centuries in driving corn mills, water pumps, etc. Like solar energy, wind power is available virtually everywhere on the Earth. Wind energy derives from the uneven heating of the atmosphere by the Sun and the speed of movement is controlled by the terrestrial thermal balance and distribution, the Earth's terrain, the Earth's rotation, vegetation, oceans and seas, etc., hence intensity is largely unpredictable. Wind speed is usually strongest over open oceans which cover over 70% of the planet because of fewer obstructions, and also because they pick up energy from warm waters. Strong wind passing through a propeller can generate considerable thrust, very much like in the aircraft, and can be used to drive turbine-alternator systems that generate electricity. If it were possible to harness just 5% of the available potential wind energy, it would provide electric power equivalent to the current global total primary energy needs. However, suitable sites are relatively few and variability is a significant challenge. Also, plants operating in the best locations often return no more than 15-30% efficiency. Wind speed is a critical feature of wind-powered systems because it determines deliverable energy, hence terrains with as few obstructions as possible are the best sites, and there are many all over the world. The major challenge is the intermittent nature which makes it hard to synchronize availability with demand for electricity. Also, good sites are often located in remote areas and on oceans, far from power demand centers and bringing the power to consumers can be expensive. Furthermore, because wind farms require substantial land areas and are noisy, resistance is building up in some developed countries which account for most of the investments in wind energy. Compared with fossil fuel-powered generating plants, capital investment is significantly higher, but a life-cycle cost analysis shows that wind-powered plants are cheaper because of the lower operating costs, since wind is free and there is no fuel to purchase. For this reason, wind energy has high value as a secure source but, in spite of the very substantial investments and progress made in the last decade or so, the combined contribution of solar and wind to the global primary energy consumption in 2018 was less than 2%.

3.2.3.4 *Hydropower*

Hydro-energy is a renewable resource because water is the primary fuel, and the global water cycle is controlled by the Sun. Virtually every country has potentially suitable hydro dam sites and, although capital investments can be very high, the technology produces electricity at a unit cost that is lower than any other power generation technology, renewable or otherwise, and produces no direct emissions. Also, impounded reservoirs offer a variety of recreational opportunities, notably, fishing and recreation. Hydropower is the leading source of renewable electricity generation globally, currently supplying over 70% of renewable electricity and 16.4% of the world's electricity from all sources. There has been a major upsurge in hydropower development globally in the past decade or so during which total installed capacity has grown by around 40%, but the rise has been concentrated in emerging countries where they not only provide electricity, but also water services and energy security. China has the world's highest installed hydrodam capacity, three times the capacity of the next country, United States. Although worldwide undeveloped potential of hydropower is huge, fewer new dams are being built in the developed countries because of environmental and ecological concerns. However, development in emerging regions remains strong, energized by the United Nations' sponsorship of hundreds of mini-hydro-dams for developing countries.

3.2.3.5 *Geothermal energy*

The Earth's crust stores an enormous amount of energy, believed to have originated largely from radioactive decay of minerals but also during the Earth's formation. The core mantle boundary which is about ten kilometers deep is believed to be about 4,000°C, hot enough to melt rocks and set up a geothermal gradient between the core and the Earth's surface. High pressures force molten rocks towards the Earth's surface, and eruptions of volcanoes and hot lava occur frequently in some regions of the world. Underground water that comes in contact with the heat becomes heated and penetrates to the Earth's surface as warm springs, and when hot enough, they heat homes, swimming pools, etc., as has been the practice from early times. In some locations the temperature could be as high as 380°C, enough to produce superheated steam which can be captured for electric power generation. It is estimated that the upper ten kilometers of the Earth's crust contains 50,000 times as much energy as found in all the World's fossil fuels. Geothermal energy is abundant in areas located around the boundaries of tectonic plates which make up the Earth's outermost shell also known as the lithosphere. Countries which are rich in geothermal energy include North and Latin American countries, African countries, Japan, Australia and Russia. Geothermal energy has been used since the ancient times as source of domestic hot water and space heating, and, currently, is being used extensively for space and industrial process heating, heating of pools and greenhouses, and water treatment in around 70 countries. An increasing number of countries are also tapping geothermal energy for power generation.

3.3. GLOBAL PRIMARY ENERGY SUPPLY (TPES)

Energy plays a key role in virtually all aspects of human development: economic well-being, health, education, safety, consumer products, etc. Increasing prosperity across all regions of the world is driving demand for even more energy and increasing pressure on all countries to expand energy infrastructures. On the other hand, energy use is degrading the environment and enhancing climate change and most countries face the dilemma of providing adequate energy in an environmentally sustainable manner, especially when local resources are limited. Fossil fuels have filled the bulk of the global primary energy demand for hundreds of years and, in view of the slow pace of phasing in lower carbon energy, the dominance of fossil fuels will likely continue for decades. The inequitable geographical distribution of fossil fuel resources and prime role in the global economy have made production and trade potent variables in geopolitics. Furthermore, it has made agreement on environmental issues associated with energy use very problematic since most countries across all economic regions of the world prioritize internal exigencies. For many, coal is the only major source of primary energy, and this explains why many new coal power plants are being commissioned or planned particularly in the emerging economies but also in some developed countries in spite of the poor environmental credentials. Energy demand is growing in all countries of the world, the primary drivers being the rapid growths in the global economy and population, and most countries are faced with the complex problem of providing adequate energy in a secure and environmentally sustainable manner. The global energy sector is being transformed rapidly: new technologies in all areas of exploration, production and utilization are making significant impacts in all sectors of energy technology, from exploration to production and end-use. New exploration technologies are moving resources to reserves, and deployments of new technologies in power generation, industrial processes, transportation and building are raising utilization efficiencies, decarbonizing energy, and lowering pollution emissions. Fossil fuel share of the global total primary energy supply (TPES) in 2017 was over 80%, with oil accounting for the largest share of 32% (Figure 3.6). Non-OECD countries accounted for around two-thirds of the demand, with Asia accounting for over one third, and China alone consuming 22%. Oil is refined into gasoline, diesel oil, jet fuel, etc., and used mainly by the transportation sector; coal, gas, nuclear fuel and renewables are used mainly for power generation; the industrial sector uses coal, gas and electric power as energy sources, and coal, oil and gas as feedstocks for the chemical and petrochemical sub-sector.

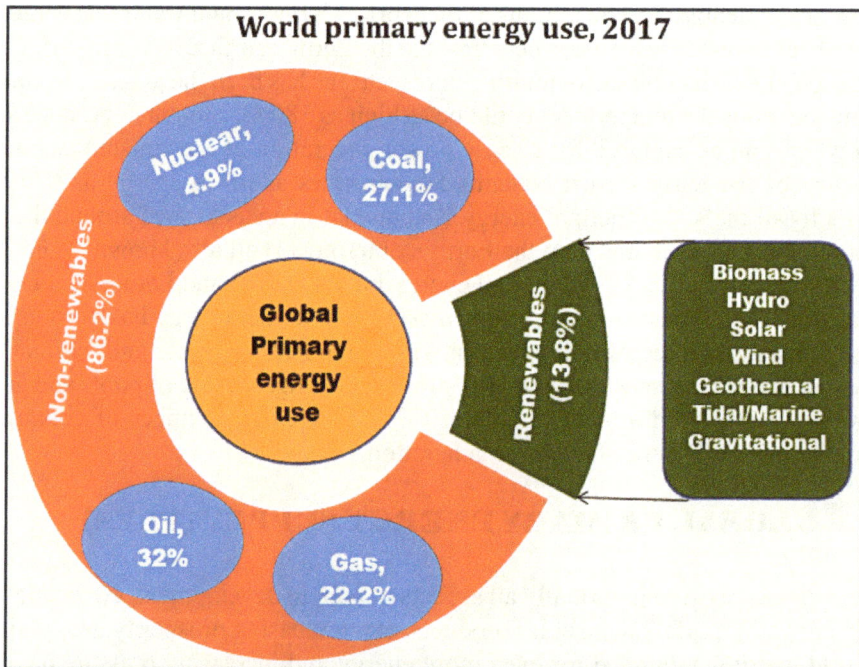

Figure 3.6a World total primary energy supply (TPES) by fuel in 2017 (13,972 Mtoe). *(Data from IEA, 2019a).*

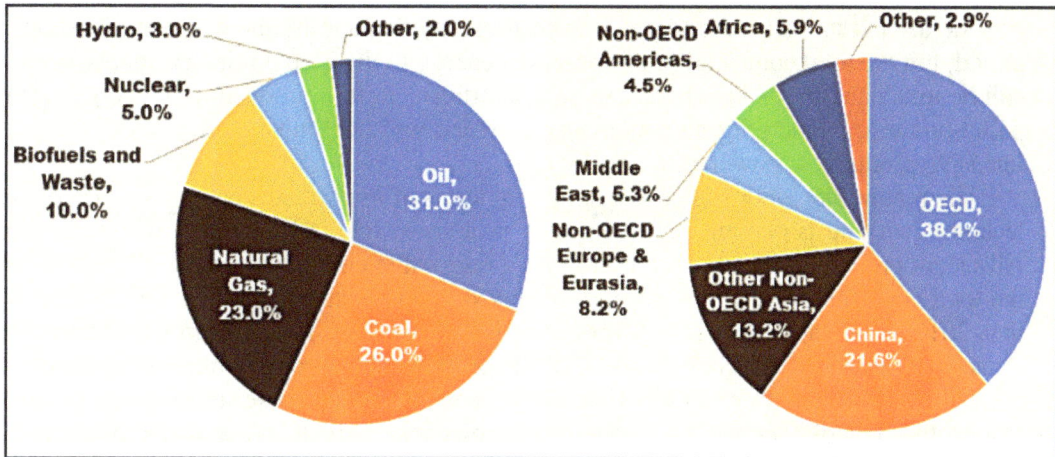

Figure 3.6b World total primary energy supply (TPES) by fuel and region in 2017 (13,972 Mtoe). *(Data from IEA, 2019a).*

Primary energy demand is projected to grow by about 35% by 2040, broadly based across all end-use sectors, with the industrial sector accounting for around half of the overall increase. Renewables will grow faster than any other primary fuel source, accounting for about 40% of the expected total growth in primary energy demand over the next two decades. Around two-thirds of the increase will be used to generate power to meet the exponentially increasing demand for access to power in the developing world. Over 90% of the primary energy end-use is for combustion to produce heat, power, and fuel internal combustion engines which drive automobiles, airplanes, ships, etc., the balance (around 6%) is used as feedstock for the chemicals and petrochemicals industry. However, non-combusted use is growing rapidly and is expected to become the largest source of fossil fuel demand growth in the next two decades in view of the rapidly rising demand for the end products across all regions of the world: industrial and pharmaceutical chemicals, polymers, fertilizers, etc. Several recent projections show that the dominance of fossil fuels in the global primary energy mix is unlikely to change very much over the next two to three decades although there will be significant changes in fuel mix across regions and economic sectors.

Economic and human development depend critically on access to a range of energy choices. The global economy is expected to at least double over the next twenty years and much of the projected growth is driven by emerging but rapidly developing economies, with China and India accounting for around half of the increase. Growth in world economy requires more energy and about 30-35% growth is projected over the next two decades. On the contrary, growth rate in demand by the developed economies has been slowing down and the trend is expected to continue in response to the adoption of more efficient technologies, and movement away from high energy intensity production (primary metals, chemicals, cement, etc.), a sub-sector now dominated by the developing countries. For decades fossil fuels have supplied most of the global demand for primary energy, accounting for over 81% of the total demand in 2017. The fossil fuel share in 1973 was 86.7%, a drop of only 5% in nearly five decades. Global climate change mitigation effort to decarbonize energy is intensifying, but all projections published recently indicate that there will be little change in the dominance of fossil fuels over the next two decades, still accounting for nearly 80% of the total primary energy supplies in 2040 and beyond. However, there will be a significant adjustment in the fuel mix and non-fossil fuels (renewables, nuclear and hydro energy) will provide around half of the increase in global energy demand by the end of the projection period, but the total contribution of non-fossil energy to the global primary energy use will still be less than 20%. Also, there will be a significant shift in production of energy-related pollutions to the emerging nations many of which will host most of the world's energy-intensive production, powered mostly by coal.

The renewable sector has grown rapidly in terms of investments and emergence of new technologies, and is playing an increasingly prominent role in heat and power generation. All recent projections show a rapid growth in renewable energy use across all regions, but the total contribution to global primary energy over the next two decades will still be less than 20% (up to 25-35% of energy for power generation), with hydroenergy accounting for around two-thirds of the contribution. All the outlooks show continuing modest growth in oil, with gas growing more rapidly than oil and coal. There are some significant differences among the outlooks, especially in fuel mix projections, which reflect differences on key assumptions, such as: the availability and cost of oil and gas supplies; the speed of deployment of new technologies; the pace of structural change in China; and the impact of energy and environmental policies on such issues as the speed of transition to lower carbon economies. However, It should be noted that all projections on energy resources, demand and supply are subject to many unpredictable variables, notably the dynamics of the global economy, global oil availability and pricing, growth rate and pattern of the transportation sector, the emergence of new technologies, unforeseen and unpredictable disruptions of global supply and demand. For example, the rapid decline in natural gas prices over the last few years due to increasingly rapid deployment of hydraulic-fracturing technologies was unforeseen just ten years ago; but it has disrupted all forecasts on global primary energy supply, stimulated investments in gas-fired power plants and slowed down investments in renewable energy development (IEA 2019c). On the other hand, declining prices of natural gas due to oversupply are having negative impacts on profitability of current operations and future investments in relatively expensive hydraulic fracking technologies. This explains why there are significant variations in nine different projections published by various organizations. The current lockdown of the global economy as a result of Covid-19 virus pandemic was unforeseen but has virtually collapsed global fossil energy demand in just weeks, providing an unprecedented opportunity for environment-friendly structural transition in the global primary energy industry.

3.3.1   Fossil fuels

Fossil fuels (oil, natural gas and coal) have supplied the bulk of global primary for centuries, accounting for over 81% of the global energy demand (roughly the same for the last four decades), and 70% of the world demand growth in 2017. Demand increased by 2.1%, more than twice the growth rate in 2016. For decades, efforts have been made to extend the global energy mix to include less carbon-intensive energy - renewables and nuclear - but the rate of deployment has been slow for many reasons, some of which have been discussed in an earlier section. All recent projections show that the situation will not change significantly over the next two decades or so, although renewables, in particular, solar and wind will play a much more prominent role in power generation. Fossil fuels drive most of the important sectors of the economy, just as the economy drives fossil fuel use and stimulates more investment in infrastructure, and recent projections indicate that the fuels will remain dominant in the foreseeable future. Oil is the world's leading fuel, accounting for around 32% of total global primary energy consumption in 2018, and the transportation sector is the primary consumer: oil filled about 90% of the fuel demand by the sector, and the prospects of viable substitutes over the next few decades are not very bright. Natural gas and coal are used mainly for power generation but gas is also used extensively in industrial processes, commercial and public services, domestic heating and cooking, while coal is the primary fuel for iron and steel, cement, chemicals and petrochemicals production. All the fossil fuels are also precursors to many non-energy products - chemicals, plastics, fertilizers, drugs, etc. Various projections indicate that, although there will be significant changes in fuel mix across regions, fossil fuels will still dominate primary energy supply in 2040, accounting for nearly 80% (Figures 3.7-3.10).

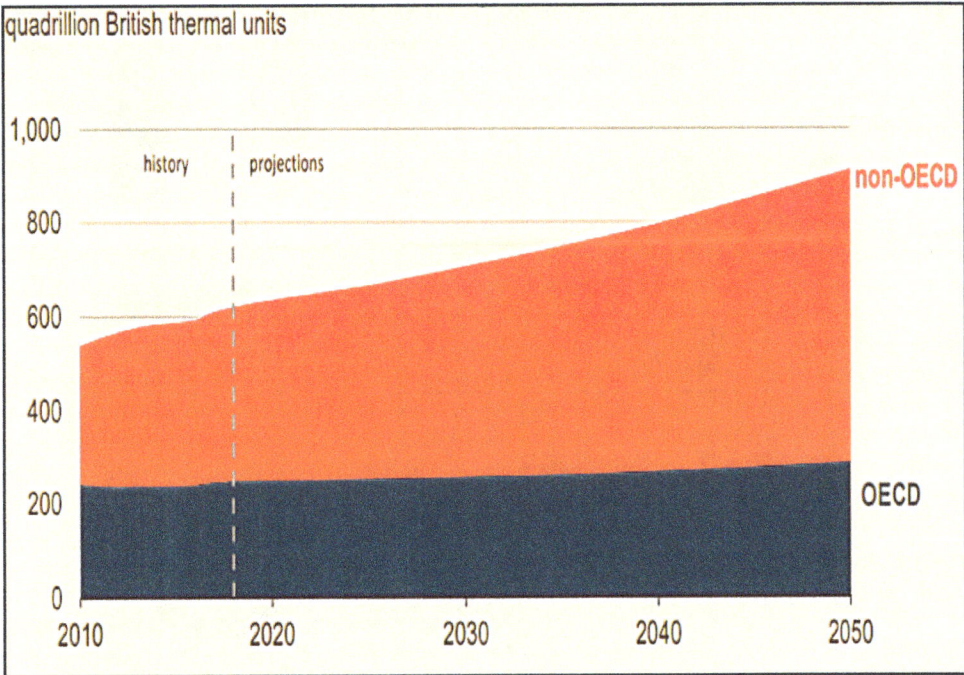

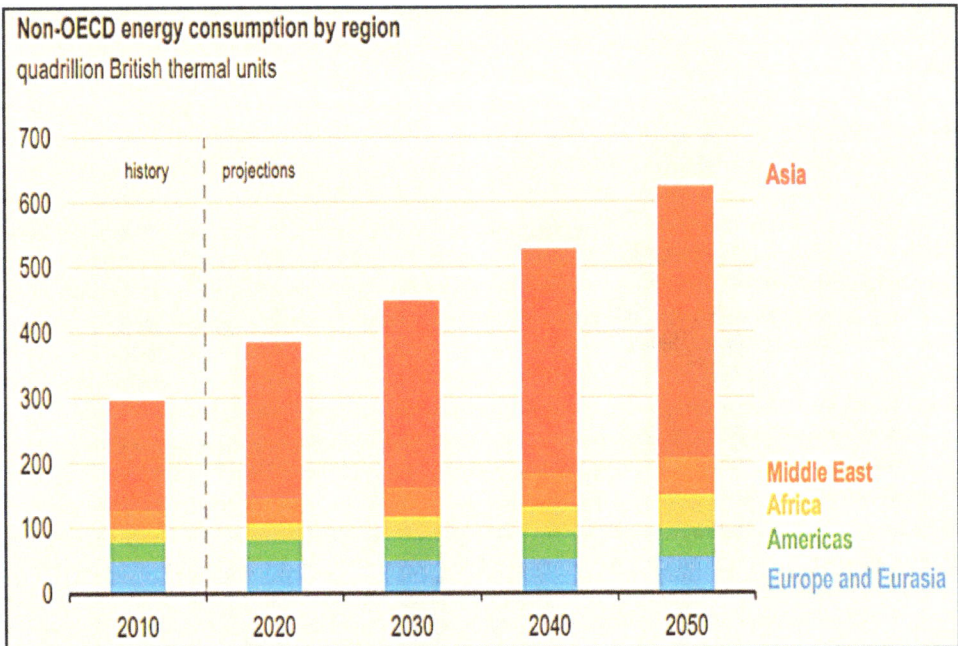

Figure 3.7 World primary energy supply projection by region *(EIA, 2019).*

Primary energy consumption by energy source, world

quadrillion British thermal units

share

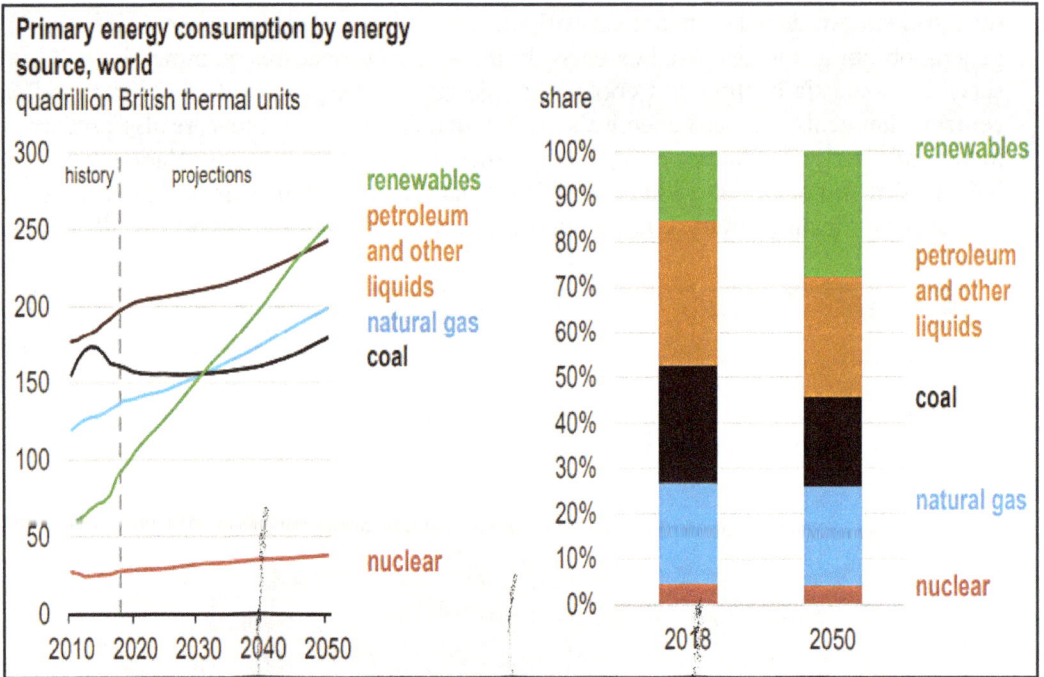

Energy consumption by sector

quadrillion British thermal units

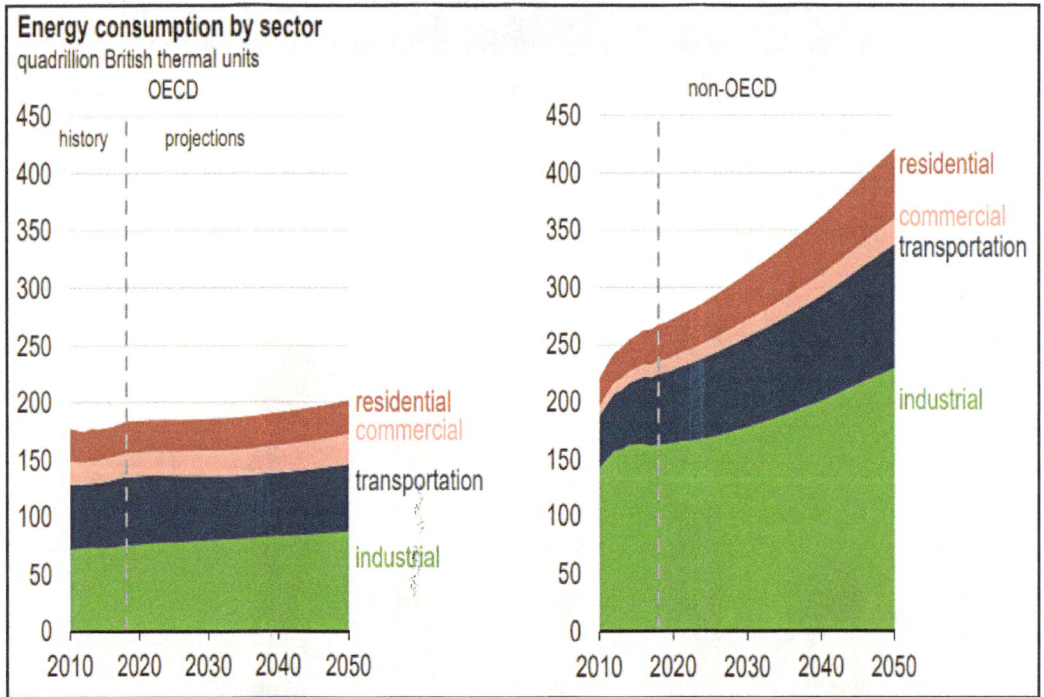

Figure 3.8 World primary energy use projection by fuel and economic sector *(EIA, 2019).*

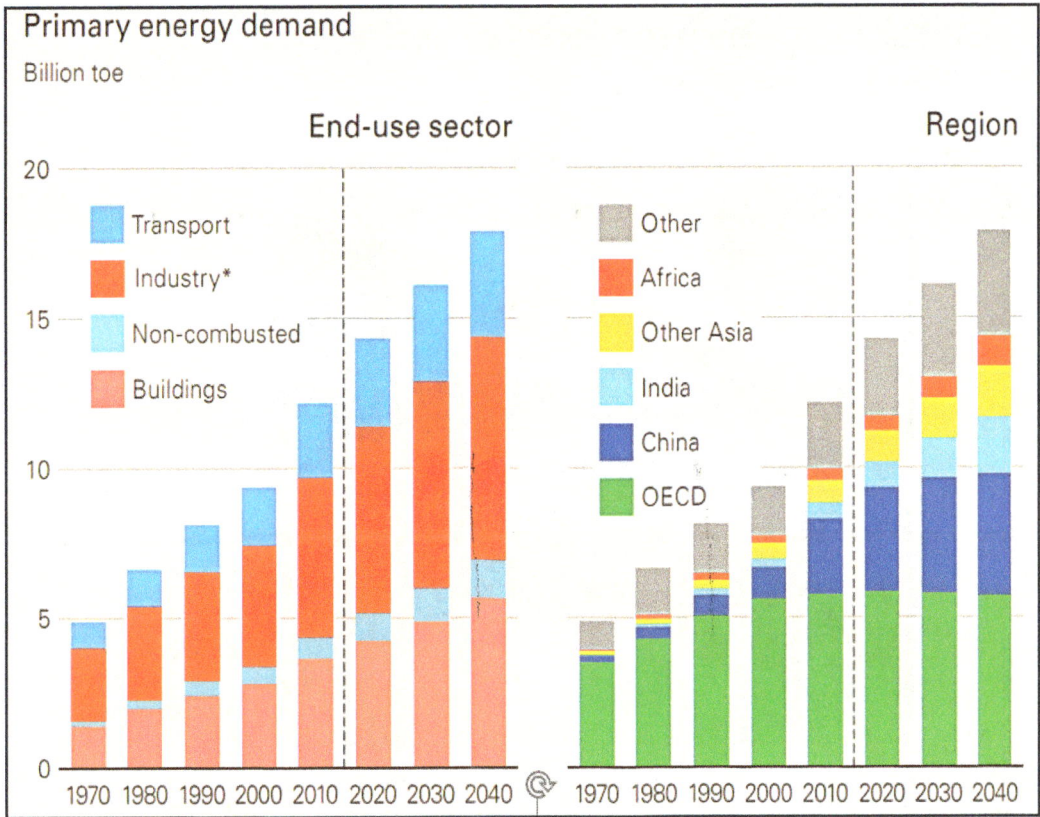

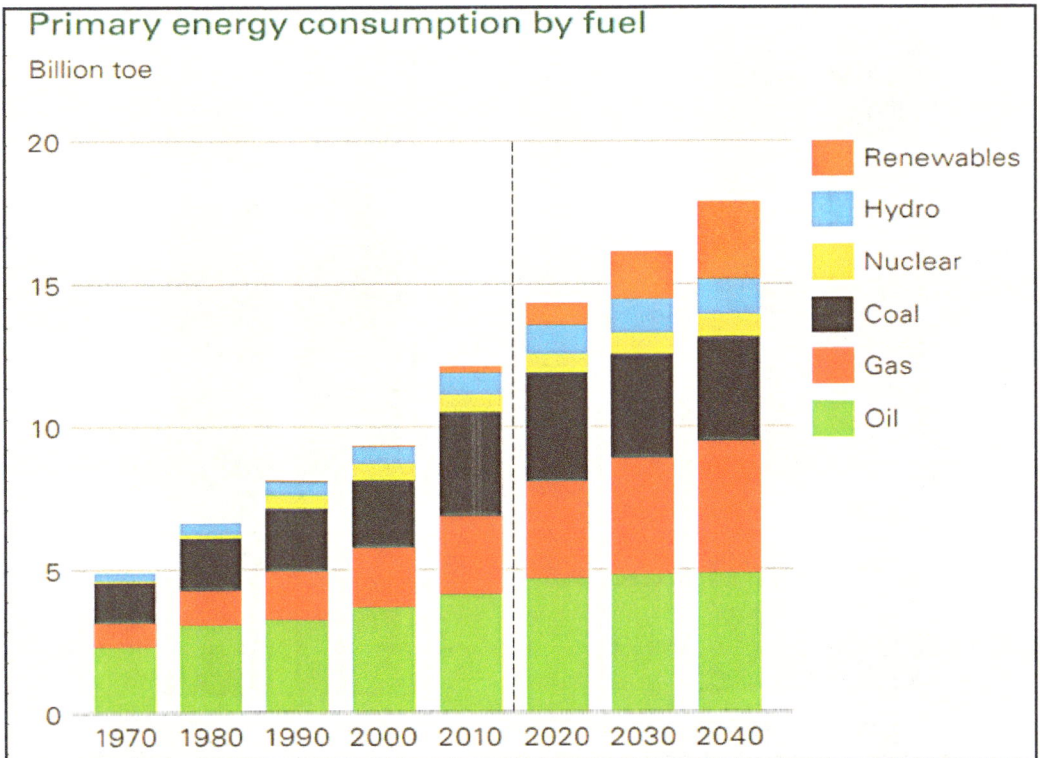

Figure 3.9 World primary energy use projection by fuel and economic sector *(BP, 2019a).*

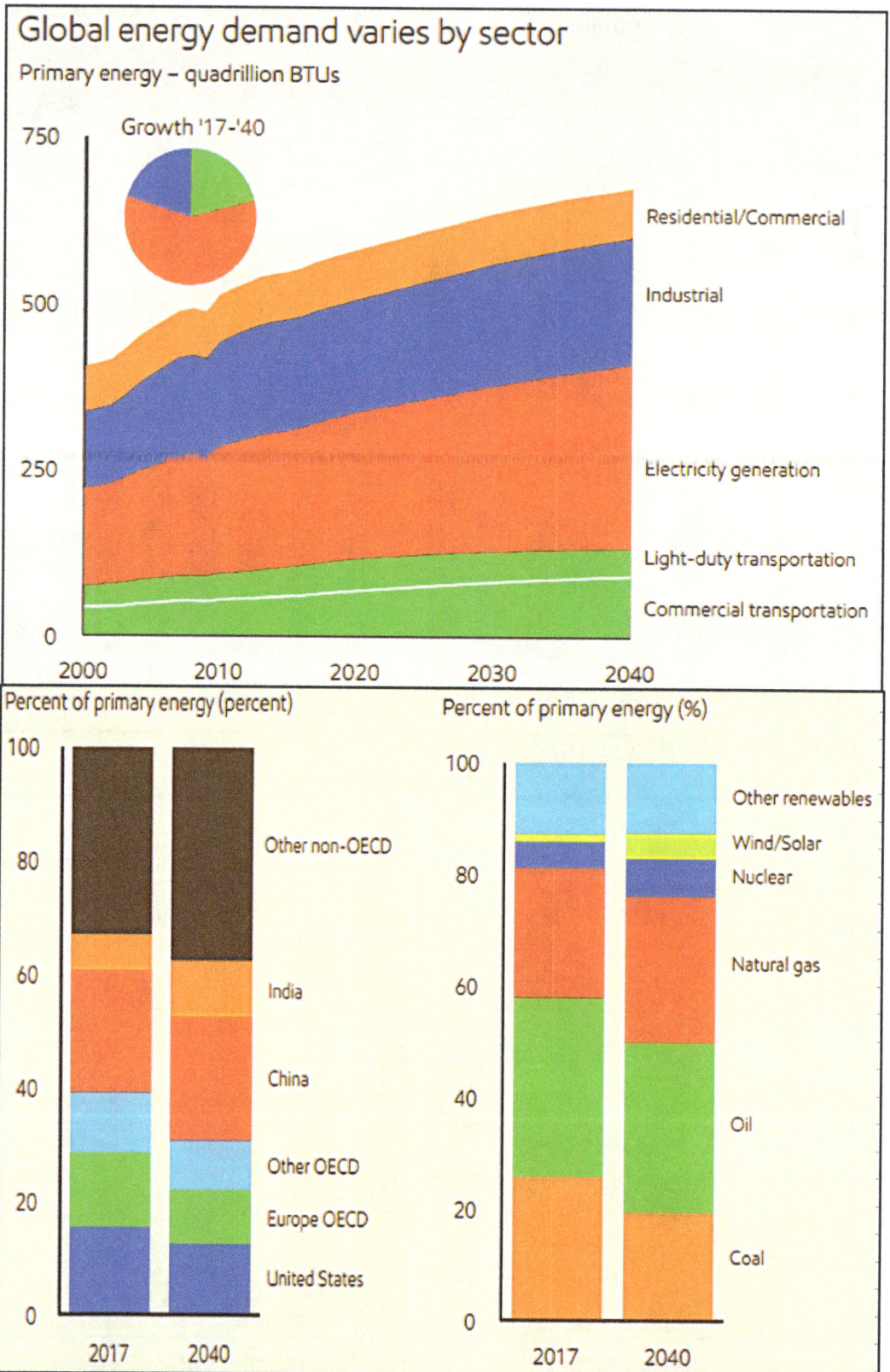

Figure 3.10   World primary energy use projection by economic sector, region and fuel
(ExxonMobil , 2019).

3.3.2   Nuclear energy

Before 1954 when the first nuclear electricity plant was commissioned in the United States, uranium was used primarily for weaponry. Over the next fifty years or so, the number of installations for power generation grew very rapidly and nuclear plants were supplying nearly 20% of the global electricity in the late nineteen eighties. Development was bolstered by the oil crisis of the early 1970s which brought energy security to the forefront of global concerns. Nuclear power generation has been dominated by the industrial world for decades, accounting for over 90% of the total nuclear electricity generation from 450 reactors in 2018. The United States has the world's largest nuclear generation capacity but its contribution to the country's power generation fuel mix is below 20%. However, nuclear power plays a more prominent role in other countries, notably France where it accounts for three quarters of power generation capacity and many other neighbouring countries that tap power from French grids.

Most of the nuclear power plants in Europe and the United States are old and nearing end of life, many are being de-commissioned prematurely, and there are no coherent plans for replacement. However, interest in nuclear electricity is growing rapidly in the emerging world: most of the 55 reactors that were under construction in 2018 were in the regions. Nuclear power share of global power generation decreased for the tenth year in a row to about 10% in 2018, yet still accounting for nearly a third of the world's low carbon electricity production. The low carbon credentials of nuclear power are being negated by increasing concerns about the competence of the world to manage nuclear disasters and nuclear waste. Almost eight decades since the first substantial use of nuclear power, the world is yet to figure out how to safely dispose of the toxic waste which is being stockpiled on temporary sites all over the world and could remain potent for centuries. Furthermore, the three major nuclear accidents in the last forty years or so have revealed serious flaws in the preparedness of even the industrial world for the management of nuclear accidents. The serious fallouts are still being dealt with in real time and the sites are still being de-commissioned and de-contaminated. Renewable energy is considered the low carbon power generation energy of the future that could replace nuclear power generation, however, in spite of policy-driven rapid expansion of the world's renewable capacity, fossil fuels, especially coal have remained dominant in global electric power generation. Also, although new renewables (wind, solar, geothermal power) have surpassed nuclear power in total installed capacity, their share of actual electricity generation is less than one third of that produced by nuclear power because of their intermittency (IAEA, 2017).

3.3.3   Renewable energy

Renewable energies are natural resources that can be converted to useful energy - solar, wind, biomass, tidal, geothermal, hydropower. They are the fastest growing primary energy sources used mainly for steam raising and power generation, meeting a quarter of global energy demand growth in 2017, with wind energy accounting for 36% of the growth in renewables-based power output (IEA, 2018). China and the United States led this highest growth ever, contributing around 50% of the increase in renewables-based electricity generation, followed by the European Union, India and Japan. In spite of the unprecedented growth rate in the last few years, renewables accounted for only 15% of the global primary energy supply and accounted for only 26% of global power generation in 2018. Around two-thirds of the share in power generation was accounted for by hydropower while the balance was shared equally between other modern renewables and traditional biomass (IEA, 2019c). The combined contribution of solar PV and wind energy to total world electricity generation was only about 6%. Most recent projections show that renewable power capacity will grow by about 50% over the next five years, led by solar PV which will account for around 60% of the expected growth while offshore wind will account for around 25%. This increase of 1,200 GW is equivalent to the total installed power capacity of the United States today. Although growth rate of renewable energy is sensitive to many variables, in particular, the fluctuating market prices of competing fossil fuels, most projections estimate that renewables will account for 25-35% of electric power generation by 2040, but some optimistic projections predict contributions of over 50% by 2030 (IRENA, 2020). However, there will be significant variations between countries, with China increasing renewable share by close to 40%. Hydroelectricity has dominated the renewable energy power generation for decades but its contribution is projected to reduce significantly due to increased use of solar and bioenergy and environmental concerns in the developed countries. Solar and wind power contributed less than 2% to the global primary energy supply mix in 2018 and only about 8% to global power generation. The intermittent nature of both resources is a key issue: the load factor is low, 15-40% compared with 75-90% for thermal plants. This means that they can only be deployed as stand-alone power systems for utilities backed by batteries, but need to be deployed as an integral part of a fossil, nuclear or geothermal energy grid system.

3.3.3.1   *Solar energy*

Solar energy is the most abundant global energy resource and is available everywhere. It can be converted to heat for use in steam raising, power generation and many other applications. If just 0.1% of the Sun's energy that reaches the Earth can be converted into heat at an efficiency as low as 10% it could supply four times the global electric power demand. Solar radiation also has indirect impact on other forms of renewable energy - wind, biomass, marine, hydro energy, etc. Solar energy may be concentrated by collectors and used to heat water and produce steam for home/industrial heating or for power generation (solar thermal systems); or converted directly into electricity by photovoltaic cells (solar PV systems). The later technology is the easier and much more common because it can be deployed in small stand-alone modules or integrated into very large plants. Rapid developments in solar PV and concentration technologies are making it possible to deploy solar electricity plants even in locations where radiation intensity is low or variable. Solar radiation is available in the appropriate intensity and for adequate duration in relatively few global locations, mostly in regions hosting emerging countries that do not have the wherewithal to harness and deploy solar-powered power generation on any massive scale. One major exception is Morocco where the world's largest solar thermal power plant was commissioned recently.

Solar energy capacity has increased by approximately 60% over the last five years, rising to 486 GW in 2018. China leads the world, with four of the world's ten largest installations. The other leaders are United Arab Emirates, India, United States and Mexico. Most of the growth in solar power deployment over the next two decades will be stand-alone and rooftop systems for small establishments, utilities, rural communities and homes across all regions of the world. The contribution of solar energy to global power generation is still very low, only 1.8% in 2018. However deployment of solar power is growing rapidly and projections indicate that the total contribution to global power generation could double by 2023. Although the cost of PV cells has dropped drastically in recent years, variability and low load factor (around 20%) remain formidable issues. However, advances in storage battery technologies could reduce this problem significantly, especially for utility deployments.

3.3.3.2 *Wind power*

Wind is available everywhere in the world and if just 1% could be effectively utilized, it could provide electricity equivalent to the capacity of all global electric power generating plants in the world today. However, intensities vary widely in many locations: there are relatively few locations where the wind is sufficiently strong and available for enough time to power electric generation plants. Wind power is useful only when the intensity and consistency are sufficient to drive a wind turbine, and the best locations are offshore mostly in remote locations where obstructions are minimal. Wind power contributed 4.5% to the total global electric power generation in 2018, although there were significant variations between countries. For many countries, wind power has become a pillar in their strategies to phase out fossil and nuclear energy in power generation. For example, wind power produced 43% and 13% respectively of Denmark's and Germany's electricity supply in 2018, and an increasing number of countries - Ireland, Portugal, Spain, Sweden, Uruguay - have reached double digit wind power share. Around 3,000 wind-powered generating plants are operating offshore in 15 countries and many more are planned or under construction. Seven countries accounted for three quarters of the total global wind power generation capacity in 2017, with China alone accounting for one-third. Like solar energy, wind energy is making a significant contribution to supply of clean, secure, reliable electricity in many countries, particularly, OECD countries. However, the contribution to the total global final energy supply is less than 1% largely because of variability and the initially large investment required.

3.2.3.3 *Hydropower*

Hydropower is available in over a hundred countries and is supplying a significant amount of electric power throughout the world. Hydropower currently accounts for around 70% of the total contribution of renewables to the total global final energy use. The global average of contribution to electric power generation is around 15% but several countries depend on the resource for over 50% of local power demand. Just five countries - China, United States, Brazil, Canada and India accounted for about half of the total global capacity in 2018. Drivers for hydropower's growth include a general increase in demand not just for electricity, but also for qualities such as reliable, clean, low-carbon and affordable power as countries seek to meet the carbon reduction goals set out in the Paris-2015 Agreement. Hydro-dams in emerging countries also serve as municipal water sources, fish farming and recreational facilities. There has also been a significant increase in pumped storage capacity in the last few years, a growing recognition of hydropower's role in enhancing the use of variable renewables (solar and wind) in more predictable power generation - surplus energy from these sources are used to pump water to higher ground, and the water is released as in a normal hydrodam to generate power when needed. Hydropower has great potential in developing countries and the United Nations is pioneering the installation of hundreds of small hydro projects in many emerging countries. However, there is increasing resistance to proliferation in developed countries due to environmental and ecological damage by large dams.

3.3.3.4 *Bioenergy*

Bioenergy refers to energy generated from various feedstocks of biological origin. fuelwood, charcoal and agricultural residues often referred to as traditional biomass have been used to generate energy from early times. However, although there has been a significant mix diversification in the last two decades or so, with the wide availability of industrial biowaste and crops which are grown specifically for conversion to energy (sugarcane, corn, soybean, sorghum, etc.), traditional biomass still dominates. Biomass can be converted into a wide range of energy sources including heat for power generation and heating, liquid and gaseous fuels. The major drawbacks that have prevented wider use are the logistics for collecting and processing biowaste and the need for massive land areas and sufficient water for the cultivation of feedstock which are also important human and animal food items. Brazil and the US lead the world in the production and utilization of biofuels from starch or sugar derived from corn, soybean, sugar cane, cassava, etc., while the main focus of Europe is on the production of biodiesel from oils and plant/animal fats (soy, rapeseed, palm oil, mustard, sunflower oils etc.) by catalyzed transesterification. The process of producing fuel from oils and animal fats involves reacting an alcohol with an ester to produce a different alcohol through an exchange of organic carbonyl carbon groups in the feedstock with organic groups of alcohol. Biodiesel produced by this process is more environment-friendly and produces 60% less pollution compared to fossil-sourced diesel oil.

The use of modern bioenergy has been growing very fast over the last five years, providing heat energy in the building and industry sectors, mostly in India, China and the European Union. Production of biofuels (mainly ethanol) has also been very strong, propelled by a surge in Brazil's ethanol production which culminated in the introduction of the Renovabio Programme in 2020, but growth in Asia has also been strong, accounting for half of the total growth. China has a very strong ethanol programme that could triple output in the next five years in order to meet future demand due to its massive rollout of 10% ethanol blending in a growing number of provinces, although coal is the main precursor. Total world output of modern biofuels is forecast to increase by about 25% over the next five years, and the United States and Brazil will lead the growth, accounting for two-thirds of total global production (IEA, 2019c).

3.3.3.5 *Geothermal energy*

The potential of geothermal energy in the global primary energy mix is enormous: it is estimated that stored thermal energy down to 3 km within continental crust is around 43×10^6 EJ, which is considerably greater than the world's total primary energy consumption (WEC, 2016). However, the contribution to the global total primary energy consumption is very small and power generation from geothermal energy accounted for less than 1% of the world's output in 2015, largely because not many countries have suitable sites but also because new technologies that cut costs are still emerging. The largest operating geothermal powerplant in the world is located in the United States, Turkey leads the world in new plant projects and accounts for half of the new global capacity additions, although Indonesia has the highest global planned capacity additions and new plants are also under construction in many other countries. Apart from power generation, geothermal energy is also used directly for domestic and industrial heating.

3.3.4 **Unconventional energy**

There are other primary energy sources apart from the traditional ones discussed above. These include shale oil, gas and bitumen, coal/biomass-based gaseous fuels, coal-bed methane, coal/gas-based liquid fuels, and hydrogen often classified as unconventional energy. Exploitation of these resources is increasing due to the development of new technologies and the impact on the global primary energy reserves, demand and supply will increase significantly over the next two decades. The highest growth will be in unconventional natural gas which is expected to account for around 60% of the projected rise in global natural gas supply, about a third of the total natural gas requirements, and nearly 90% of North America gas production by 2040. Shale oil is being processed into petroleum products, shale (natural) gas is being produced by hydraulic fracking, and exploitation of coal bed methane for power generation is increasing.

3.3.4.1.   *Shale oil and gas*

Substantial gas and some oil resources are trapped in petroleum-bearing formations of low permeability: in rocks (caprocks), shale or tight sands deep beneath the Earth and seabeds, and are not recoverable by conventional techniques, especially when the gas may be dispersed in the rock rather than occurring in a concentrated underground reservoir. Fine-grained sedimentary rock may contain significant amounts of a solid mixture of organic chemical compounds (kerogen) from which liquid (shale oil) and gaseous (shale gas) fuels can be extracted. Unconventional oil recovery accounts for about 30% of the global recoverable oil reserves and oil shale contains three to four times as much oil as conventional crude oil reserves which are projected at around 1.2 trillion barrels (WEC, 2016). About 600 shale oil deposits have been identified in 33 countries on all continents of the world but less than fifty have been proved to be economically viable and moved to reserves. Current estimates put the global resources of shale oil (also known at tight oil or shale-hosted oil) at around 6,050 billion barrels, around four times the size of the world's conventional crude oil resources. The United States alone holds about 80% of the total global proven reserves, with China, Russia, Israel, Jordan and DR Congo holding most of the balance. It is believed that current estimates for oil shale resources are rather conservative, as many deposits have not been adequately quantified and many others have not even been identified. It should be noted also that only a small fraction of the resources has been proven and moved to reserves - shale oil, that is economically extractable using currently available technologies. There is little doubt that new technologies will improve access in the near future.

Global resources of shale gas are estimated at around 0.2 trillion cubic meters, enough to fill the global demand for natural gas for over sixty years at the current rate of consumption. Again, many potential resources are either not yet identified or evaluated. Six countries (U.S., China, Argentina, Algeria, Canada and Mexico) hold around two-thirds of the technically recoverable global shale gas resources but only three countries - the United States, Australia and China - are actively exploiting shale gas deposits currently. Production of shale oil and gas requires special technology known as hydraulic fracturing: vertical wells are drilled to depths of up to several thousand meters and may include horizontal sections. Shale rocks are fractured by pumping high-pressure fluid, typically water, proppant (sand, ceramic pellets, other small incompressible particles) and proprietary chemical additives. The high-pressure fluid fractures the rocks while the proppants hold the fractures open to allow gas or oil flow. Hydraulic fracturing technology was developed in the late 1940s and has been in use since the 1980s, initially for fracturing coal seams to release methane, primarily for safety reasons, since coalbed methane is both poisonous and a major cause of coal mine explosions (Figure 3.11). The fracking technology has been further developed and deployed primarily in the United States and Canada, and can produce large volumes of shale oil and gas at relatively low cost, although still significantly higher than traditional methods. Two other countries - China and Argentina complete the list of the only four countries in the world currently exploiting shale oil and gas commercially. In 2017, shale oil provided 50% of total U.S. crude oil production and shale gas provided 60% of total natural gas production. As a result of the rapid and wide deployment of fracturing technology, the country has emerged as the world's largest oil producer of oil and gas, and the largest consumer. However, production of unconventional oil and gas raises new environmental problems and issues: fluids employed in cracking up caprocks and coalbeds are potential groundwater contaminants and fracking is believed to induce geological instability which may increase the incidence or intensity of earth tremors and earthquakes. The fracking process is controversial worldwide: France and Germany have banned fracking, so have many states in the United States and increases in tremors are raising issues in Europe. Furthermore, fracking is significantly more expensive than conventional technologies and a major drop in global oil and gas prices can quickly collapse the industry.

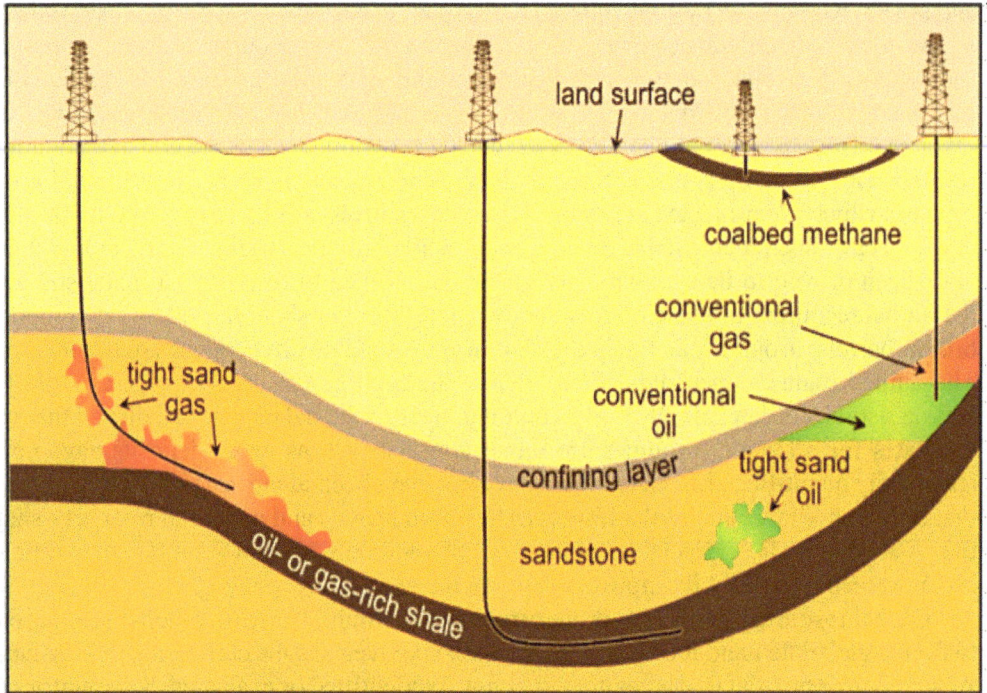

Figure 3.11 Unconventional production of coalbed methane, shale oil and gas by hydraulic fracturing. *(Wikipedia, 2018).*

3.3.4.2.   *Bitumen/Oil sand/tarsand*

Bitumen (oil sand/tar sand) refers to naturally occurring thick oil trapped in loose sandstone and clay. The texture varies from highly viscous oil to solid mass, depending on the depth of the deposit. Deposits have been found in many countries all over the world and while some are less than a hundred meters deep, most deposits are located deeper underground. Global bitumen resources are estimated at more than 2 trillion barrels although much of it is yet to be moved to reserves. Canada holds the largest deposit in the world and utilizes the most advanced production technologies which include complex mining and in-situ oil recovery. Currently, tar sands represent about 40% of the country's oil production. Venezuela is the only other country in the world currently exploiting bitumen in commercial quantities, although on a much smaller scale. Complex technologies are required to separate the oil from the sand and clay, and refine it to the quality required for processing by regular refineries.

3.3.4.3. *Coal-bed methane (CBM)*

Methane is the major constituent of natural gas and is normally associated with petroleum deposits. However, the gas also occurs in association with coal deposits, often referred to as unconventional natural gas (Figure 3.11). During coalification, large quantities of methane-rich gas are generated and stored within the coal on internal surfaces. Methane occurs wherever large coal deposits are found and is released into the atmosphere regularly during coal mining for safety reasons. Some coalbeds are vented through boreholes before mining, but in the majority, the methane is diluted to below its explosive limit of around 5%. The gas is not only poisonous but also potentially explosive, and a cause of many mining accidents. Also, methane is a potent anthropogenic gas and releasing the gas into the atmosphere causes major pollution problems. Some mines burn the methane on site but this also releases carbon dioxide (another potent anthropogenic gas) into the atmosphere. In recent years, coalbed methane has gained prominence as a potential energy resource. Because coal has such a large internal surface area, it can store very large volumes of methane-rich gas, six or seven times as much gas as a conventional natural gas reservoir of equal rock volume can hold (USGS, 2000). In addition, much of the coal, and thus much of the methane, lies at shallow depths, making wells easy to drill and inexpensive to complete. With greater depth, increased pressure closes fractures (cleats) in the coal, which reduces permeability and the ability of the gas to move through and out of the coal. Exploration costs for coal-bed methane are low, and the wells are cost-effective to drill. Unlike natural gas from most petroleum deposits, coalbed methane contains very little of the heavier hydrocarbons such as propane and butane which need to be removed from associated natural gas before use. In a conventional oil or gas reservoir, for example, gas lies on top of oil which, in turn, lies on top of water. An oil or gas well draws only from the petroleum that is extracted without producing a large volume of water, but water permeates coal beds, and its pressure traps methane within the coal. The methane is in a near-liquid state, lining the inside of pores within the coal (called the matrix). The open fractures in the coal (called the cleats) can also contain free gas or can be saturated with water. Coalbed methane can be recovered from underground coal before, during, or after mining operations. It can also be extracted from unmineable coal seams that are too deep, too thin, or of poor or inconsistent quality.

Coalbed methane resources are commensurate with the world's conventional gas resources. Methane concentration in the mixture of coalbed natural gases reaches 95 to 98% but carbon dioxide and ethane may also be present in significant quantities reaching around 50% and 10% respectively. Although coal deposits abound all over the world and all contain methane, not all deposits are suitable for coalbed methane exploitation and not every type of coal deposit is suitable for methane production. Thus, long-flaming brown coal fields are featured with low methane content and anthracite coal is characterized with high gas content but, it cannot be recovered economically due to high density and very low permeability of the deposit. Mature coals (bituminous) that fall somewhere in between the brown coals and the anthracite coals are attributable to the most favorable ones for methane production. Another important determinant of the economic potential of coalbed methane is the cost of competing primary energy resources, in particular, natural gas. Active drilling for coal bed methane began in the 1980s, mostly in the U.S. but spread rapidly to many other countries and the global CBM market is expanding rapidly. North America holds the largest share, followed by Europe and Asia Pacific. However, many emerging countries are also developing and exploiting resources, in particular, China, India and Indonesia. Application areas are diverse - power generation, heat for industrial processes, transportation, and steam-raising for commercial and residential use.

3.3.4.4   *Synthetic gas and liquid fuels*

Coal, and biomass can be converted to more efficient, lower carbon fuel including synthetic oil gas, synthetic natural gas, hydrogen, gasoline, diesel oil, and even ethanol which is mixed with gasoline. Also, natural gas can be steam-reformed to produce cleaner, higher-value hydrogen fuel. The technologies are well developed and have been around for decades. While syngas produced from coal, natural gas and biomass is fueling many power plants all over the world, proliferation of coal/gas/biomass-to-fuels has been slow because the economics is closely tied to global oil pricing which has been unstable for many years. Conversion of coal to transportation fuels has been the main source of liquid energy in South Africa for many decades and China also operates a commercial plant, with many more being planned or under construction. The main advantage of coal-to-fuels from the environmental point of view is that most of the pollutants emitted at the gasification stage can be captured at source relatively easily because they are at high pressure, thus greatly reducing the life-cycle carbon footprint of the synthetic fuels significantly.

3.3.4.5   *Flared fuels*

Combustible gas/liquid fuel is released during normal or unplanned operation in many industrial processes - oil and gas extraction, chemical plants, coal mining, landfills - and is usually burned off. The flaring process releases anthropogenic gases and particulate matter into the atmosphere and is a major environmental issue. Flaring in the oil and gas industry is perhaps the most prominent in terms of atmospheric pollution but also because enormous potentially valuable fuel is being wasted. The World Bank reports that between 150 to 170 billion cubic meters (BCM) of gases are flared or vented annually with no energy benefit and significant environmental damage, an amount valued at about $ 30.6 billion, equivalent to one-quarter of the United States' or 30% of the European Union's gas consumption annually. Russia leads the world in gas flaring and, with nine other countries, (mostly developing countries with low access to electricity), account for around three-quarters of the total global flaring (Figure 3.12).

The current amount of flaring releases around 350-400 million metric tons of CO_2eq into the atmosphere, in addition to toxic particulate matter. This level of emissions, combined with similar levels in midstream operations, (leakages, routine venting during handling and transportation), account for around 20% of total greenhouse gas emissions from natural gas systems (the balance of 80% is accounted to combustion processes). Flaring gas wastes a valuable energy resource that could be used to support economic growth and progress; and it contributes to environmental pollution by releasing millions of metric tons of CO_2.eq to the atmosphere. This has prompted a World Bank initiative of "Zero Routine Flaring by 2030." For decades, global efforts to reduce flaring had not been very successful primarily because global prices of conventional oil and gas were high and there was little incentive to invest in expensive associated gas gathering and utilization infrastructures. However, the drastic fall of crude oil prices in the last decade, coupled with increasingly stringent environmental regulations, have repositioned associated gas as a potentially valuable primary fuel. It is interesting to note that a significant amount of flaring is practiced in Africa which also has one of the lowest rankings in global modern energy access, especially electricity. It is estimated that flared gas by oil and gas companies in Africa could supply around 50% of the continent's electricity needs (Andersen *et al.*, 2012).

**Top ten gas flaring countries in 2018
(Billion cubic meters)**

Russia	21.3
Iraq	17.8
Iran	17.3
United States	14.1
Algeria	9
Venezuela	8.2
Nigeria	7.4
Libya	4.7
Mexico	3.9
Angola	2.8

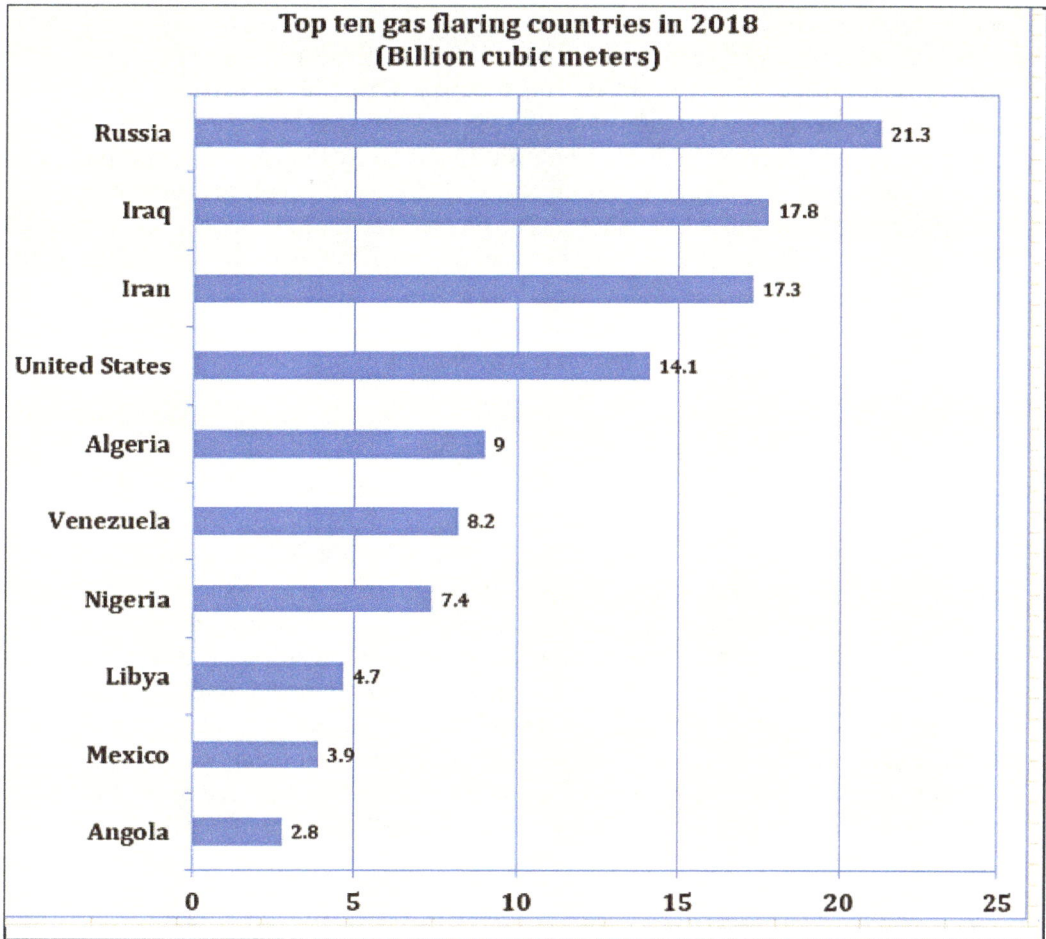

Figure 3.12 Top seven gas flaring countries in the world in 2018 *(NOAA, 2019).*

3.3.4.6 *Hydrogen energy*

Hydrogen is the simplest element in the chemical chart and the most abundant in the universe. Because the gas is very reactive, it is never found in pure state but is always combined with other elements. Water is a combination of hydrogen and oxygen and many organic compounds contain hydrocarbons which are compounds of hydrogen and carbon - gasoline, natural gas, coal, methanol, propane, etc. Hydrogen is a versatile energy resource which can be produced from many sources including sea water, coal, natural gas, coalbed methane, and biomass. It can be burned in oxygen to produce clean energy that powers gas turbines for electric power generation, and in the propulsion of spacecraft. It is used to power fuel cells and also has the potential to be mass-produced and commercialized for passenger vehicles and aircraft. Hydrogen energy is considered renewable because it can be produced from water and biomass. It is also considered the fuel of the future because of its low carbon intensity (the only emissions from hydrogen combustion are heat and water) and high versatility because it can be deployed in many ways: power generation, fuel cell-powered transportation, chemicals production, etc. However, production of hydrogen from any source is considered inefficient because more energy is always required to produce the gas than can be retrieved from it in end use (Zehner, 2012). Furthermore, electricity required to produce the gas will likely come from generating plants powered mainly by fossil fuels but also renewables or nuclear energy, hence hydrogen has a significant carbon footprint on a lifecycle basis. Also, handling hydrogen is extremely dangerous and this could remain a significant constraint to widespread use as an energy source. There are many precursors to hydrogen production but the most common is natural gas which is reformed by reacting with high pressure steam at high temperature in the presence of a catalyst. The product carbon monoxide is reacted with high pressure steam at lower temperature to produce more hydrogen in accordance with the equations below.

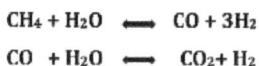

$$CH_4 + H_2O \rightleftharpoons CO + 3H_2$$
$$CO + H_2O \rightleftharpoons CO_2 + H_2$$

Synthetic gas (syngas) produced from coal or biomass contains mainly hydrogen and carbon monoxide and can also be steam-reformed to produce pure hydrogen for power generation (clean coal technologies). Research on new sources of hydrogen is intensive and the focus is on the electrolysis of water. The process involves application of an electric current to separate water into its components: hydrogen and oxygen. This process has numerous advantages: water is abundant; electricity can be produced from wind or solar powered generators and the hydrogen produced can be liquefied, stored or transported for use as required, thus resolving the issue of variability of renewable power supply. Liquid hydrogen is a very versatile fuel and has powered space rockets for decades and potentials for decarbonizing transportation are high. Many power plants feature hydrogen-powered gas turbines; and electric vehicles powered by hydrogen-fueled solar cells are already on the roads in some developed countries.

3.3.4.7   *Fusion energy*

The Sun produces energy by fusing two atomic hydrogen nuclei under very high temperatures and pressures to form a heavier nucleus with the release of high energy, a small part of which reaches the Earth and other planets. Fusion reactions occur when two or more atomic nuclei come close enough for long enough to enable the nuclear force pulling them together overcome the electrostatic force pushing them apart, thus fusing them into heavier nuclei with the release or enormous energy. Global efforts to simulate these fusion reactions using isotopes of hydrogen such as deuterium and tritium, and harness the energy released for useful purposes have intensified over the last three decades or so but progress has been extremely slow due to problems of initiating and sustaining steady-state reactions and safely capturing and utilizing the enormous energy released, for example in powering steam turbines that generate electricity. Fusion reactors are much less dangerous than the current fission reactors and the hydrogen isotopes are abundant in the Earth's oceans, unlike fission reactors which require radioactive uranium fuels. Most of the 200 or so research fusion reactors in many countries have shut down because of problems with the production of net energy (more energy produced than consumed), the extremely high-precision systems control required, and lack of adequate materials that can withstand such extreme temperatures and pressures. However, the International Thermonuclear Experimental Reactor located in France is making significant progress and could be switched on in the next five years. Fusion is considered one of the world's most promising potential sources of abundant, low-carbon energy and active research focusing on different aspects of the problem is ongoing in many countries around the world.

Chapter 4
World final energy consumption

4.1. INTRODUCTION

Primary energy is used in a variety of ways: Oil is transformed into a variety of fuels by refining; coal, natural gas, biomass, nuclear power are transformed into electricity by power plants; coal, natural gas and biomass are used directly in heating; coal, natural gas and biomass are transformed into biofuels; and coal, natural gas, oil are feedstocks for the production of a wide variety of chemicals and petrochemicals. The final consumers are industry, transportation, buildings, and a few others. Global primary energy demand is driven largely by economic growth, which in turn is driven by the rising population and increasing prosperity, particularly in the developing world. However, the overall growth in energy demand is moderated significantly by declining energy intensity (energy used per unit of GDP). Less than 70% of the global total primary energy supply actually is available as end-use energy, the balance being used in conversion processes or lost to inefficiency across the whole spectrum of energy production and utilization. Global final energy consumption (TFC) is mixed and includes both primary and transformed fuels - refined transportation fuels, natural gas, oil, coal, biofuels. Primary energy is used in four key sectors of the global economy: electric power generation, industry, transport, and buildings. Oil accounted for the largest share (41%) of TFC in 2017, followed by electricity (19%) (Figures 4.1 to 4.3).

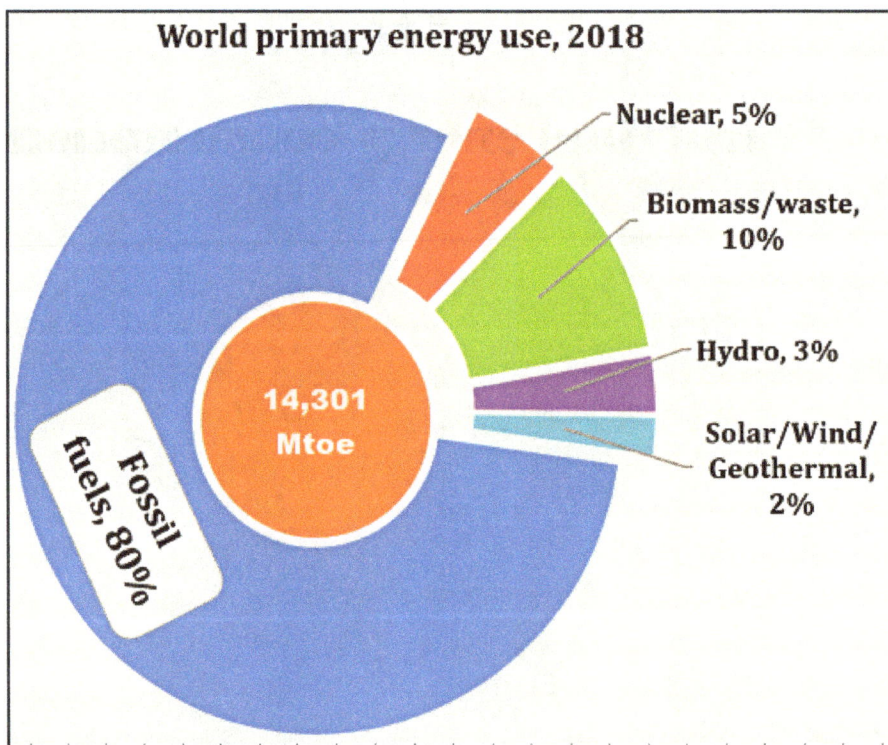

Figure 4.1a World primary energy supply (TPES), 2018. *(IEA, 2019c).*

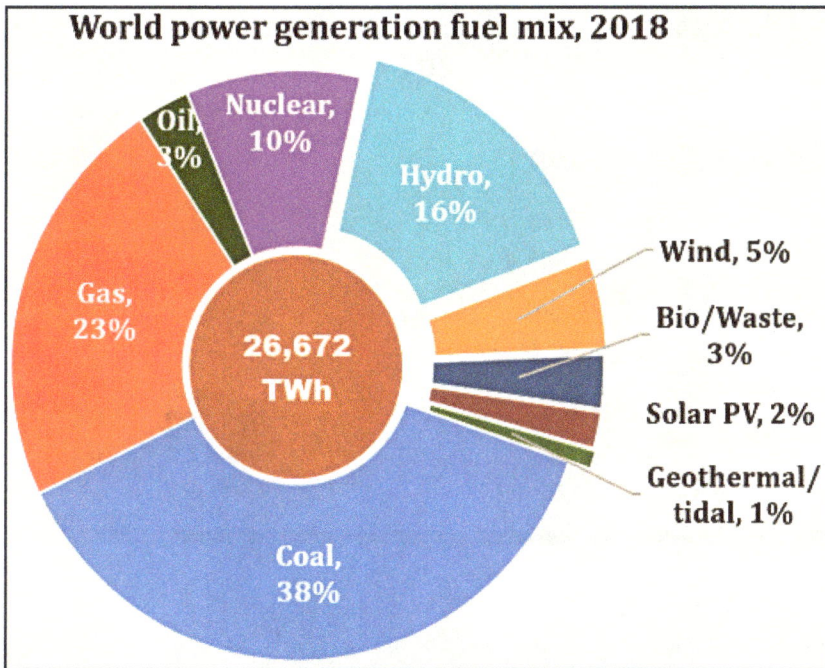

Figure 4.1b   World power generation fuel mix, 2018. *(IEA, 2019d).*

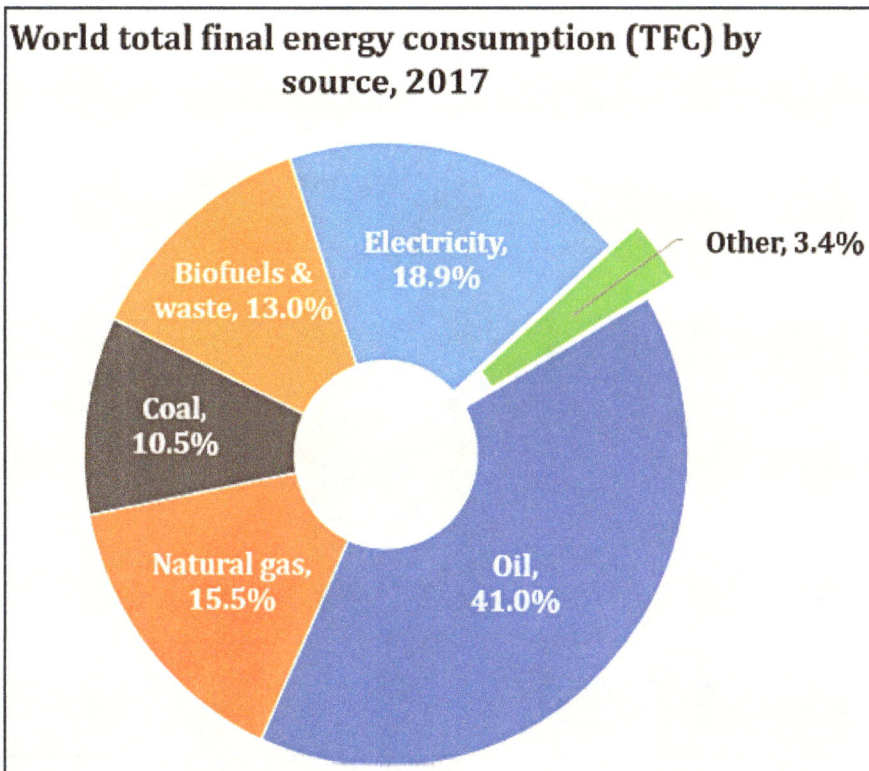

Figure 4.2a   World total final energy consumption (TFC) by fuel in 2017 (9,717 Mtoe). *(Data from IEA, 2019a).*

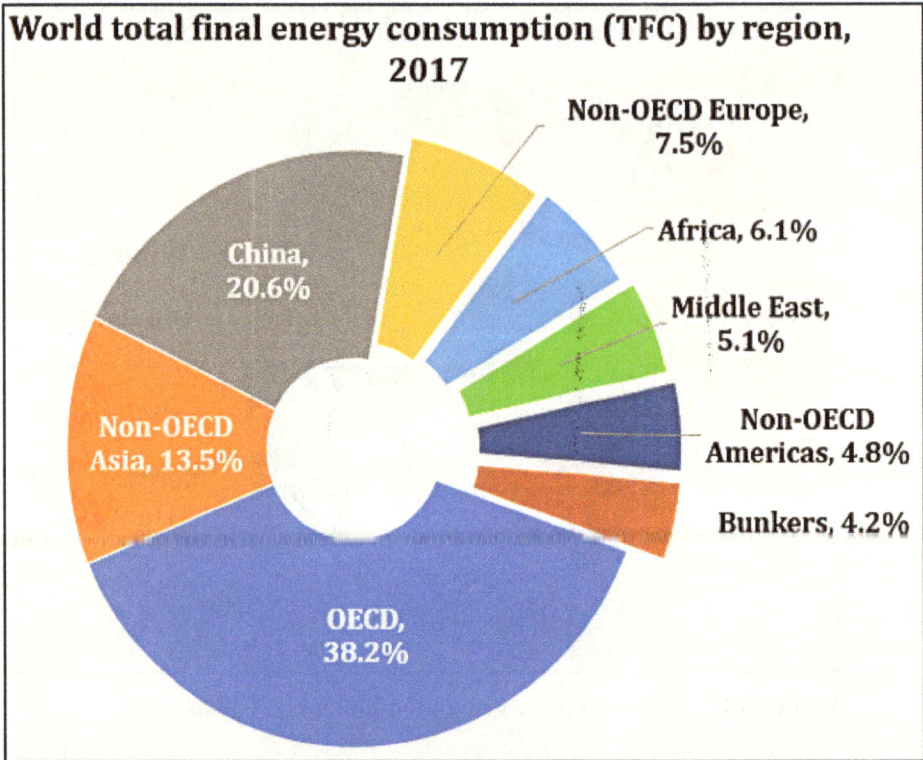

Figure 4.2b   World total final consumption (TFC) by region in 2017 (9,717 Mtoe). *(Data from IEA, 2019a).*

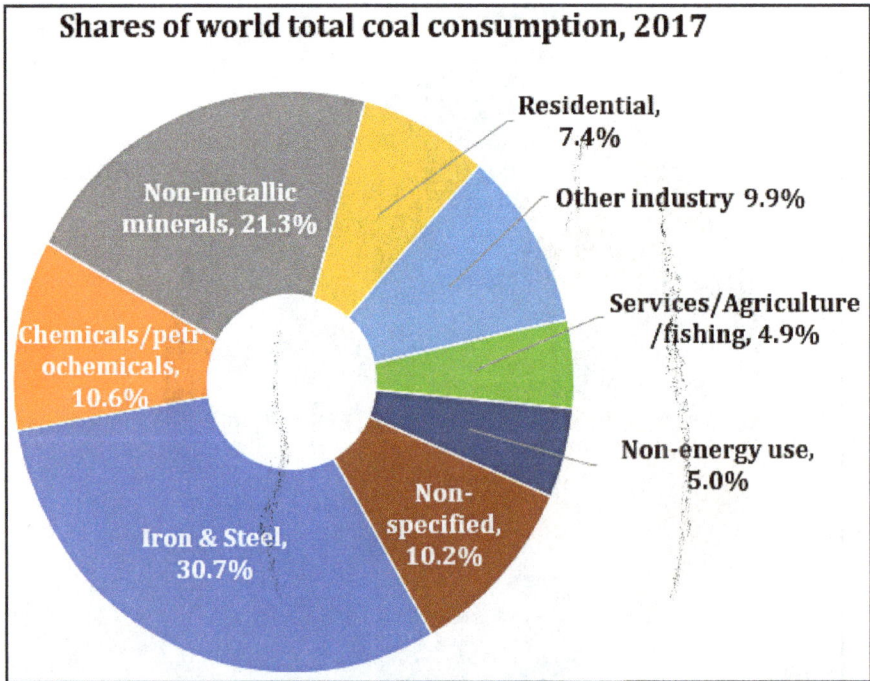

Figure 4.3a   World total final coal consumption (TFC) by economic sector in 2017 (1,020 Mtoe). *(Data from IEA, 2019a).*

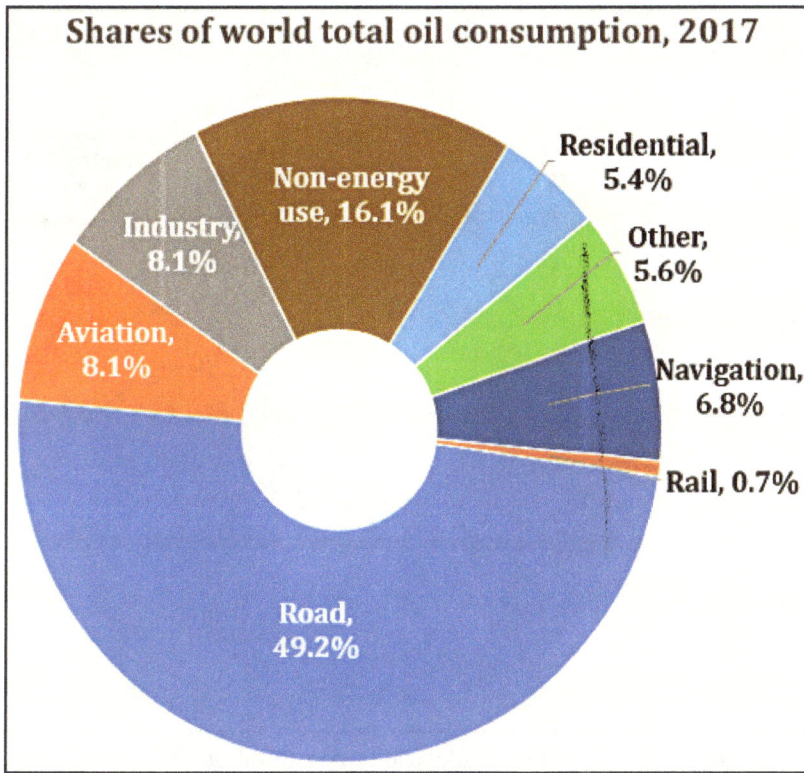

Figure 4.3b World total final oil consumption (TFC) by economic sector in 2017 (3,985 Mt). *(Data from IEA, 2019a).*

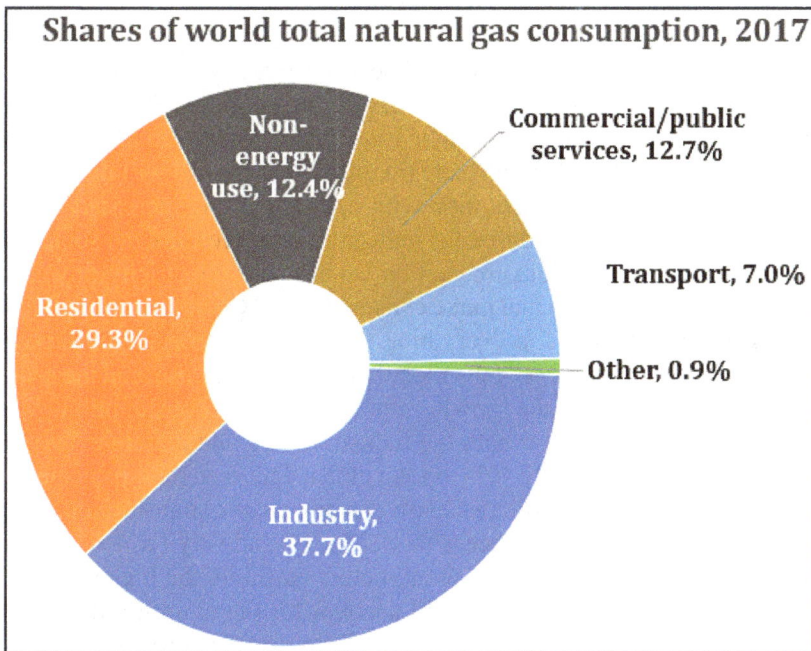

Figure 4.3c World total final natural gas consumption (TFC) by economic sector in 2017 (1,502 Mtoe). *(Data from IEA, 2019a).*

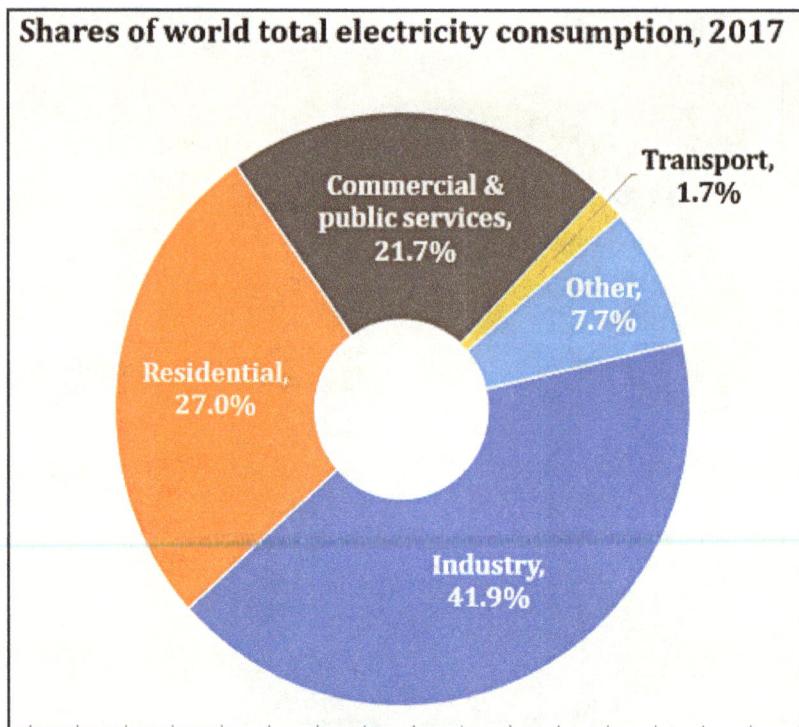

Figure 4.3d World total electricity consumption (TFC) by economic sector in 2017 (1,838 Mtoe). *(Data from IEA, 2019a).*

4.2. POWER GENERATION

Electric power is not a primary energy because it is produced by converting a primary energy source, and any of all available primary sources is potentially suitable for conversion although fossil fuels still account for by far the largest share of the total global electric power generation. However share of fossil fuels (mainly oil and coal) will drop significantly over the next two decades, due to substitution with natural gas and increased use of renewables. Electricity is the prime energy source for industry, commerce and the social sector - homes, hospitals, educational institutions, recreation centers, etc. Electricity powers all aspects of human development - factories that make consumer goods, commercial buildings, heat, light, air conditioning, the Internet and everything that connects to it. Electric power generation currently accounts for about 42% of the global primary energy use, and the sector is projected to account for around 70% of the growth in primary energy consumption over the next two decades, with comparable rise in both the industrializing and mature economies. Presently, electric power accounts for nearly 40% of total global delivered (end-use) energy and demand is expected to rise as new technologies evolve, many more countries industrialize, and more people in emerging nations gain access to electricity. Industry (which includes agriculture and other minor sectors) and the buildings sectors consume the highest proportion of final energy use (including electricity) and will account for the majority of the growth over the next two decades (Figures 4.4 to 4.11). In 2018, fossil fuels provided 65% of the primary energy used for power generation, and demand by the non-OECD countries has risen dramatically over the last few decades, accounting for around 55% in 2017. Several projections indicate that demand will rise by 60-70% by 2040 and, although increases are expected from all parts of the world, the bulk of the rise (85%) will come from non-OECD countries. Also, fuel mix for power generation will change significantly globally as well as across regions. Industry accounts for about half of global electricity usage while the other half comes from commercial and residential usage. All reviewed projections agree that the share of coal in power generation will decline from about 40% to around 30% by 2040, due to increasing shares of natural gas, renewables and nuclear energy.

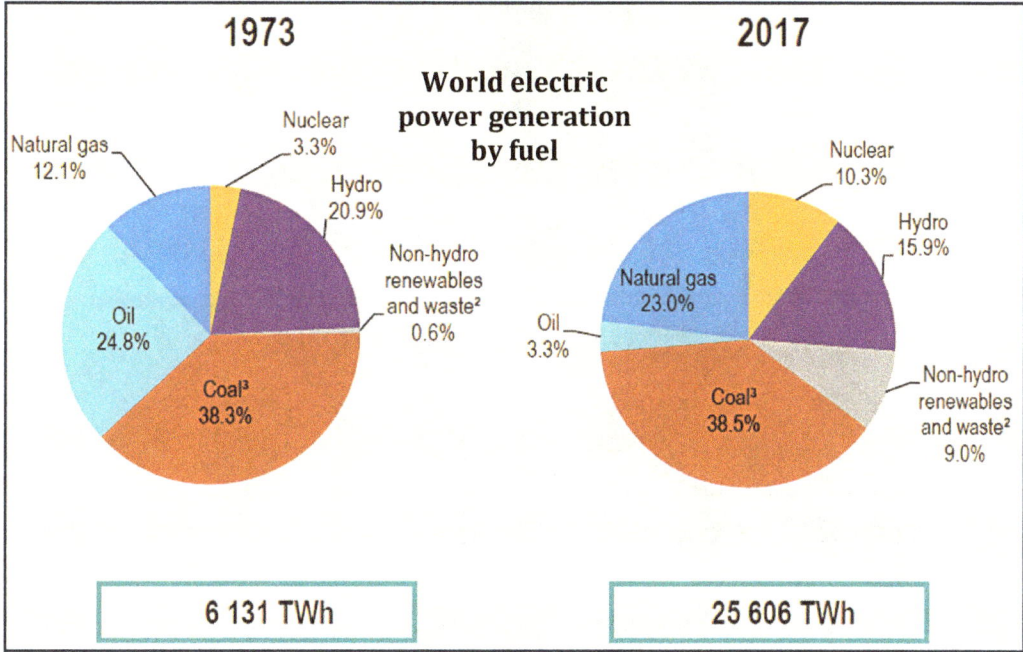

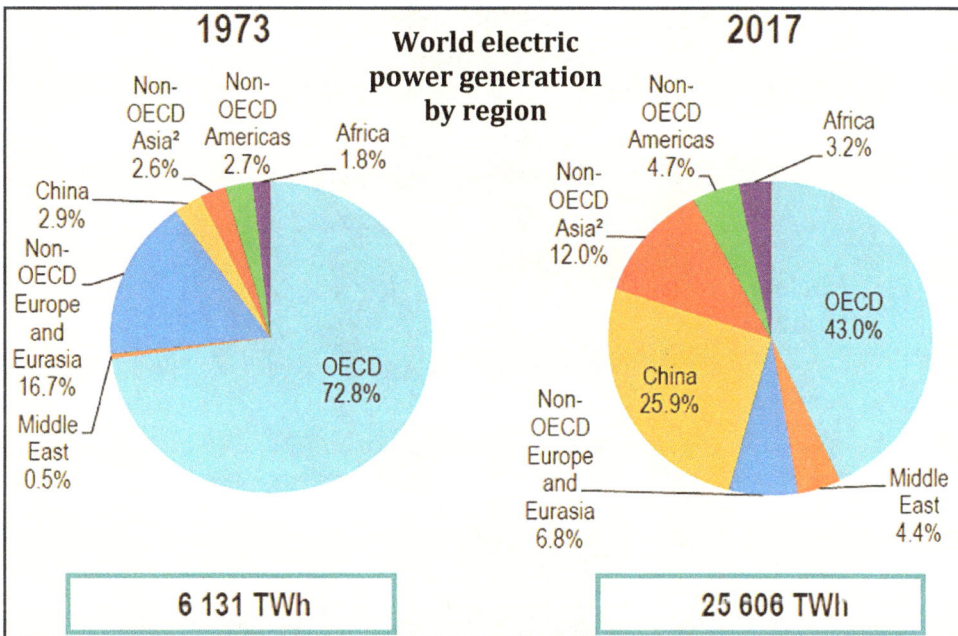

Figure 4.4 World electric power generation by fuel and region in 2017 (*IEA, 2019a*).

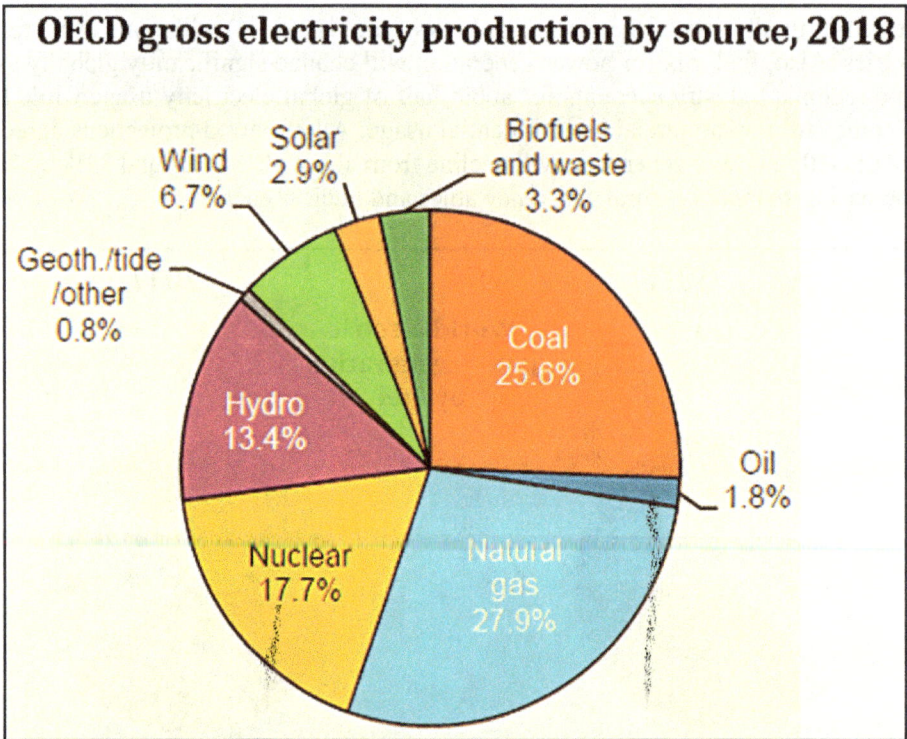

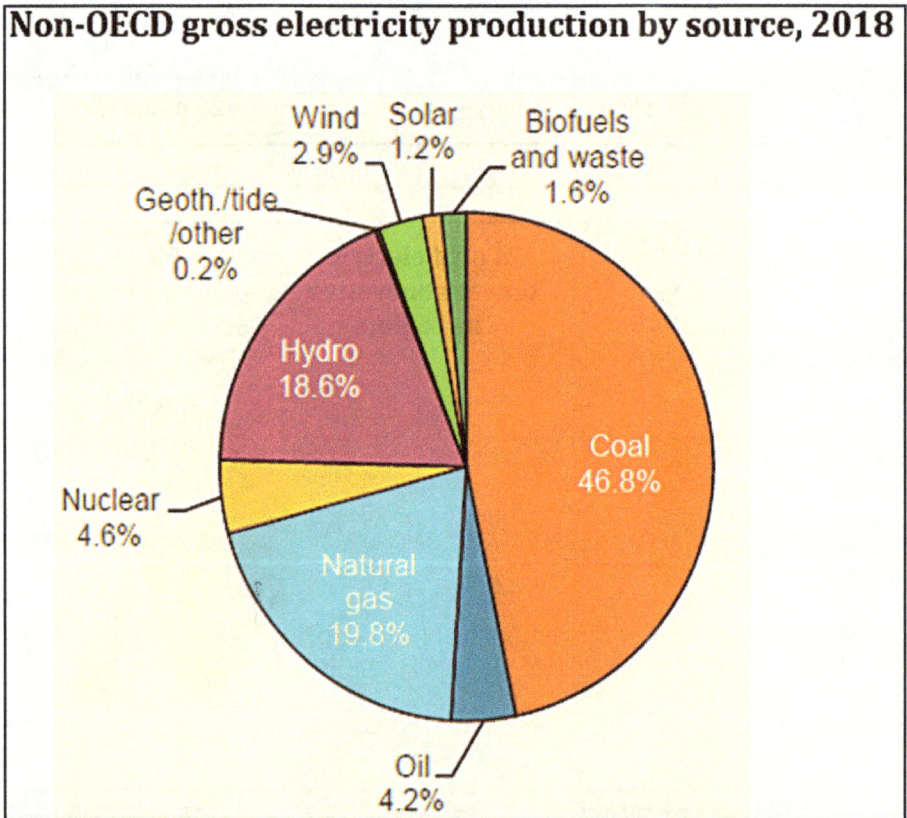

Figure 4.5a   Fuel mix for electricity production by source and region
IEA, 2019f; 2019g).

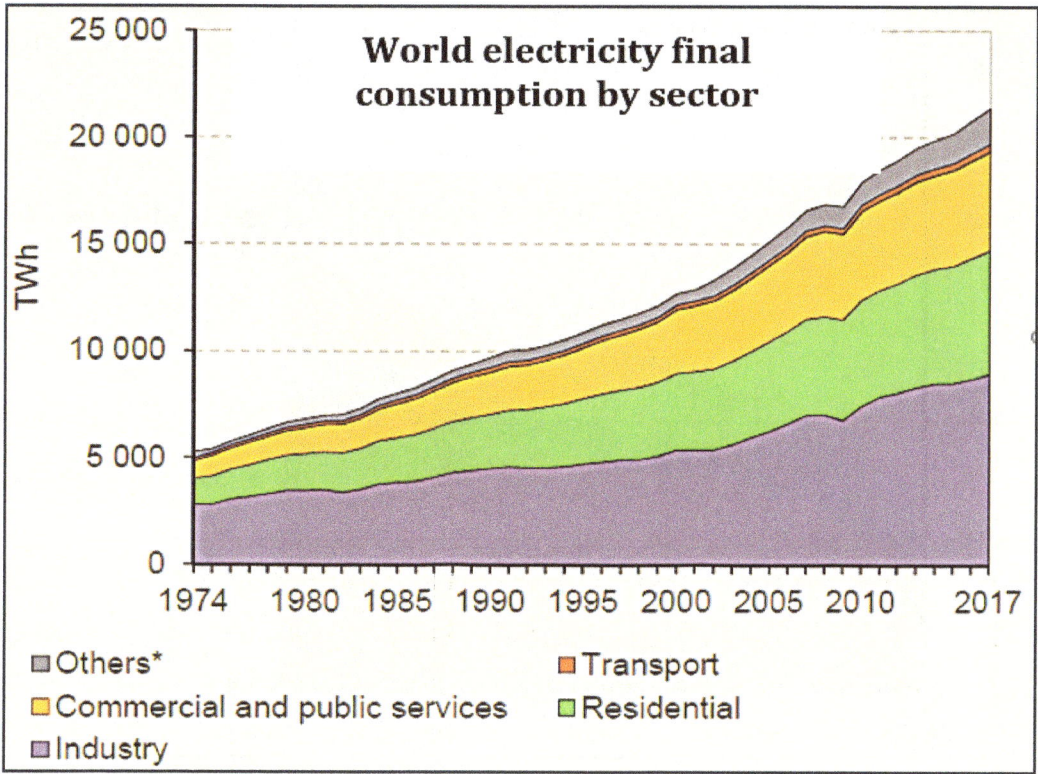

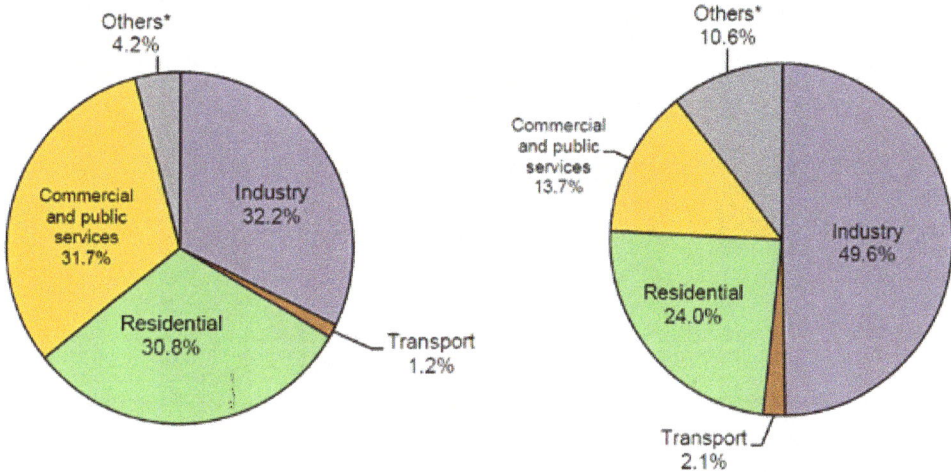

Figure 4.5b World final electricity consumption by economic sector and region. (*IEA, 2019f; 2019g*).

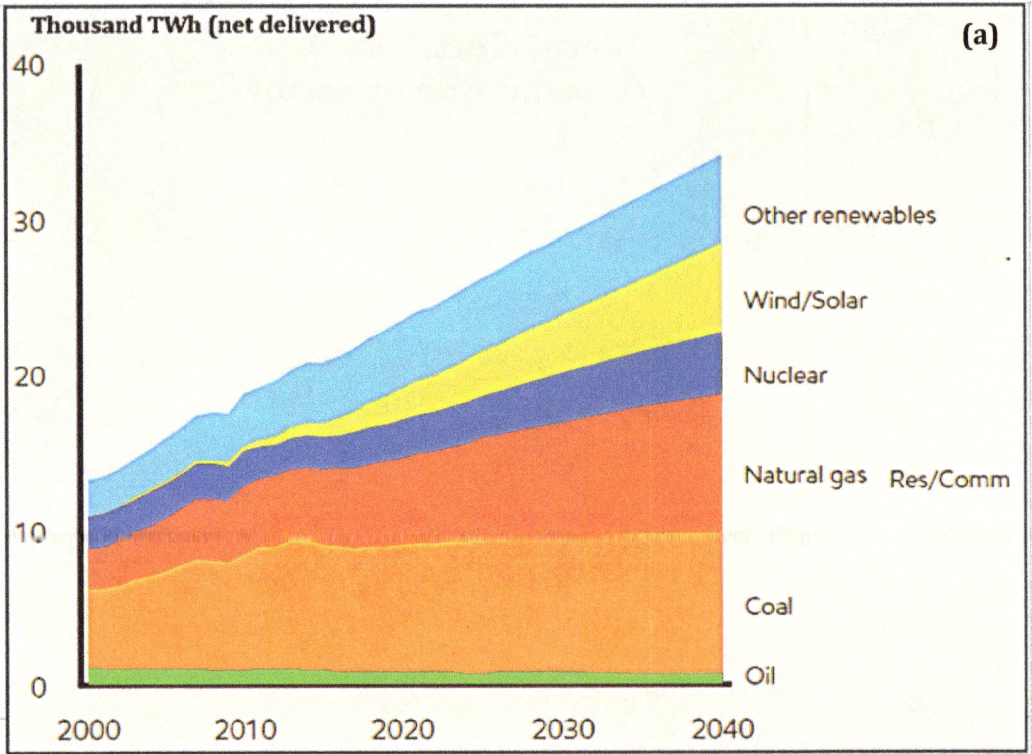

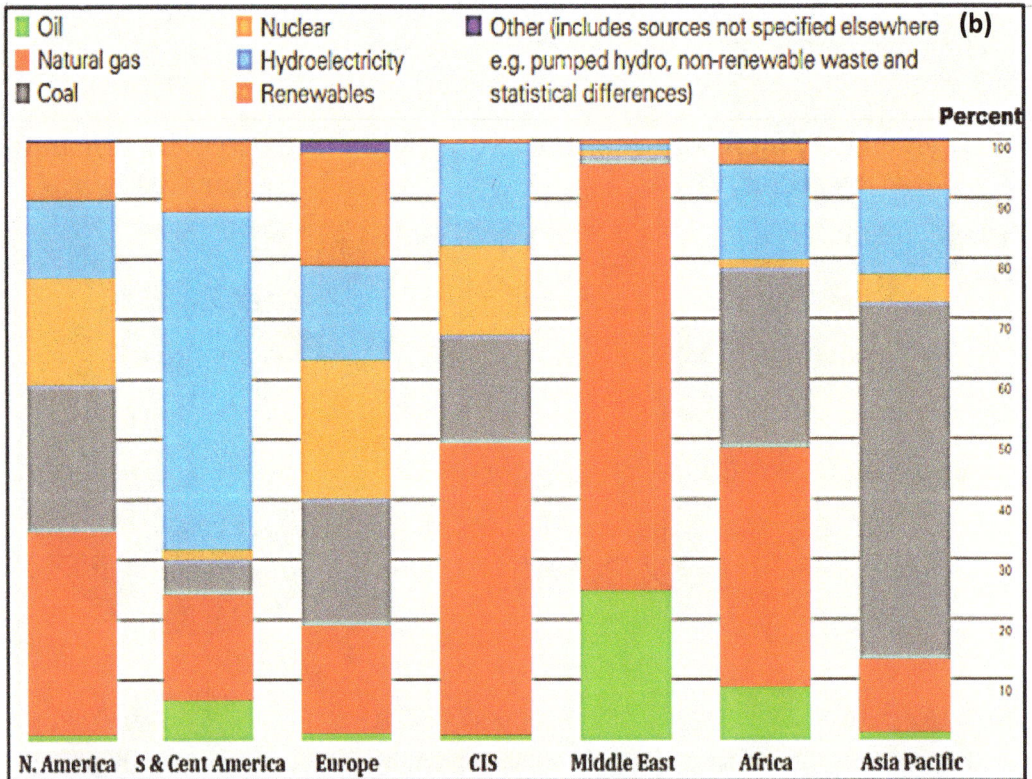

Figure 4.6 (a) Outlook of electricity genertion by fuel (b) Electricity generation by fuel and region in 2018 *(BP, 2019b).*

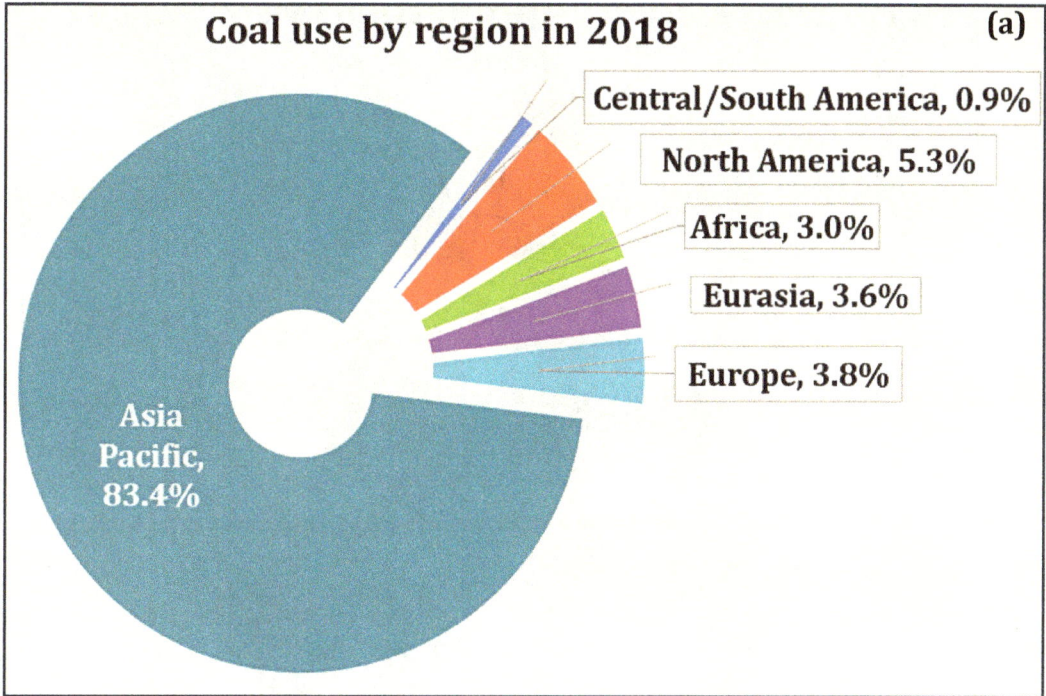

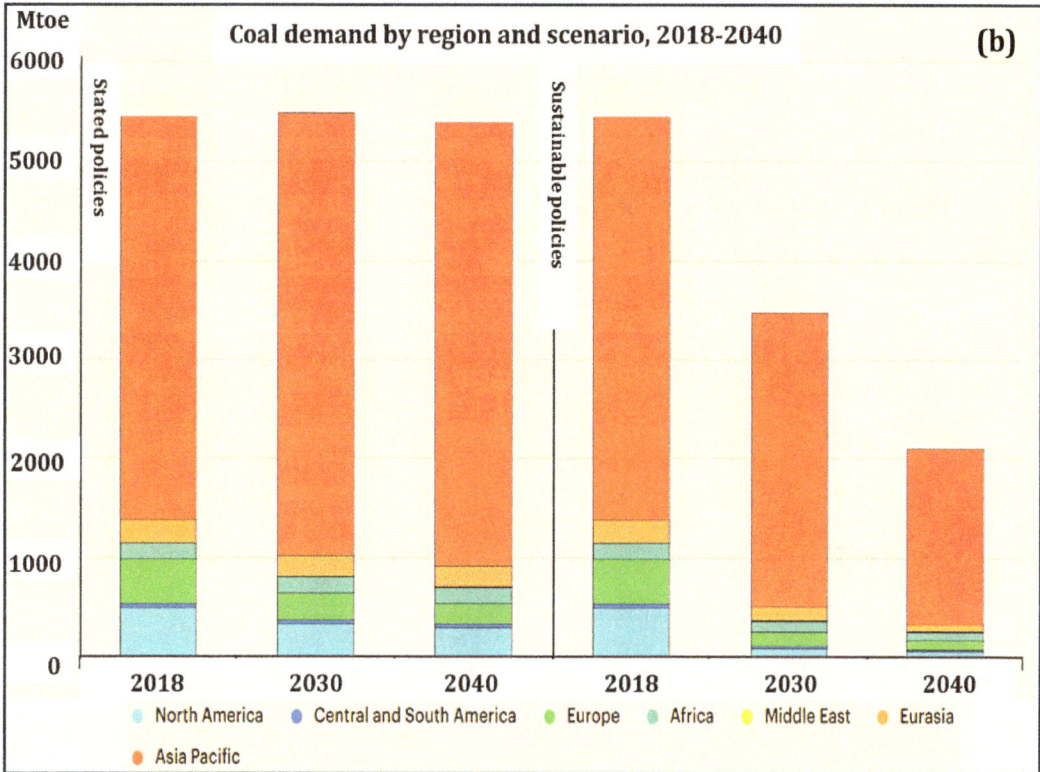

Figure 4.7 (a) Coal consumption by region (b) Projected demand for coal by region and scenario (*IEA, 2019a; 2019g*).

World gross electricity production by source, 2018

(a)

Wind
4.4%

Solar
1.8%

Biofuels
and waste
2.3%

Geoth./tide
/other
0.5%

Hydro
16.3%

Coal
38.3%

Nuclear
10.2%

Natural
gas
22.9%

Oil
3.3%

(b) **Major coal users for power generation**

Kosovo	98%
Botswana	97%
South Africa	93%
Mongolia	93%
Poland	81%
India	76%
Serbia	73%
Kazakhstan	72%
China	71%
Hong Kong	66%
Australia	63%

0%   20%   40%   60%   80%   100%

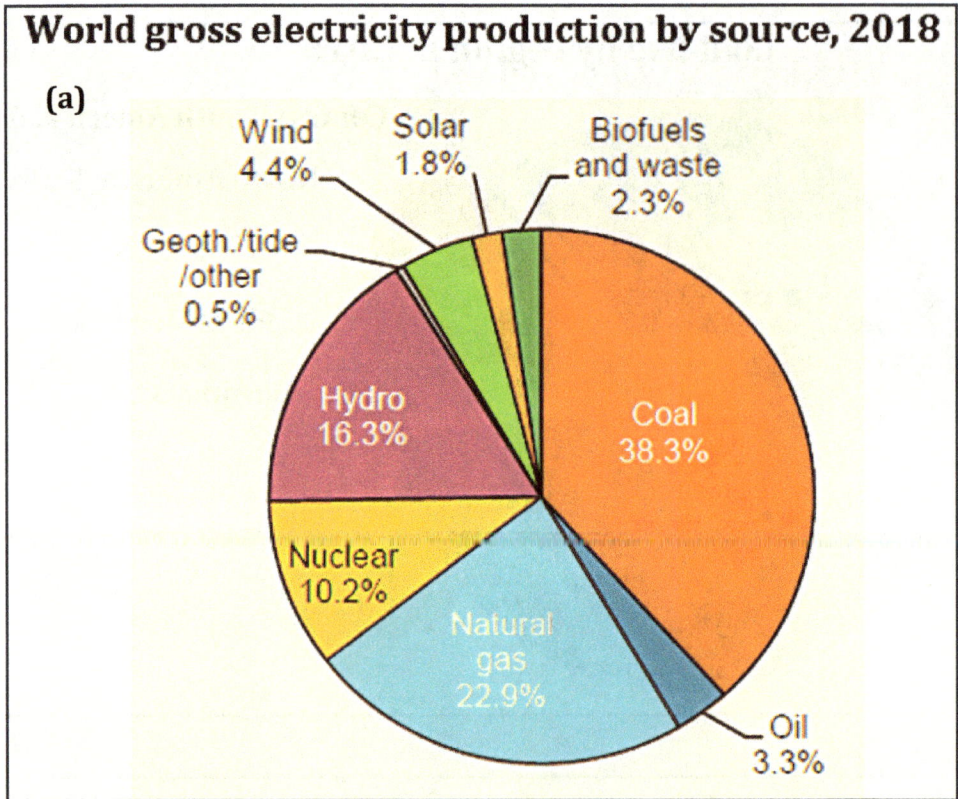

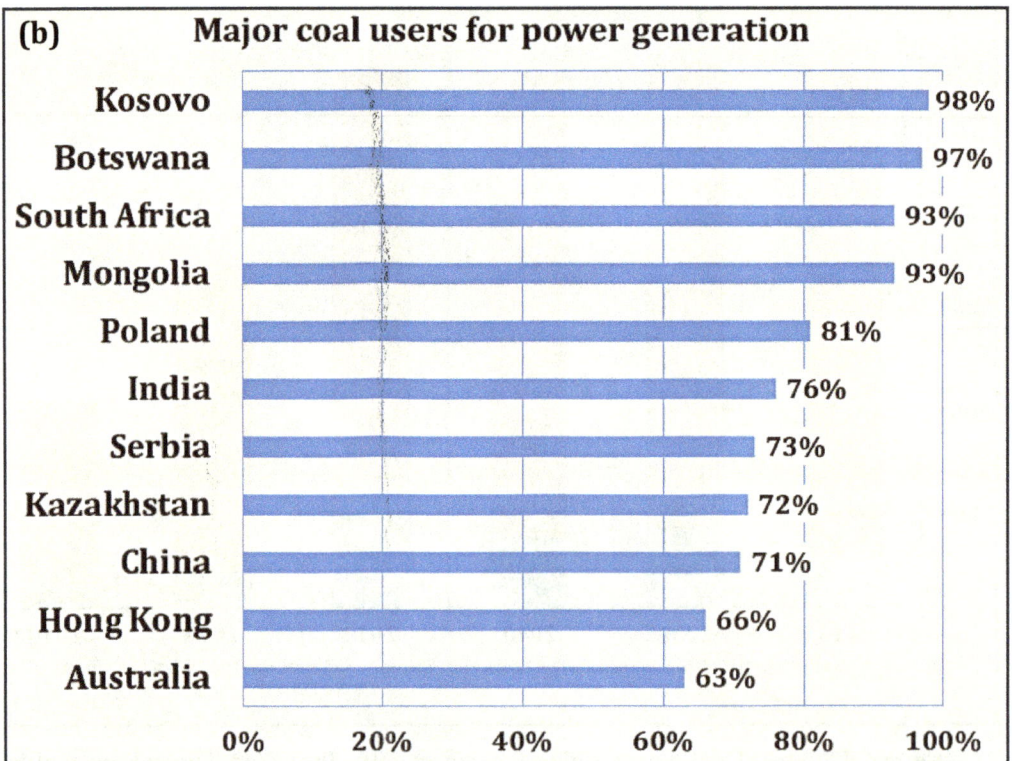

Figure 4.8 (a) World electricity production by source (b) Countries with the highest dependence on coal for power generation in 2018. *(IEA, 2019a; 2019f).*

Countries with the highest coal-fired power generation capacity, 2018

Country	TWH
Turkey	74
Kazakhstan	77
Ukraine	80
United Kingdom	97
Indonesia	110
Taiwan	120
Poland	130
Australia	152
Russia	156
South Korea	211
South Africa	237
Germany	263
Japan	299
India	868
USA	1711
China	4090

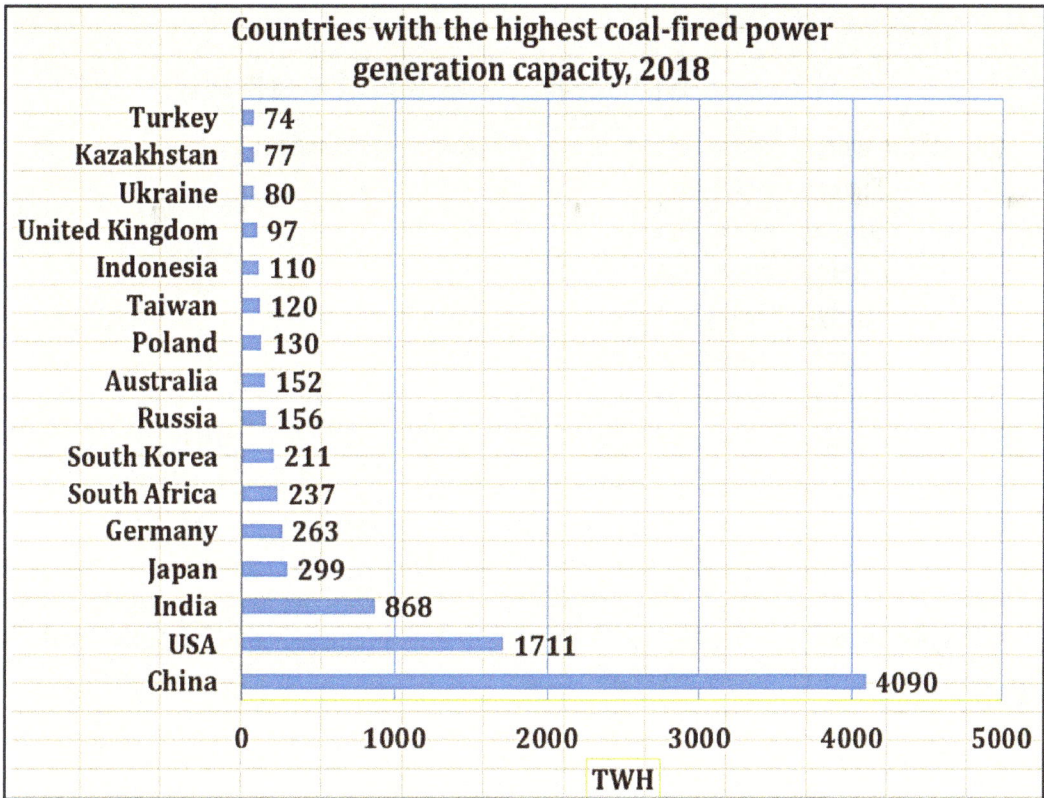

Figure 4.9 Countries with the highest coal power generation capacity in 2018 (Data from *Coal, 2019 Electricity Information, iea.org*).

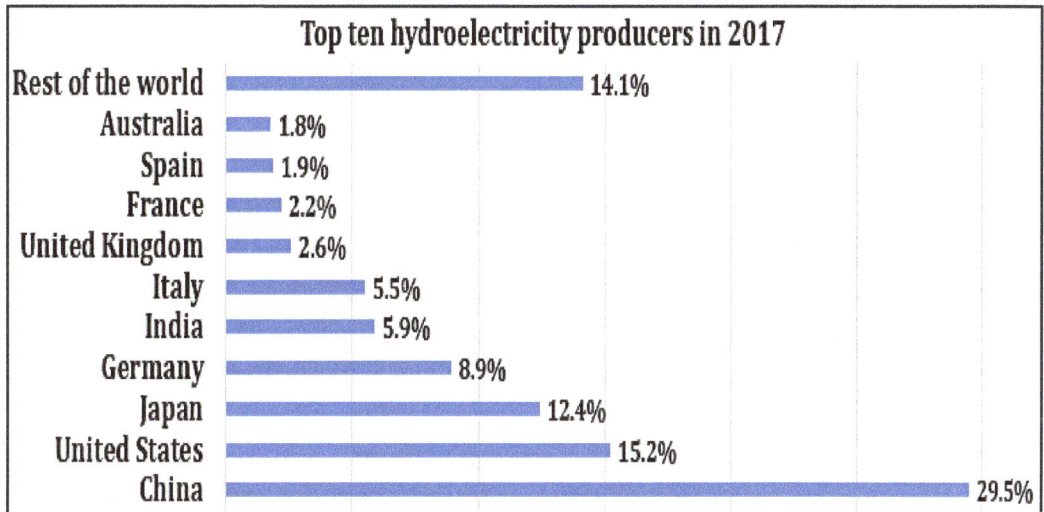

Top ten hydroelectricity producers in 2017

Country	%
Rest of the world	14.1%
Australia	1.8%
Spain	1.9%
France	2.2%
United Kingdom	2.6%
Italy	5.5%
India	5.9%
Germany	8.9%
Japan	12.4%
United States	15.2%
China	29.5%

Figure 4.10a Top ten renewable hydro-electricity producers in the world in 2017. *(Data from IEA 2019a).*

Top ten wind electricity producers in 2017

Country	%
Rest of the world	18.1%
Turkey	1.6%
France	2.2%
Canada	2.6%
Brazil	3.8%
Spain	4.4%
United Kingdom	4.4%
India	4.5%
Germany	9.4%
United States	22.8%
China	26.2%

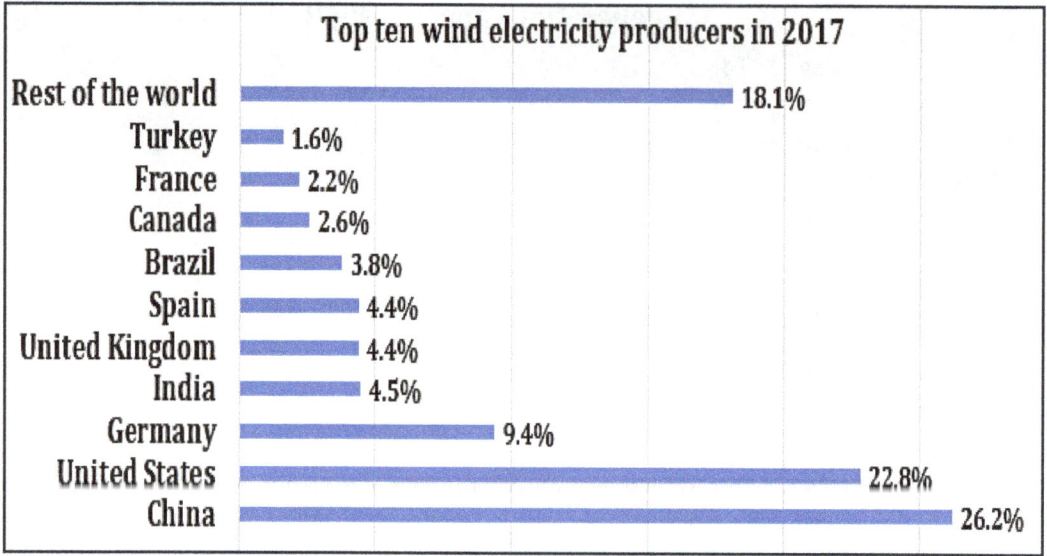

Figure 4.10b Top ten renewable wind electricity producers in the world in 2017 *(Data from IEA 2019 Keyworldstatistics).*

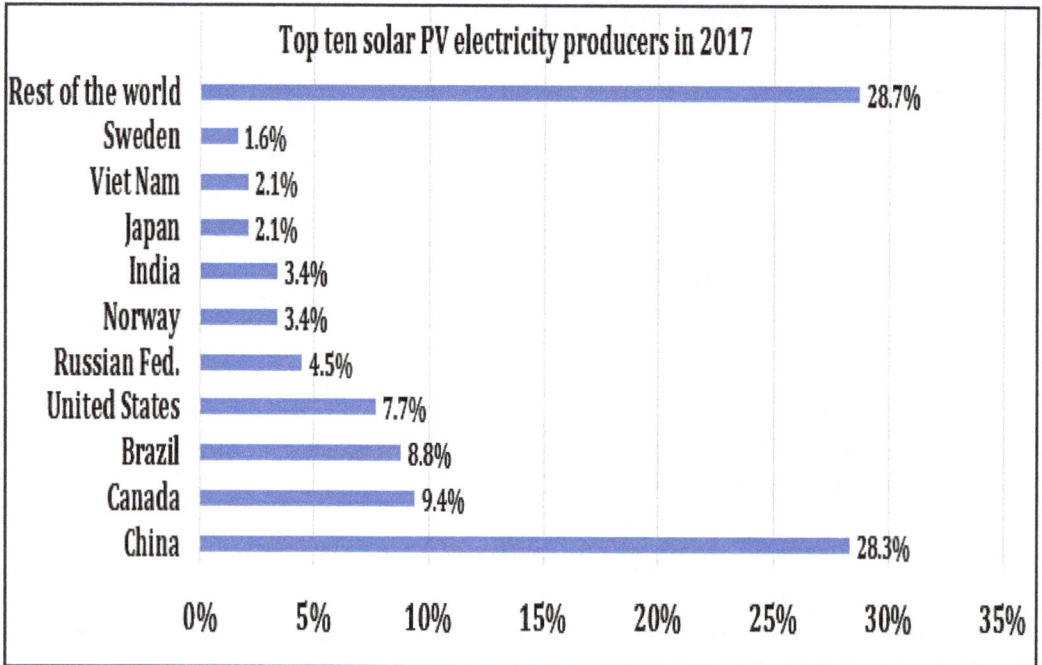

Top ten solar PV electricity producers in 2017

Country	%
Rest of the world	28.7%
Sweden	1.6%
Viet Nam	2.1%
Japan	2.1%
India	3.4%
Norway	3.4%
Russian Fed.	4.5%
United States	7.7%
Brazil	8.8%
Canada	9.4%
China	28.3%

0%   5%   10%   15%   20%   25%   30%   35%

Figure 4.10c Top ten renewable solar PV electricity producers in the world in 2017. *(Data from IEA 2019a).*

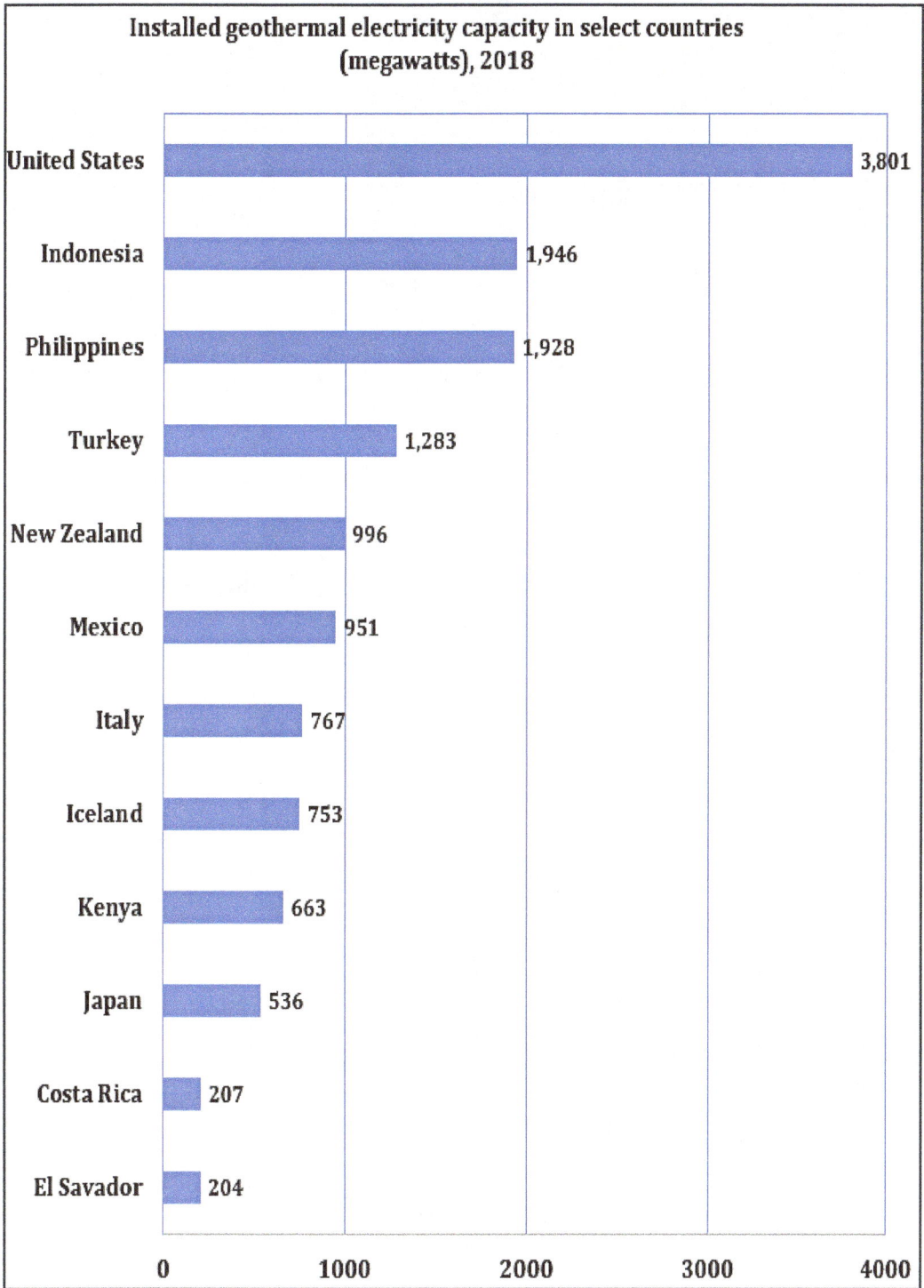

Installed geothermal electricity capacity in select countries (megawatts), 2018

United States	3,801
Indonesia	1,946
Philippines	1,928
Turkey	1,283
New Zealand	996
Mexico	951
Italy	767
Iceland	753
Kenya	663
Japan	536
Costa Rica	207
El Savador	204

Figure 4.10d Top ten renewable geothermal electricity producers in the world in 2017 *(Data from IEA 2019a).*

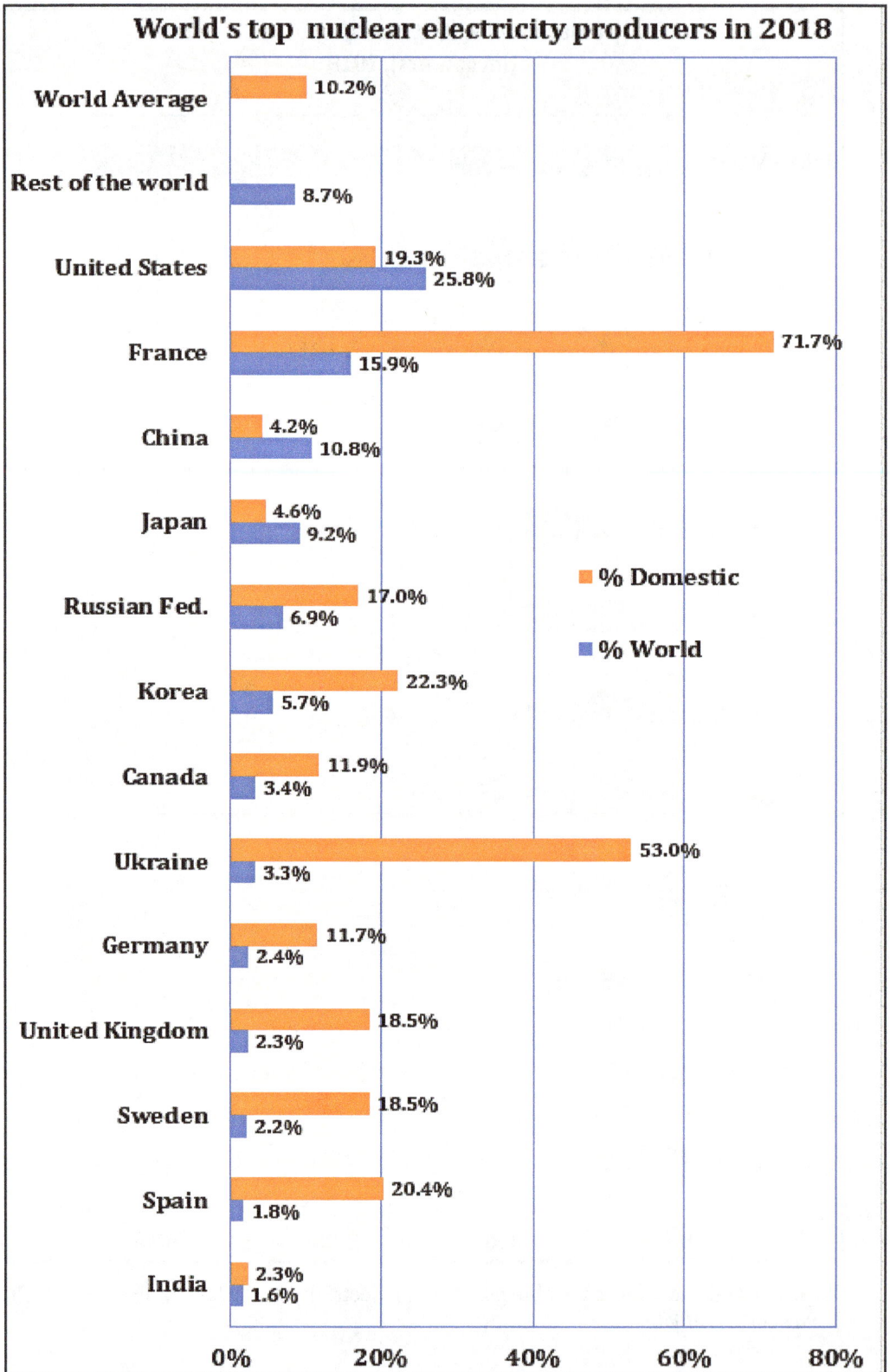

Figure 4.11 World's top producers of nuclear electricity in 2018.
(IAEA, 2019).

Coal has been the major fuel for power generation in decades and is the main fuel for many industrial processes: iron and steel, cement, chemicals and petrochemicals. Coal and natural gas are also feedstocks for the production of chemicals and petrochemicals. However, coal is also the most carbon-intensive fuel, accounting for over 40% of total emissions from fossil fuels. Increasing efforts to decarbonize fuel use by increasing the contribution of renewables, emerging technologies for producing cheaper gas, economic forces and strong policy disincentives are beginning to reduce the dominance of coal in power generation overall but many countries (mostly emerging nations) are increasingly meeting their growing energy demands with cheap, readily available coal. Over two thousand new coal power plants are currently under construction or planned worldwide. Furthermore, most of the coal-intensive industries of the world are now located in the emerging regions, in particular, Asia Pacific which accounts for over 80% of the total annual world demand for coal. On the contrary, developed economies are slowing down on the use of coal and saving fuel by improving efficiencies across fossil fuel production and use in industry and transportation. Many coal-fired power plants are being de-commissioned or converted to more efficient, increasingly competitive, and less carbon-intensive gas-fired plants. Coal is predominantly an indigenous fuel, mined and used in the same country, and only hard coals are traded internationally. Despite international action to reduce dependence on coal for heat and power generation, consumption has remained around 40% for the last 40 years, fueling 40-98% power generation even in some developed countries. Although the share of coal in primary energy demand and in electricity generation slowly continues to decrease, it still remains the world's largest source of electricity and the second-largest source of primary energy. China leads the world in coal-fired power output, followed by USA.

In spite of the very poor environmental credentials, coal has remained prominent in the global primary energy mix for nearly three centuries, still accounting for the second largest share after oil, and the largest source of electricity in 2019. However, the role of coal has continued to decline slowly over the last few years. But, after several consecutive years of slower growth rate, coal share increased again over the last two years by around 1% per year, although significantly lower than the typical 4.5% growth rate per year from 2000-2010. The rise over the last two years is believed to have been partly triggered by the unusually cold weather in Europe and North America over the period, but also due to increasing shift of demand and use to Asia Pacific that accounted for around 40% of the global power generation in 2018, with coal as the primary fuel. The region also produced over 60% of the total global output of coal, and projections show that the dominance of coal use in the emerging world is unlikely to decrease significantly over the next two decades.

Technologies for gasifying coal for power generation have been available for decades but have gained impetus only in the last decade or so as an effective way of reducing pollution, and many new coal-fired power plants across all regions will likely adopt this technology. The key advantage is that most of the toxic emissions can be captured at source and sequestrated. Coal is still the primary energy for the production of iron and steel, and 15-20% of the annual global consumption is accounted for by the sector. There is currently no feasible replacement for coal in the blast furnace and coal-fired direct reduction process, both of which produce around 75% of global steel requirements, but an alternative gas-based direct reduction technology is gaining ground, especially in developing countries which have access to cheap local gas resources. The main reason for the continued dominance of fossil fuels well into the foreseeable future is the lack of viable replacements in many applications. In spite of massive investments in renewable energy sources and the projected future rapid growth, the main contribution to global primary energy demand over the next twenty years or so will be less than 20%, and power generation will account for most of the use, with a share of 24-35%.

Hydroenergy and traditional biofuels will account for around 75% of this share and the combined contribution of solar and wind energy will still be around 10%.

Energy-related emissions make up three-quarters of global GHG emissions, and the power sector shows the largest growth. There is a consensus on the increasing urgency to decarbonize the sector, and promoting greater use of low-carbon renewables and nuclear energy is considered a top priority. Stronger policy support will likely continue to promote renewable energy across all regions of the world, in particular, solar and wind energy while most of the growth of hydropower will likely occur in the emerging world. There has been a major global shift of energy investments from fossil fuels towards low carbon technologies in the past few years, mostly in renewables and energy efficiency initiatives which accounted for over 95% of total global energy sector investments in low-carbon power in 2016. However, the scope for switch to lower carbon fuels may be limited in emerging countries which will account for most of the projected rise in electricity demand in the next two decades, with Asia accounting for around 60% of the growth. Many emerging countries will likely commission new coal-fired power plants, coal being the only locally available, affordable and relatively secure primary energy resource. Furthermore, some OECD countries: Australia, United States, Poland, Germany, currently depend on coal for 45% - 86% of electric power generation.

All recent projections agree that fossil fuels will still account for nearly 80% of the global primary energy consumption in 2040, that the share of power generation in total end-use primary energy will be around 50%, and that coal will continue to dominate the power generation fuel mix in most of the countries in the non-OECD regions. Although electric power demand is expected to rise by up to 70% by 2040, accounting for 55% of the world's energy demand growth, the share of power generation in total primary energy use will remain virtually unchanged, at around 42%. This is due to the expected significant improvement in efficiencies of existing and new generating plants. However, as electricity use rises, the types of energy used to generate it will diversify globally and regionally, led by natural gas, renewables, and, possibly, nuclear energy. Renewables (including hydropower) are the fastest-growing sources of power generation, increasing at an average rate of 2.8% per year. However, although non-hydro renewables are the fastest growing power generation fuels, they are starting from a very low base of just 7% in 2018, and the most optimistic projections estimate the contribution of renewables to power generation at no more than about 20-35% by 2040, around the same as coal, with wind and solar accounting for up to a third (IEA, 2017; EIA, 2017; IRENA, 2017).

Nuclear power, as a low carbon energy source has significant potential to contribute to the global effort to reduce the carbon footprint of electricity, but the future contribution is uncertain because most of the generating capacity is located in OECD countries , most plants are old and nearing their lifespan, and there are no clear policies on de-commissioning or replacement. However, Japan which shut down 52 of its 54 power plants after the Fukushima accident in 2011 has started to reactivate some, and it is expected that many more will come on stream over the next two decades. Also, some non-OECD countries already have nuclear power plants (China, India), and around 60 reactors are either under construction or at the planning stage in the region. Several projections show that global capacity in nuclear electric power generation will grow, mostly in non-OECD regions, and the contribution of nuclear energy to power generation will rise from its current 10-11% to 17-18% by 2040, based on the assumption that the current strong growth of capacity deployment in non-OECD countries will be sustained. However, the International Atomic Agency projects a global decline, even if all currently planned capacities become operational, unless many more countries especially in the OECD region decide to include nuclear electricity in their fuel mixes.

Production of biofuels (mainly ethanol and biodiesel) from sugarcane, corn, soybean, sorghum, and other organic matter has been rising, although sustainability has been questioned in recent years in connection with food-versus-fuel priority in land use. According to the United Nations Food and Agriculture Organization (FAO, 2018), about a third of the world population is currently hungry or malnourished. Biofuel products are used primarily in transportation, although a small proportion (biowaste) is also used for power generation, accounting for around 2% of global electricity output. The development of renewable energy, with the exception of hydro-energy has been slow despite the extensive availability of renewable resources, due to the intermittent nature of these resources and dependence on weather by most of them. For example, draught affects water levels for hydropower generation and cultivation of crops for biomass energy. The bulk of renewable energy, in particular, biomass, solar and wind (variable renewable energy, VRE) is used for electric power generation and the intermittent nature makes it difficult to integrate the power produced with national electricity grids. Also, the very low utilization capacities of solar and wind energy (20% and 30% respectively) are major constraints to their combined share in global power generation, which was only about 7% in 2017, projected to grow to about 10-12% in 2040, while the total contribution of all renewables including hydropower will rise to only about 25%. In the last two decades or so, efforts have been made to develop bi-fuel hybrid power plants which combine variable wind, solar, biomass or tidal power with a fossil fuel. Some hybrid solar-gas plants are already in commercial operation in some developed economies, in particular, the United States: renewable energy is deployed whenever available and the intermittent gaps are filled by natural gas. Developments in pumped storage are also creating new options of using solar/wind electricity when available to pump water to higher ground for later use in hydropower generation. Advances in battery storage technologies and increasing availability of used electric vehicle batteries which still have around 80% power storage capacity will also greatly improve the status of solar utility deployment.

Technology advances, particularly those that unlock efficiencies, and falling costs in recent years have been driving the adoption of renewable energy, particularly in the developing world, with the power sector leading the way. Weighted average levelized cost of PV solar and off-shore wind power generation now fall within the fossil fuel range, and projections show that both will become even more competitive over the next few years. Global investments have also been increasing, much of it in the developing world. In 2015, developing countries attracted the majority of renewable energy investments, with China alone accounting for about one-third of the global total (IRENA, 2017). Most of the investments in recent years have been in PV solar and wind, which together accounted for about 90% of total global investments. Solar PV is particularly suitable for utility power generation, serving small off-grid communities, commercial units, or household power generation, enabling consumers to produce power for their own needs and feed surplus energy into the grid. Utility scale installations are becoming increasingly competitive with new fossil-fuel power generation, even without subsidies. Renewables contribution to global primary energy use increased by 4% in 2018, accounting for almost one-quarter of demand growth and the power sector led the gains, with renewables-based electricity generation increasing at its fastest pace this decade. Solar PV, hydropower, and wind each accounted for about a third of the growth, with modern bioenergy accounting for most of the rest. Renewables covered almost 45% of the world's electricity generation growth, now accounting for over 25% of global power output, second after coal. China accounted for over 40% of the growth in renewable-based electricity generation, followed by Europe, which accounted for 25%. The United States and India combined contributed another 13% (IEA, 2019c).

Many developed countries are also providing attractive incentives to promote investments in solar and wind power generation, with very positive results. Renewables accounted for about 12% of the United States total energy consumption in 2017, with hydropower contributing only 2%. A combination of reductions in technology costs and implementation of policies that encourage the use of renewables at the state level (renewable portfolio standards) and at the federal level (production and investment tax credits) has been driving down the costs of renewables technologies, in particular, wind and solar photovoltaic, supporting their expanded adoption. Wind and solar generation will lead the growth in power generation fuel use by the country over the next two to three decades, accounting for about 64% of the total electric generation growth. The increase in wind and solar generation will lead the growth, accounting for around 94% of the total growth in renewable energy use (EIA, 2018). Many other countries in Europe also have strong policies that are promoting the use of renewables. It should be noted however that growth of renewable use is very sensitive to the cost of alternative fuels, in particular, natural gas. Lower natural gas prices always impact negatively on renewable power investments, as has happened in the last two years or so. Furthermore, although growth in solar and wind energy deployment has been very strong and is projected to rise even faster in the coming years, the actual contribution to renewable power generation will remain low, probably not more than 10-15% over the next two decades or so.

Bioenergy falls into two categories: biomass and biofuels. Biomass (traditional bioenergy) comprises largely wood, animal waste, agro-waste, all of which can be combusted to generate heat or biogas for home heating or raising steam for electric power production. About three-quarters of the world's renewable energy use involves bioenergy, with more than half of that consisting of traditional biomass use. Bioenergy accounted for about 10% of total final energy consumption and 1.4% of global power generation in 2017. Bioenergy has significant potential, particularly in populous emerging nations because biomass can be directly burned for heating and power generation or it can be converted into valuable liquid fuels which are much less polluting than fossil fuels. Investment in modern bioenergy has been strong in the last decade and world capacity has been rising by around 7% per year. In 2018, production grew at its fastest pace for five years, propelled by a surge in Brazil's ethanol output. Overall, Asia accounts for half of the growth, as its ambitious biofuel mandates aimed at reinforcing energy security also boost demand for agricultural commodities and improve air quality.

Geothermal power is used to generate steam for heating or producing electricity. About 100 countries have geothermal potential but only seventy are currently exploiting the resource for heating while twenty six are generating electricity. The United States accounts for about 30% of the global installed capacity and countries generating more than 15% of their electricity from geothermal sources include El Salvador, Kenya, the Philippines, Iceland, New Zealand, and Costa Rica. It is estimated that less than 10% of the global potential has been tapped so far (Figure 4.10d).

4.3.   END-USE (DELIVERED) ENERGY CONSUMPTION

Three sectors (industry, transport and building) use both primary and converted fuels. Industry and the building sectors use a variety of fuels, mainly coal, natural gas, and electricity while most of the products of oil - gasoline, diesel, jet fuel - are used by the transport sector. Natural gas and electricity account for less than 5% of the transport sector energy use. However, industry also uses oil, coal and natural gas as raw materials for chemicals and petrochemicals production, and bioenergy features significantly in the household energy mix Figure 4.12).

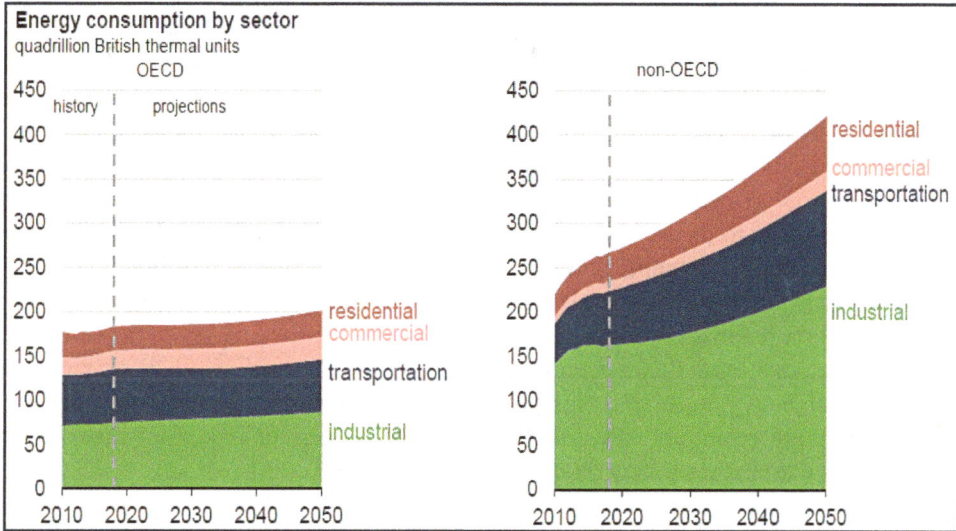

Figure 4.12a Projection of primary energy consumption by sector and region. *(EIA,, 2019).*

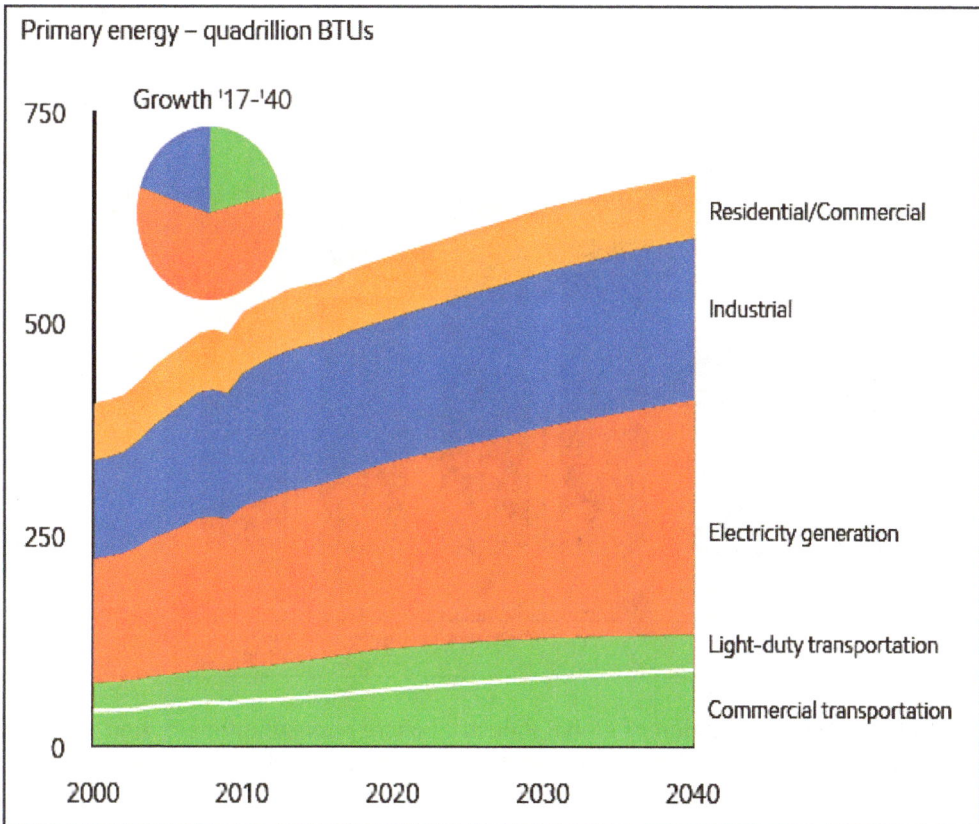

Figure 4.12b Projection of primary energy consumption by sector. *(ExxonMobil, 2019).*

4.3.1.   **Industry sector**

Modern economy depends critically on the industrial sector (including infrastructure and agriculture) for survival and growth. The sector is the largest user of energy globally, accounting for around half of the primary energy demand and 50% of electric power consumption. Areas of utilization include mining, manufacture of machinery, automobiles, production of consumer goods, primary metals, chemicals, paper, cement, metals polymers, agriculture, pharmaceuticals, food, etc. Industry also uses energy both as fuel and as feedstock for the production of chemicals, petrochemicals, lubricants, waxes, asphalt, etc. Oil, natural gas and electricity dominate the industrial fuel mix, and will contribute about one-third of industrial energy growth over the next two decades (Figure 4.13). The projected rise in oil share is due mainly to its use as a chemical feedstock. Coal plays a key role in the production of iron and steel, cement and other non-metallic manufacturing, chemicals and petrochemicals, heat and steam raising, and the emerging economies will account for most of the growth in demand. Non-OECD regions currently consume seven times as much coal as OECD regions, and this scenario is expected to continue in view of the continuing movement of heavy industrial manufacturing to the emerging nations. Fuel switching opportunities are relatively rare for these industries, and, in any case, many of the host countries have no access to any other cost-effective fuel.

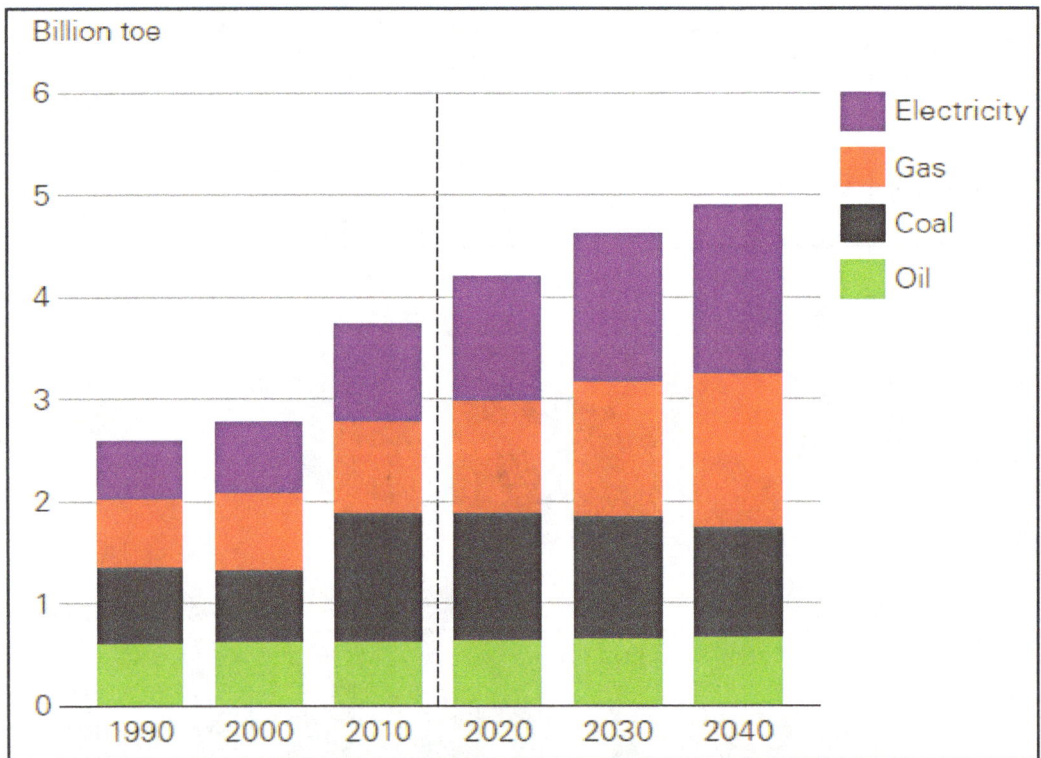

Figure 4.13a Projection of global industrial energy consumption by fuel *(BP, 2019).*

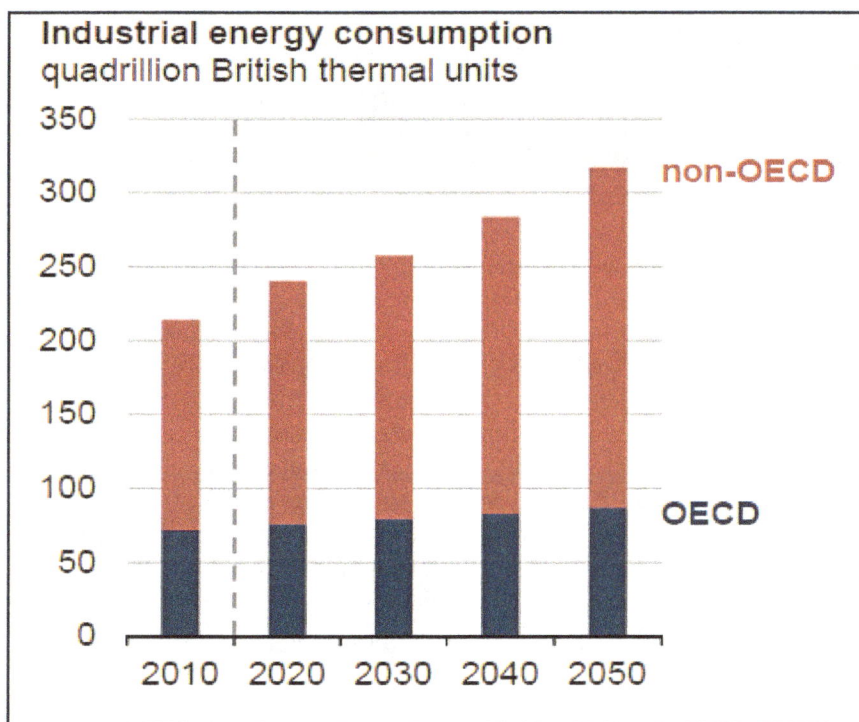

Figure 4.13b Projection of global industrial energy consumption by region *(EIA,, 2019).*

With the expected growths in the global economy, population, human development and consumer demand, energy demand by the industrial sector is expected to grow by about 25-30% by 2040, but much of the growth will be in the emerging countries in view of the expected rapid growth in industrialization, especially the heavy industries such as iron and steel, chemicals, petrochemicals and cement production. Demand for fossil fuels as feedstocks for chemicals production has been rising and most facilities are in the emerging economies. Chemicals are the building blocks for a very wide variety of consumer products: fertilizers, adhesives, lubricants, cosmetics and personal care products, textiles, plastics, paints, medical devices, plastics, pharmaceuticals, automobile, airplane and computer components, and many other industrial and home goods. The primary products industries are energy-intensive and offer relatively small scope for efficiency improvement. Energy demand projection in the industrial sector has been based on the assumption that the current global upward trend in industrial manufacturing efficiencies and adoption of new technologies which translate to lower energy intensity will be sustained. While this is likely in OECD countries and China where energy intensity in the industrial sector has improved markedly over the last two decades, other emerging economies may not be able to afford the added cost of deploying newer, more energy efficient technologies.

The overall projected rise in energy demand in the industrial sector by 2040 is around 30% but there will be significant variations in the sub-sectors: while the growth in the iron and steel and cement industries will be about 25%, energy demand by the chemical industry sector will be double, around 50%, largely because of the strong growth in demand for products in the emerging regions. A significant shift in the fuel mix is expected by 2040, particularly in the heavy industry sub-sector. Coal use will fall from over 20% to around 15%, natural gas will rise to about 25% and oil use for energy will decline. China accounted for the bulk of the growth in industrial energy consumption in the last two decades but a significant shift in the regional dynamics is expected, and India will account for the highest growth rate over the next two decades. About 10% of current oil and gas demand by industry is used in non-energy applications, mainly as feedstock for the chemical, petrochemical, and other industries. Around 6% of the global demand of coal, 16% of oil and 12.5% of natural gas were used as raw materials for non-energy products in 2016 - chemicals, fertilizers, polymers, waxes, lubricants, bitumen, etc. The demand for these products has been rising, consistent with rising prosperity across all regions. The chemicals sector (which uses energy both as fuel and feedstock) is emerging as a major energy consumer in the industrial sector of the economy, particularly in many emerging markets where demand for chemical products is growing exponentially, outpacing GDP growth. Furthermore, the global economy is becoming increasingly dependent on these sources for the supply of vital primary materials and components sourced mainly from the emerging regions.

Growth of non-combusted fossil fuel demand has been very strong, and is projected to rise almost twice as fast as the rate for other sub-sectors of industry, rising by around 45% from current level by 2040. Oil will account for nearly two-thirds of the growth in the non-combusted use of energy, with natural gas providing much of the remainder. The chemicals/petrochemicals industry is also the fastest growing energy user in the industrial sector, and has been migrating to many emerging nations with competitive advantage like low-cost feedstocks or increasing local demand. Developing Asia Pacific will account for the largest growth, driven by population size and strong economic growth. By 2040, growth of the sub-sector is expected to more than double in the Middle East, stimulated by access to low-cost oil and gas feedstocks, while China is also expanding its chemicals industry which uses coal both as fuel and chemical feedstock. Most of the products of the chemicals industries which are vital inputs for many industrial products and consumer goods - naphtha, methanol, ethanol, fertilizers, plastics, pharmaceuticals, and many other products are traded globally as intermediate products. For example, China supplies around 40% of global demand for methanol, and more than a quarter of the world's total plastic consumption, both produced mainly from coal. The chemicals sector of industry offers limited scope for efficiency gains, especially when operating in emerging countries, and has become the most polluting sub-sector of the global industry.

4.3.2. Transportation sector

The transport sector currently accounts for 21% of global primary energy demand. The global car fleet in 2015 was about 0.9 billion and is expected to double to around 2 billion by 2040 (BP 2018, Statistica 2020). Almost all the growth is in emerging markets where rapid economic growth, urbanization and rising incomes boost car ownership. Car ownership in non-OECD countries will triple over the same period, from 0.4 billion to 1.2 billion and a similar growth pattern in the commercial vehicle sector is also projected. Oil will remain the world's primary fuel in the foreseeable future, driven by demand for transportation and by the chemicals industry. Oil accounted for about 32% of the global primary energy in 2018 and there will be no significant change in 2040 - around 32%. About 94% was used for transportation in 2018 but some reduction is expected in 2040 (possibly down to 85%) due to increasing efficiency and the expected rise in the use of electric cars, biofuels, and natural gas in transportation (Figure 4.14).

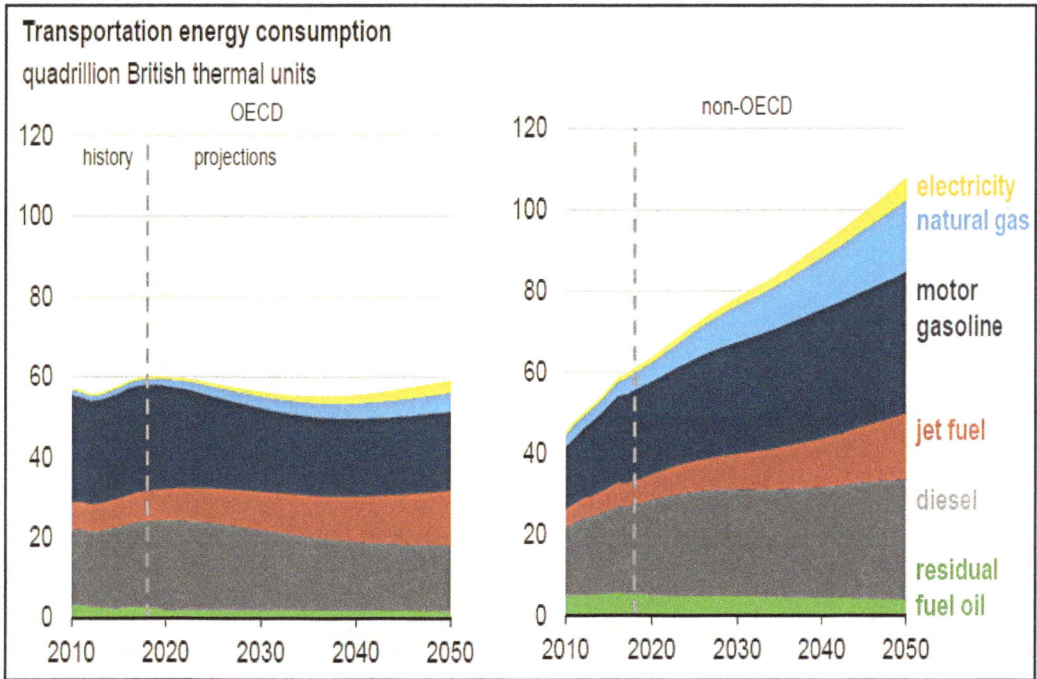

Figure 4.14a Projection of transportation energy consumption by sub-sector and region in 2018. *(EIA, 2019).*

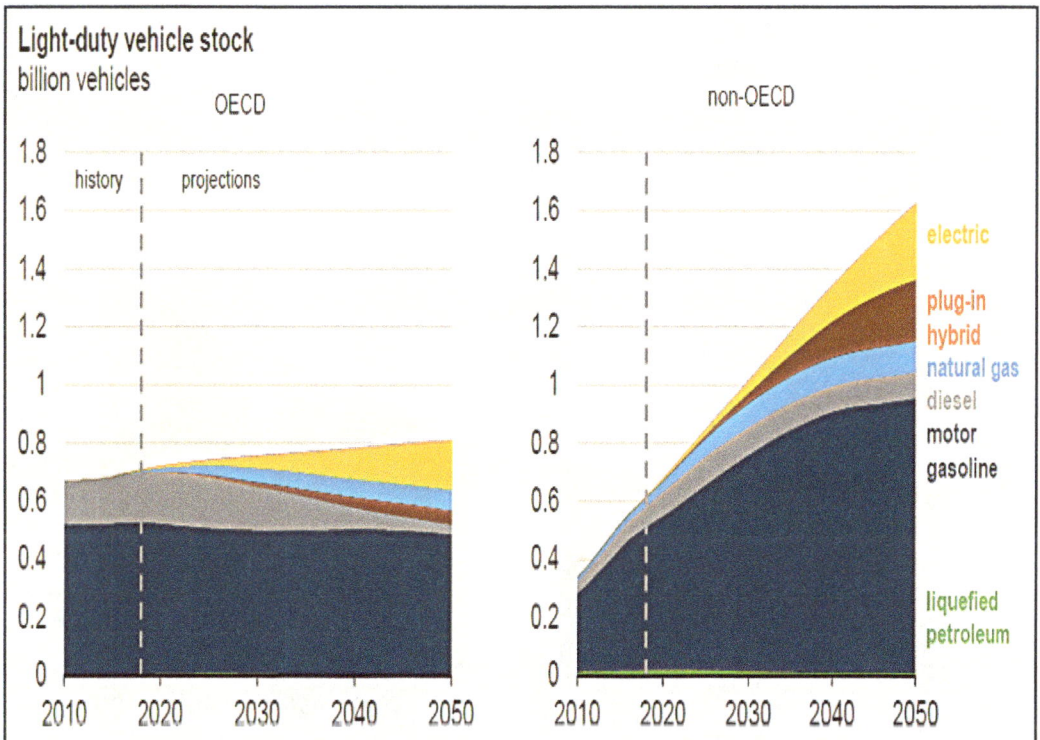

Figure 4.14b Projection of light-duty transportation energy consumption by fuel in 2018. *(EIA, 2019).*

Passenger transportation energy consumption
quadrillion British thermal units

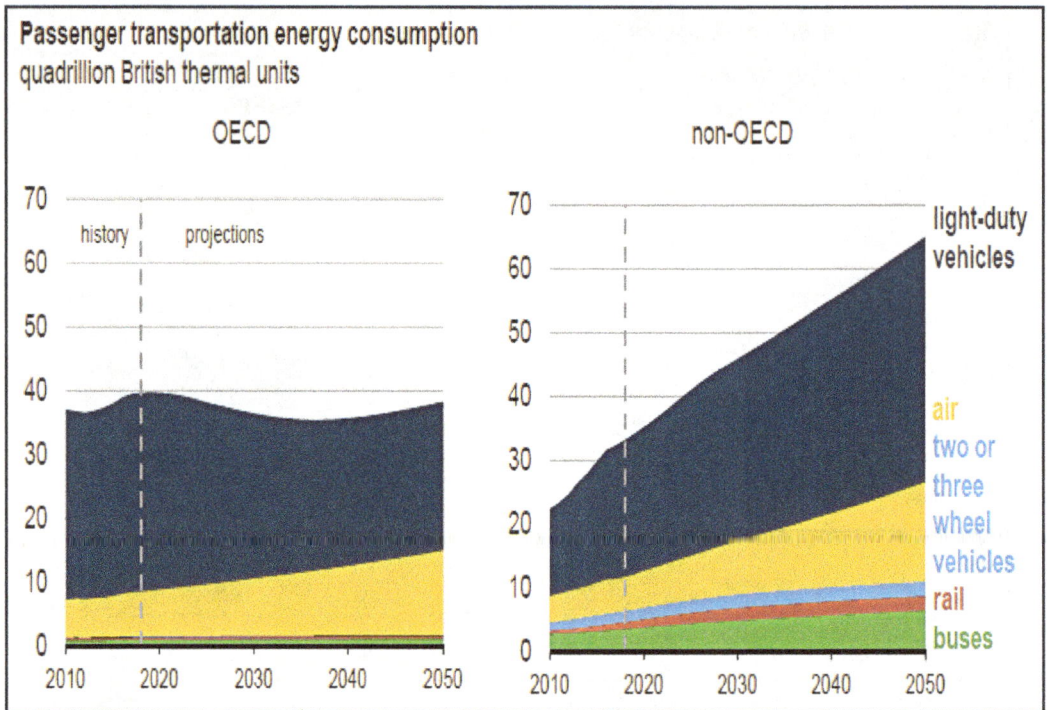

Figure 4.14c   Projection of passenger transportation energy consumption by type in 2018 *(EIA, 2019)*.

Freight transportation energy consumption
quadrillion British thermal units

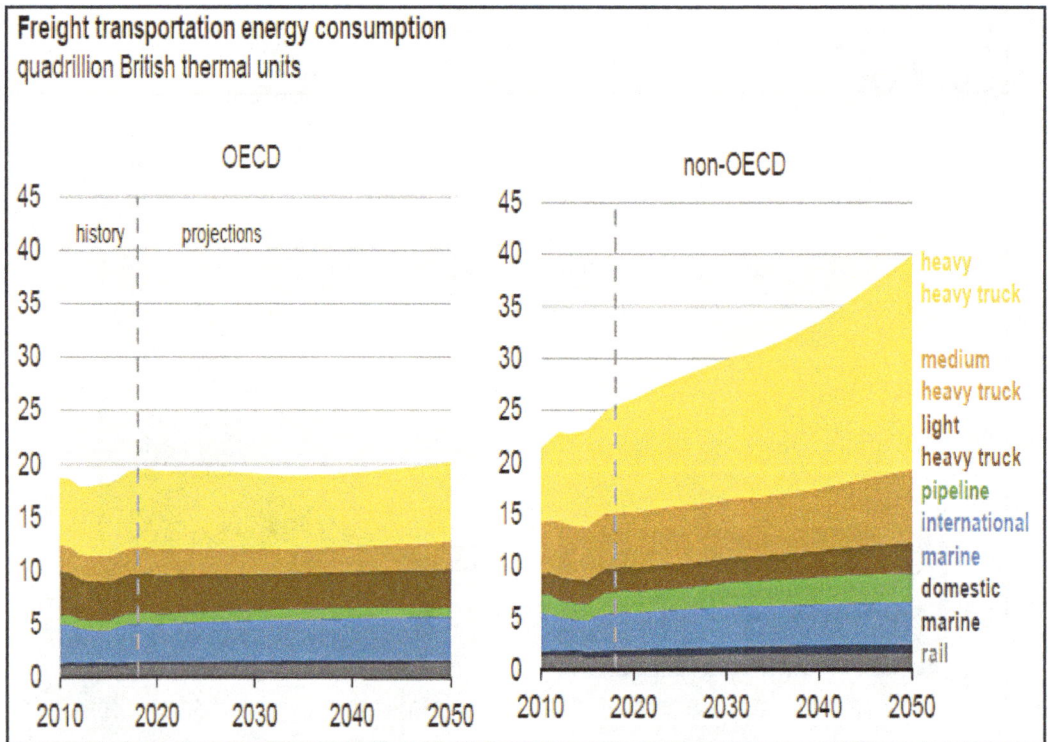

Figure 4.14d   Projection of freight transportation energy consumption by type *(EIA, 2019)*.

The demand for diesel is expected to grow by about 30% by 2040 to meet the projected increase in trucking and marine transportation, while jet fuel will rise by about 50%. Fossil fuel-sourced petrol, diesel, gas, and aviation fuel remain dominant in the transportation sector but increased conversion of biomass to gasoline and diesel will help to reduce demand to some extent over the next two decades or so. Fuel efficiency of transportation vehicles has been rising and the trend is expected to continue: the average kilometers covered per liter of fuel is expected to double by 2040, and this should reduce demand for oil significantly. However, the population of vehicles is growing exponentially, expected to double by 2035-2040, and the bulk of the growth will be in the emerging world where there is little policy-inspired fuel efficiency drive. This is expected to push up the overall demand for transportation fuel by about 30% although the demand in OECD countries will remain flat or decline.

The main components of the oil demand are gasoline, diesel oil and aviation fuel, gasoline being in the highest demand. Diesel accounts for around 35% of the total energy used for transportation but demand is growing rapidly and is expected to surpass gasoline by 2040. The main driver is the growth in heavy-duty truck transportation which accounts for about 80% of the total demand for diesel oil consumption. Global energy demand for heavy-duty vehicles is expected to increase by about 45% by 2040 and about 85% of the growth will be in emerging economies where economic activities are increasing most rapidly, but growth in the developed countries will also be strong because of the rising trend in e-commerce. Light-duty vehicles and marine transportation account for the remaining 20%. Clearly, diesel oil will remain predominant in the heavy-duty transportation sector in the foreseeable future. Some heavy trucks and buses are designed to use compressed natural gas but the share of the fuel is only about 2% of the total transportation fuel demand, although this is likely to rise to about 5% by 2040. The main problem is the cost of gas-powered trucks which is significantly higher than the cost of a comparable diesel truck but, depending on local availability and pricing of gas, there could be significant savings in fuel cost. Some city bus and heavy truck fleets are running already on liquefied natural gas in some OECD countries.

Fuel use by the air planes, ships and trains is growing rapidly from about 50% of the amount used by light-duty vehicles in 2014 to about 85% in 2040 (ExxonMobil, 2016). Over 90% of the demand by the sector will be met by oil and strong growth in demand is expected in spite of increasing fuel efficiency, largely because the aviation industry is growing rapidly in response to the increasing air travel worldwide, and there is as yet no viable substitute for jet fuel. Modern bioenergy technologies produce liquid fuels that are renewable substitutes for gasoline and diesel from a wide range of starch crops that can be fermented to produce alcohols – grains, sugarcane, vegetables. etc. Bioethanol is perhaps the most advanced and is being blended with gasoline in many advanced countries and Brazil, the world leader in biofuel technology has the largest fleet of vehicles running partially or fully on bioethanol. Other valuable fuels notably biodiesel and aviation fuel are also being produced. Although jet fuel from biomass is beginning to emerge as jet fuel mix in some developed markets, consumption remains very low: biofuel contribution to jet fuel use in 2018 was less than 0.1%. In addition to biofuels, renewable electricity is projected to provide around 10% of renewable energy in transport by 2024, most of which will be in China because of the current large investments in ethanol production largely from coal. Growth in Brazil and the United States has been strong and both will still account for about two-thirds of the total world output of biofuels in 2024 (IEA 2019c). Biofuels will make the most significant impact on transportation but, starting from a very low base, the maximum contribution to global transportation and aviation energy needs to grow from about 3% in 2018 to around 9% by 2030 under the International Energy Agency Sustainable Development Scenario (SDS).

The electric car market has been growing, be it slowly over the past decade or so. There are three types of electric vehicles: Battery-electric (BEV) which is powered solely by a battery charged from mains electricity; Plug-in Hybrid (PHEV) which features both a battery and an internal combustion engine powered by petrol or diesel; Extended range EV (E-REV), a version of PHEVs, powered by a battery but with a petrol or diesel powered generator on board. A battery-electric vehicle runs purely on mains-charged battery and has a typical range of up to 150 kilometers between charges, depending on driving style and use of accessories (heating/air conditioning and electronic equipment etc.); the PHEVs have a range of around 45 kilometers on pure electricity but once the battery is depleted, the vehicle can switch to hybrid mode and run on gasoline/diesel, delivering 400-450 kilometers. The E-REV is a variant of the PHEV, and is powered by a battery with a petrol or diesel powered alternator on board that recharges the battery continuously, delivering a total range of up to 450 kilometers. Plug-in hybrids are currently more popular compared with plug-in vehicles which rely entirely on batteries that have to be recharged frequently, but they are more expensive.

The main advantages of electric cars are the relatively lower fuel consumption (around 30% fuel economy compared with gasoline powered vehicles) and very low tailpipe carbon footprint because water is the only exhaust emission. Also, EVs are particularly well-suited to urban driving, one of the main sources of ambient pollution. Apart from being environment-friendly, plug-in cars offer a number of other benefits compared with conventional vehicles. Depending on the relative costs of gasoline and electricity, specific energy cost (energy cost per kilometer) for BEVs could be only a quarter of the cost of a similar internal combustion engine (ICE) vehicle, especially when owners have access to low off-peak electricity tariffs; there are fewer mechanical components in an electric vehicle when compared with conventional vehicles, which often results in lower servicing and maintenance costs; and there are many policy incentives which include subsidies on initial purchase cost of eligible EVs and cost of installation of charge points; lower taxes; and even free parking in some urban areas. However, there are also several issues with electric vehicles, associated mainly with batteries and access to charging bays. Batteries of electric cars are huge and expensive, accounting for up to 40% of the manufacturing energy and costs of an electric vehicle; it takes 6-8 hours to fully charge a lead-acid battery which delivers only a few hundred kilometers between charges; there are few commercial battery charging bays compared with conventional gasoline stations; and, crucially, consumer interest is low and government policies supporting EV ownership are generally weak and shifting in many countries. However, several more efficient batteries, notably lithium-ion batteries with fast recharge cycles are becoming the choice for all but the low-cost electric vehicles and costs are declining rapidly.

The penetration of electric vehicles into the global auto market has been very slow: there were about 5 million electric vehicles in the global fleet (about 0.2% of the total) in 2018, with 45% in China, 24% in Europe and 22% in the United States (Figure 4.15). Several recent projections on EV market penetration show that the global population of electric vehicles compared with conventional autos will remain low in the foreseeable future. The most optimistic outlooks project that electric vehicle population concentrated within passenger cars, light duty vehicles and public buses will grow to about 350 million (of which around 300 million are passenger cars) accounting for about 15% of the total global vehicle fleet in 2040. One optimistic projection expects that around 14 million EVs (around 26% of the projected global sales of new vehicles) could be sold in 2020 mainly in China, Japan, United States, and Western Europe. About 3 million will be fully electric cars and range extenders, while the balance of 11 million will be hybrids (BCG, 2018). The global electric vehicle outlook published recently by the International Energy Agency (IEA, 2019) shows that the world electric car fleet in 2018 was about 5 million, up by around 2 million from the previous year, projected to rise to between 130-250 million by 2030. The projection was based on historical analysis and all current and anticipated electric vehicle development initiatives.

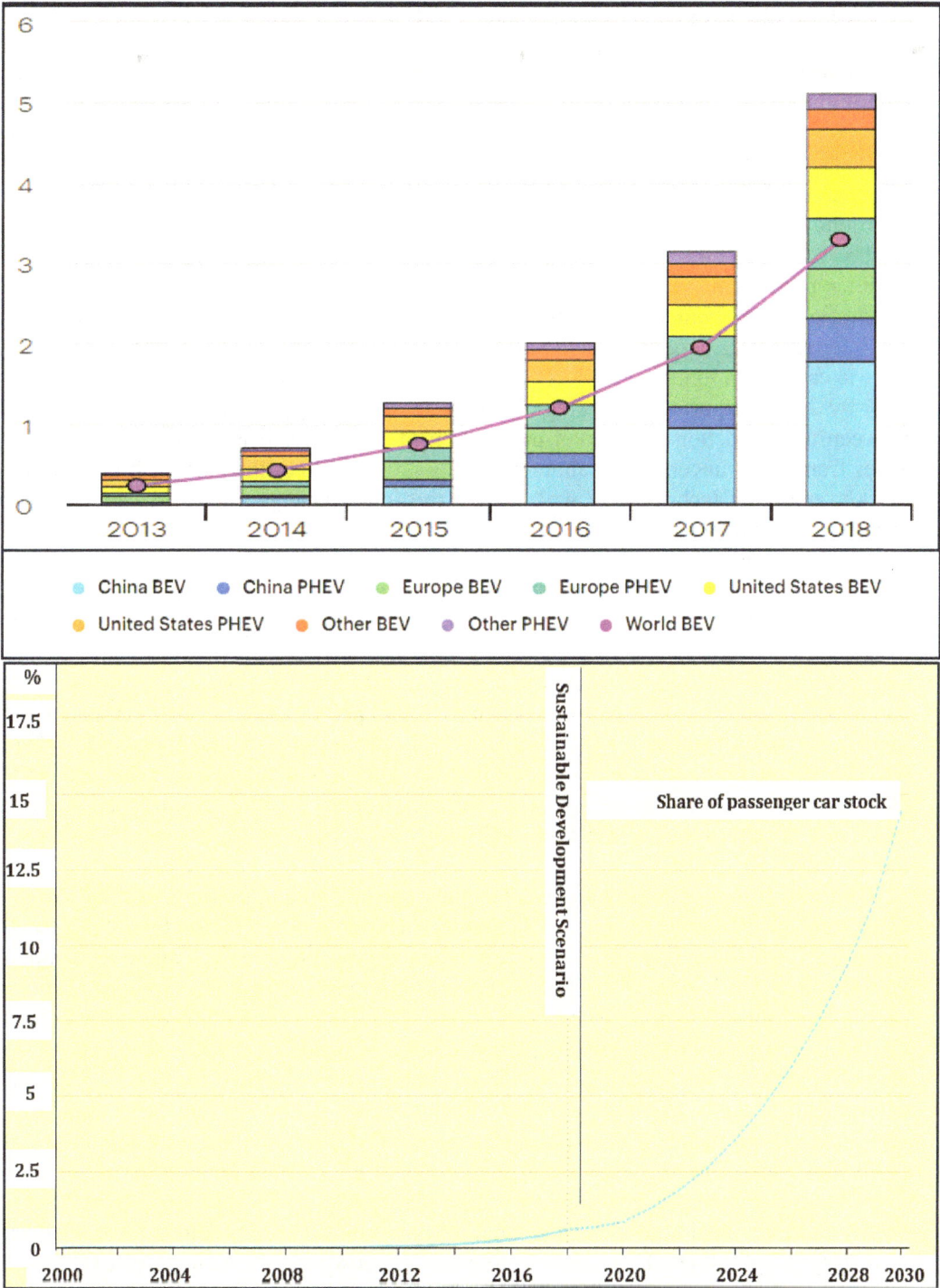

Figure 4.15 (a) Electric vehicle market penetration, 2013-2018 (b) Sustainable Development Scenario for electric vehicles *(IEA, 2019e).*

Electric cars and light commercial vehicles have dominated the market so far but there has also been substantial growth in the global stock of two-wheelers, electric buses and medium electric truck sales. China remains the world's largest electric car market, accounting for nearly half of the total stock, followed by Europe and the United States. This growth rate is phenomenal but the total EV population was still less that 3% of the global vehicle fleet in 2018. A combination of technological innovations particularly in battery performance and cost, and strong, resilient policy support instruments could push EV vehicle sales to around 23 million in 2030 and the total global stock to 130-150 million, but this would still be less than 10% of the projected 1.8-2 billion vehicle population. A more optimistic EV30@30 initiative (with a combined pledge of 30% market share of EV vehicles by 2030) projects sales and stock at 43 million and 250 million respectively. BP (2019) projects a steady rise to around 350 million by 2040, with roughly equal proportions of PHEVs and BEVs, but electric vehicles will still account for much less than a fifth of the expected 2 billion cars. Another projection by ExxonMobil (2018) is less optimistic and expects a total market penetration to be about 160 million in 2040, with BEVs accounting for around 60%. Under the International Energy Agency Sustainable Development Scenario, the penetration of electric cars needs to grow from its current less than 1% of the total global vehicle fleet to 15% by 2030, requiring an average growth of 30% per year. The significant difference between the projections of market penetration of electric vehicles in the various analyses arises from many uncertainties about the likely growth rate of the market. It is unclear how quickly emerging technologies will resolve the formidable problems of battery cost and paucity of private/public low-speed/high-speed charging bays, and battery swapping bays. Also, a faster pace of fuel efficiency improvement of conventional engines, a slow-down in the global economy resulting in weak oil demand and low oil prices in the local or global markets could slow down EV market penetration, as has been happening in the United States for some time. Furthermore there is clear preference for PHEVs which use both electricity and gasoline in the United States and several countries in Europe. A typical internal combustion engine car will consume around 3 liters per 100 km, in 2040 compared with 5 liters today and 7 liters in 2000 (BP, 2018). Only about 2-4% of global electricity demand will come from the transportation sector in 2040 and oil will still account for around 85% of total fuel demand by the sector, down from the current 94-95%. Natural gas, electricity, and a mix of other types of fuels are each projected to account for around 5% of transport fuel by 2040 (BP, 2018).

The mostly inefficient two-wheeler population in China still dominates the global EV market but the estimated average global emissions of 41 million metric tons CO_2eq in 2018 was nearly 50% lower compared to an equivalent internal combustion engine fleet, although the electricity for battery charging still comes from carbon intensive sources. The lower carbon footprint of electric vehicles (on well-to-wheel basis) ultimately depends on the fuel mix for power generation: while the carbon advantage would be significant in Europe due to major progress in decarbonizing power generation, the   positive impact of EVs on the environment could be minimal in China which hosts around half of the global population of EVs but generates most of its electricity (around 65%) from coal, the most highly-polluting fossil fuel. Furthermore, a substantial proportion of EVs would be hybrids which still use fossil fuels. For example, although more than two-thirds of EVs sold globally in 2018 were battery electric vehicles (BEV), almost half of the electric vehicles sold in the United States were Plug-in hybrids (PHEVs), driven largely by the downward trend in the price of gasoline. Also, about three-quarters of electric vehicles sold in Japan and several European countries in the same year were PHEVs. There are several major constraints to the growth of the EV market, notably the need for an electric vehicle to be lightweight whereas the average vehicle weight has more than doubled over the last twenty years or so; consumer apathy; inadequate and shifting policy support; and lack of adequate number of fast charging bays. Although there were as many battery charging infrastructures as there were EVs in 2018, most were private, low-charging systems which require 6-8 hours recharge time. This effectively excludes the large number of people who live in apartments or homes without a garage or front parking lot, and accentuates fears of being stranded on trips due to discharged batteries.

The hydrogen-powered car (FCEV) is a new variant of the electric vehicle which is becoming a reality, with several automobile manufacturers bringing out models. The car is powered by a fuel cell which generates electricity from liquid hydrogen that can be produced from many sources. The main advantage is that there is no direct ambient tailpipe pollution which is a major source of urban smog and risks to human health and other ecosystems since the only emissions are heat and water. Also, hydrogen vehicles do not require battery charging, and hydrogen refilling stations could produce hydrogen from water on site, using solar power when available, and store as liquid hydrogen, thus eliminating the need for bulk transportation of highly inflammable liquid hydrogen. Furthermore, any pollution associated with hydrogen production (which can be substantial) can be contained at source. One major disadvantage is that hydrogen production by any method requires enormous electricity, hence hydrogen fuel is environmentally sustainable only if the electricity comes, not from fossil fuels but from renewables or nuclear energy. Hydrogen for transportation can be produced from many sources - fossil fuels (natural gas, coal, oil), biomass, or water which is the most environment-friendly source. It is abundant, renewable, and cheap, but production is power-intensive. Solar energy is particularly promising because it can be used to produce liquid hydrogen when available and the product can be stored or transported easily like conventional liquid fuel. This takes care of the prominent disadvantage of solar power intermittency. It also means that both the electricity and liquid hydrogen can be produced at filling station complexes thereby eliminating transportation of highly inflammable liquefied gas over long distances.

A typical hydrogen car currently on the market delivers 400-450 kilometers on a tankful of liquid hydrogen, comparable with a regular petrol/diesel car, and higher than most mainstream electric cars. Some developed countries are already promoting hydrogen cars and developing hydrogen filling stations. The London Metropolitan Police Service in the United Kingdom recently took delivery of eleven Toyota Mirai hydrogen cars. It is the world's largest fleet of hydrogen fuel cell police cars, and the Force plans to grow the fleet to 550 by 2020. The current five hydrogen filling stations in London are adequate for now, but success of the hydrogen car will depend on an aggressive policy-driven proliferation of filling stations, particularly in cities where the positive impact on the environment would be highest. Also, the cost of a hydrogen car would be a major deterrent for quite a while - a hydrogen-powered car costs around three times that of a comparable internal combustion engine car, although they have similar fuel efficiency. The global population of fuel cell electric vehicles (FCEV) was only about 11,000 in 2018, with half of them in the state of California, United States and another 25% or so in Japan. While FCEVs do not need charging bays, the number of hydrogen refueling stations is still severely limited. Furthermore, the risk of explosion is still very high because the hydrogen is stored under high pressure in dedicated tanks - there were several serious accidents in Europe in recent months. Nevertheless the future prospects for FCEVs are bright: in addition to passenger cars, fuel cells are featured in various other transport applications including fuel cell buses, heavy trucks, trains, ships, spacecrafts.

Current global discourse on climate change which has aroused keen public interest in contributing to global mitigation efforts on energy-environmental sustainability could continue to push interest in EVs among the car-buying public as well as policymakers.

Furthermore, increasing prominence of environmental issues in global discourse and politics, policy incentives for EV ownership (and disincentives for purchase of ICE vehicles, in particular, diesel cars), particularly in North America and Europe are beginning to stimulate consumer interest. Apart from subsidies on cost of purchase of eligible EVs and provision of home, street and workplace charging facilities, conventional ICEs are facing stiff tax regimes which are already discouraging many from investing in new cars particularly in Europe. However, there has been no dramatic rise in the sale of electric vehicles either, indicating that people may be confused and are simply delaying investment in new vehicles for now: EV sales in Europe in 2019 accounted for only 3% of the total European vehicle sales. The potential for reducing lower atmospheric (tropospheric) pollution which casts toxic smog over urban areas round the world and accounts for numerous health issues is particularly high, and many features of EVs make them particularly suitable for urban transportation. There has been growing worldwide optimism that electric vehicles (EV) will virtually take over the current role gasoline-powered light passenger transportation by 2040, and for good reason: there is no pollution at point of use, although they could still have a substantial carbon footprint depending on the source of electricity. In fact, some European countries have recently announced plans to ban the sale of conventional gasoline/diesel cars by 2040. However, none of the projections reviewed supports this optimism, based on the following facts.

- A fully charged lead-acid battery delivers less than 500 kilometers and the average recharge time is around 6-8 hours depending battery type and size; facilities are mostly private and are slow chargers. Lack of adequate public battery recharging/swapping infrastructure will remain a major constraint to EV proliferation, especially for people who live in apartments or park in the streets. However, battery technologies are on a fast track and new lithium-ion batteries that can be fully charged within an hour are coming on the market, but they are very expensive and require special charging bays which are also expensive. Policies for providing fast public charging facilities or proliferating commercial battery swapping facilities are currently inadequate in most countries of the world. Proliferation of commercial battery swapping bays could greatly resolve the issue of being stranded due to flat batteries because such facilities would enable drivers swap batteries in minutes and move on. Pilot projects are emerging in some countries, notably Israel. However, batteries need to be standardized for such schemes to be successful, currently they are not. Lithium-ion batteries have memory issues and energy retention declines with each charge and the battery has to be replaced when energy retention capability falls below around 80% (around 150,000-200000 kilometers). Replacement cost currently would be about half the cost of the car but there is considerable optimism that battery costs will decline rapidly over the next few years.

- Conventional gasoline cars are significantly cheaper than equivalent EV cars, fuel-efficiency is rising, maintenance costs are falling, and attractive features of EVs, like start-stop engines are beginning show up in conventional cars, creating doubt whether consumer preference, (currently heavily in favor of conventional vehicles and hybrids which combine ICE and EV features) will change very much over the projection period. Conventional cars are backed by over a hundred years of innovative research and development experience, which places electric vehicles at a significant disadvantage. Although many auto manufacturers already have EV models, most still consider ICE vehicle manufacture the business core. Furthermore, in China and the United States which accounted for nearly 75% of global EV sales in 2018, consumer preference for heavier ICE vehicles such as SUVs is evident whereas EVs need to be as light as possible. Furthermore, China has pulled back on EV policy incentives and the United States is in the process of doing same. Most recent projections show that refined petroleum and other liquid fuels will remain dominant, supplying 90-95% of global transportation fuels in 2040, and conventional liquid fuel-powered vehicles will remain the most popular, especially with low global oil prices, rising transportation efficiencies, and rapidly growing commercial, marine and air transportation. The most optimistic projections expect the population of electric vehicles to grow substantially, making up 15-35% of the projected 1.8-2 billion vehicles in 2040, but this would depend critically on upgrading current policy supports. EVs will penetrate mainly the light-duty vehicle sub-sector which accounts for less than half of transportation energy, and most of the growth will be in OECD countries while emerging countries will account for most of the rise in transportation energy demand. The currently low access to modern energy and electricity will be a major disincentive to EV penetration in these regions, with the exception of China which currently accounts for the world's largest sales of EVs but sources two-thirds of its electricity from high-carbon fuels.

4.3.3. Building sector (Residential and Commercial)

Building sector primary energy demand which currently accounts for about 29% of the global total is projected to grow to nearly a third by 2040, faster than the growth in demand by the industrial and transport sectors. The total number of households worldwide will increase by around 40% by 2040 and over 90% of the growth in building energy use will be in the emerging world. The relatively warm climate in most of these regions means a low demand for space heating while the majority of household energy demand will be driven by the need for space cooling. Also, the growing urbanization, prosperity and rapid upward social mobility will drive up ownership of electrical appliances and building sector energy demand. (Figures 4.16).

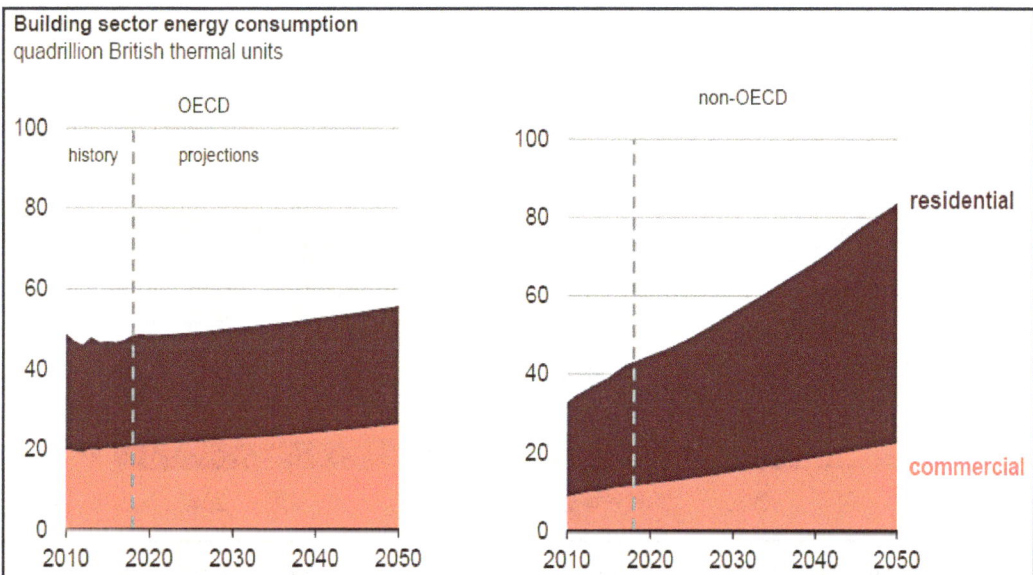

Figure 4.16a Projection of building sector energy consumption by region and sub-sector *(EIA, 2019).*

Residential energy consumption per person
million British thermal units per person

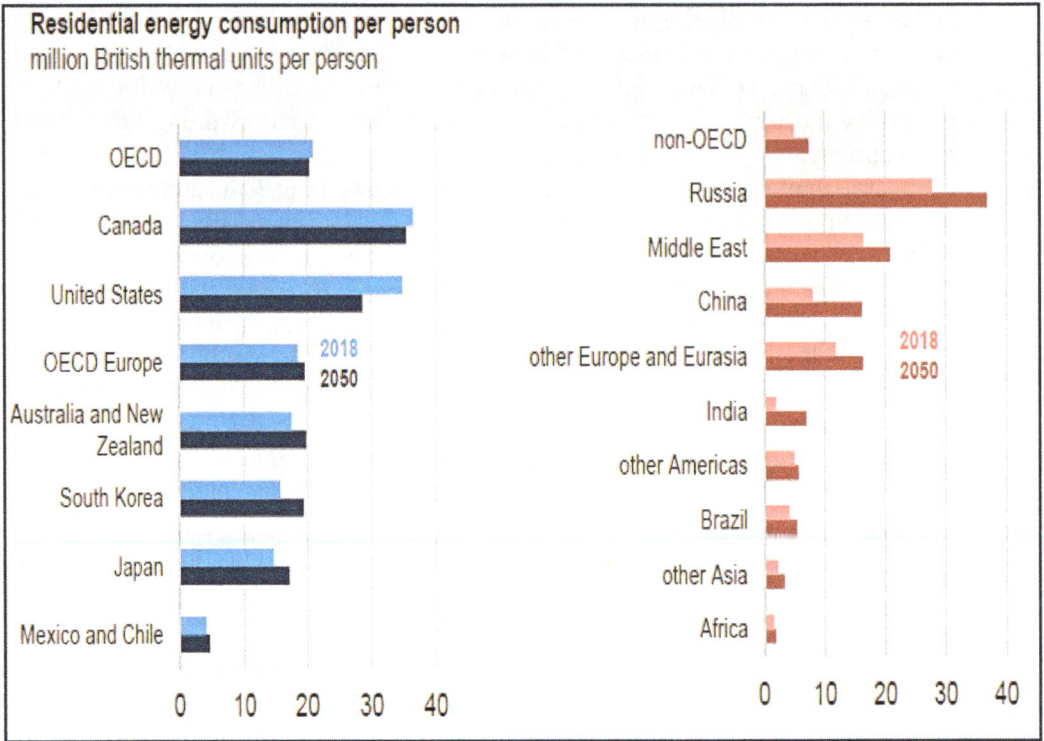

Figure 4.16b   Per capita projection of building sector energy consumption by country and region *(EIA, 2019).*

Residential sector energy consumption by fuel
quadrillion British thermal units

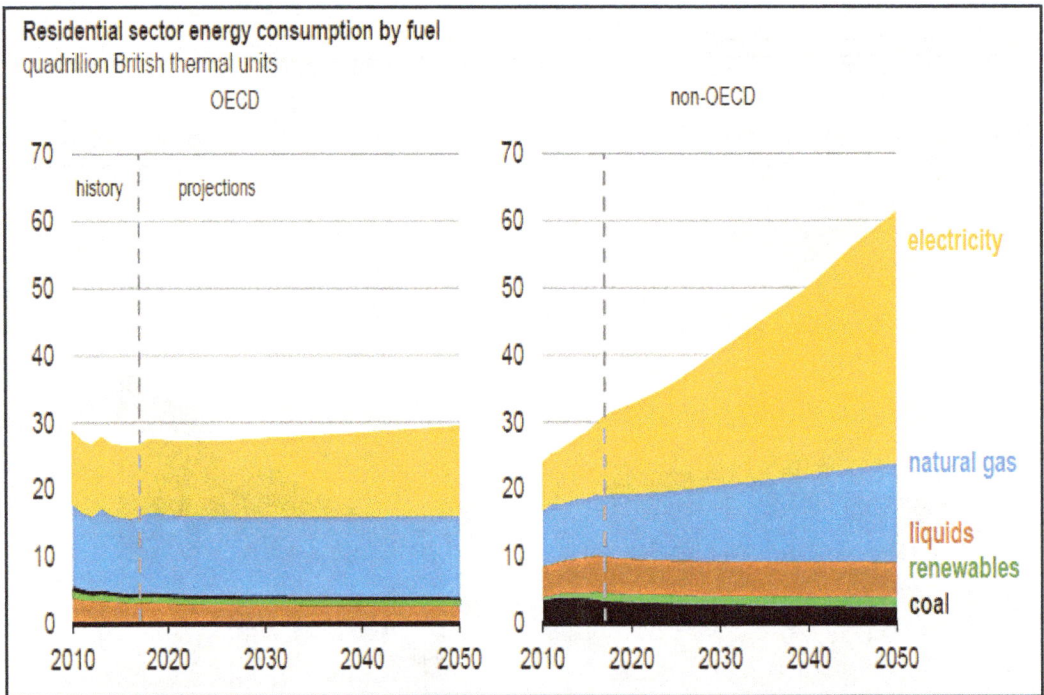

Figure 4.16c   Projection of building sector energy consumption by energy and region *(EIA, 2019).*

The energy use per household has been declining and the trend is expected to continue to 2040, due to the increasing energy-efficient homes, appliances, etc. Household energy demand will rise by only a quarter, nearly all of it accounted for by the emerging countries. The increase should be around 100% but for the mitigation of rising energy efficiencies in building construction, fittings and appliances across all regions of the world. About 90% of this demand growth will be met by electricity which will emerge as the preferred household energy. The average worldwide household electricity use will rise by about 30% by 2040, again nearly all of it in emerging non-OECD nations where increase is expected to double. Residential, commercial and public places account for about 15% of the global primary energy use and half of the global electricity demand for heating, lighting, refrigeration, air conditioning, cooking etc. Biomass currently dominates the energy mix in the sector, accounting for about 40% followed closely by electricity. The bulk of the biomass consumption currently is in the emerging world and comprises mainly wood and charcoal. The share is expected to decline to about 30% due to rapid urbanization and more people entering the middle class and gaining access to more modern fuels - electricity, natural gas, liquefied petroleum gas (LPG), solar energy.

China and Africa will each account for 30% of the increase in commercial and residential energy demand. Use of natural gas will rise by 25%, demand for biomass should peak around 2020 and then start to decline, but oil and coal use have been declining and the trend is expected to be sustained. The building sector is dominated currently by electricity and natural gas, accounting for around 75% of total energy use. Residential electricity use will grow by around 70-75%, accounting for nearly all the growth in total energy demand from 2016 to 2040, while natural gas consumption rises by 20%, but there will be significant regional/country differences in terms of fuel mix. Growth in non-OECD countries will be around 150%, and as high as 250% in India and Africa, filled mostly by fossil fuel-based power generation, although use per household will remain low. China and India will account for a quarter of the world's total building sector electricity demand in 2040. Most developing nations have relatively warm climate, hence the need for space heating is small. Instead, the majority of demand is driven by the need for space cooling/refrigeration, and electricity is the most efficient source of energy for meeting these demands. Household energy is being modernized all over the world, with coal being replaced gradually by natural gas for space heating, promoted largely by the increasing global availability of liquefied natural gas, which is being delivered by truck and marine transportation from source to all parts of the world. This resolves the perennial problem of the need for proximity to a natural gas supply pipeline network. Progress is slower in the developing world, although China and India have strong policies for promoting the use of natural gas and liquefied petroleum gas in household heating and cooking. Other regions of Asia and sub-Saharan Africa will still depend largely on traditional energy for household needs over the next two decades.

4.4. OUTLOOK ON GLOBAL ENERGY USE

Many independent and partisan organizations publish outlooks over two to three decades on global energy demand and use, based on historical antecedents, and various assumptions on the future pathways of the main determinants of energy demand: global economy, population growth and demographics, global energy market dynamics, emerging technological innovations, urbanization, emerging policies, etc., all of which are subject to significant uncertainties. The energy industry is in transition, shaped by growing global economy and population, rising prosperity and urbanization, new technologies, government policies, social preferences, challenges of producing more energy with less carbon emissions, and the net effect on future energy demand and consumption is impossible to predict today. This is evident from significant changes which are often made in projections and forecasts from year to year by the same organizations, and the evolving strategies of exploring different potential scenarios. Nevertheless, even though there may be differences in specifics, there is general consensus on the likely growth pathways of global energy demand and energy-related emissions. This section presents a summary of in-depth evaluation of projections and related analyses published by many independent and partisan organizations in the last five years, with a view to identifying the most likely scenario of the global energy demand, utilization and growth of associated emissions over the next two to three decades. There are some differences in the projections but there is also a fair consensus on expected trends of the critical energy variables and potential pathways to a sustainable environment. Many events none of which can be predicted accurately will shape the global energy demand outlook: economic and population growth dynamics, prices of options, potential investments, new technologies, national priorities and policies, changing dynamics of fuel mix, changing consumer preferences, unpredictable severe disruption of global energy demand/supply system, etc. In effect, any energy projection should be regarded as intelligent speculation based on currently available facts. Nevertheless, it is useful to identify the likely future direction in the context of trends over the last two decades, and various known and anticipated technological and policy innovations that could impact on global trends in energy supply and demand.

4.4.1.   Energy demand growth

The global trend towards rising income and population growth greatly stimulates energy demand rise, but that growth is moderated significantly by increasing energy use efficiency as indicated by energy intensity (energy consumed per dollar GDP). The expanding middle class in Asia and Africa will also make a major impact on energy demand and consumption over the next two decades or so. Global primary energy demand will grow about 25-30% by 2040, driven largely by the growing economy which is projected to double over the same period. Shift in world demographics, another major driver of energy use, over the period will also be significant, with a projected 25% growth to around 9.1 billion by 2040. Rising prosperity and urbanization should push up energy demand at roughly the same rate as the economic growth, but emerging, more efficient technologies, particularly those that impact on energy production and use have played a major role in the gradual de-coupling of economic and energy demand growth over the last decade, and the mitigation is expected to continue at an even faster rate. Without this development, energy demand would probably more than double by 2040. Projected improvements in efficiencies across the energy chain from production and distribution to utilization should fully de-couple economic growth and energy demand over the next two or three decades. Emerging nations (non-OECD) will account for much of the growth in both the economy and population, and for about 70% of the global total primary energy demand, with China alone accounting for about 20%. Increasing urbanization is projected to continue worldwide, with around 2 billion additional people likely to live in urban centers by 2040. Much of this urbanization will occur in Africa where the urban population is projected to grow by around 600 million, about a third of the global increase. However, the impact on the region's energy consumption and intensity will depend on the extent to which this urbanization facilitates increased levels of industrialization and prosperity.

Emerging economies will account for most of the growth in global primary energy demand as a result of the projected strong economic, population, and transportation growth in the region. While total demand is projected to rise by nearly 40% in the region, overall consumption in OECD countries will actually fall by around 3% due to increasing transportation energy use efficiency. Much of the growth in the global transportation energy demand will be accounted for by Asia (China, 36%, India, 142% growth from 2015 level).

4.4.2. Future of fossil fuels

Recent technology developments particularly in production of unconventional primary energy (shale, bitumen, coalbed methane) and liquefied natural gas production are changing radically the pattern of global energy supply and diversifying trade in fossil fuels. All recent projections show that fossil fuels will continue to provide the bulk of primary energy demand in 2040, accounting for around 78% compared with about 82% currently. However, there will be a significant shift in mix, with natural gas and renewable energy (in particular, hydro, solar and wind) playing a more prominent role in power generation at the expense of coal.

In spite of the anticipated growth in global electric vehicle population, oil will continue to dominate transportation in the foreseeable future, simply because there is no viable option. Electricity will account for less than 2% of global transportation energy in 2040, and natural gas (used mostly by heavy trucks and buses) will also account for only 2-4%. Oil has been and will remain the world's leading primary energy resource, and global demand will rise steadily by about 20% in 2040 compared with today's consumption level. Around two-thirds of the oil is used by the transportation sector (road, 49.2%; aviation, 8.1%; navigation, 6.8%; rail, 0.7% in 2017) but non-energy use, in particular, the petrochemicals industry that converts oil and gas to high-premium chemicals, fertilizers, and polymers, is rising (16.1% in 2017). The transport sector consumes most of the world's liquid fuels and the share will remain flat over the next two decades or so. The sector will also account for two thirds of the projected growth in liquid fuel demand, shared almost equally by all its sub-sectors - cars, trucks, ships, trains, airplanes.

Natural gas is gaining increasing prominence in the global primary energy scene across all regions of the world, primarily because of its lower carbon foot print compared with coal, but also due to relative ease of production and transportation by pipeline or shipment in liquefied form. All reviewed projections show that gas consumption worldwide will grow by 43-45% over the next twenty years or so, both in OECD and non-OECD countries, but more than twice as fast in the latter because of the expanding industrial sectors and electricity demand, and the region will account for around 60% of the total global demand on natural gas in 2040. All regions are finding natural gas increasingly attractive for the power and industrial sectors, both of which will account for about 75% of the projected increase in total consumption by 2040. This has greatly stimulated extensive investments in the development of international gas pipeline networks and liquefied natural gas production and transportation. Natural gas when available at competitive prices is the first choice for many new power plants all over the world because of its low capital costs, favorable heat efficiency and emissions rates, and relatively low fuel cost. Also, demand is growing in the industrial sector, in particular, chemicals, refining and primary metals production. Natural gas use by heavy trucks, buses and ships has been rising in recent years and demand is expected to increase by around 43% by 2040, be it from a very low starting base. Developments in shale oil production (hydraulic fracturing) technologies have opened up access to previously unrecoverable shale gas, led by the United States and China which will account for 20% and 18% of the projected growth respectively. Around 50% of the natural gas output by the United States already comes from shale and tight gas deposits, moving the country to the world's largest producer of oil and gas, and projections show that these sources will account for around 70% of the country's gas production in 2040. China is also aggressively developing shale resources which could account for around 50% of the country's natural gas output in 2040. Many other countries also have shale gas deposits which are under development, in particular, Canada which is expected to source most of its future gas output from the rich deposits from several regions in British Columbia and Alberta. Production of conventional natural gas will also grow due to increased output by Russia and the expansion of its international pipeline networks in Europe and Asia. Demand for liquefied natural gas (LNG) is growing in every region particularly in countries that have no deposits and are not near any natural gas pipeline networks. World LNG trade has increased by about 10% per year in the last two years and stood at 350 million metric tons in 2018, projected at around 600 million metric tons in 2040. Qatar led five other countries (Australia, Malaysia, United States, Nigeria, and Russia), accounting for around 75% of the total global exports in 2018, but all the leading producers and most of the fourteen remaining exporters are expanding their capacities, and many other countries are developing their resources. However, projections show that existing and planned new capacities will not be sufficient to meet the global demand for LNG in 2040 (IGU, 2019).

Most of coal used worldwide is for power generation, industrial processes (iron and steel production, cement production, conversion to premium chemicals and petrochemicals), and home heating. Over the last two decades or so, coal energy-intensive industrial production of primary materials has been moving to emerging countries and the trend is expected to continue. Worldwide coal use will not change very much over the next two decades or so because projected decrease in China and the United States (the two leading global consumers of coal) will be offset by growth in India and other developing nations (Africa, the Middle East, and other non-OECD Asia). Although the rate of growth of coal use by China has been declining over the last few years due largely to increasing use of solar and nuclear energy for power generation, but also because of transition to less energy-intensive manufacturing, overall demand is expected to continue to rise, and the country will still remain the world's largest consumer of coal in 2040 because new, more efficient coal power plants are being commissioned or planned and use of coal as feedstock for chemicals/petrochemicals production is rising. India's coal use has continued to grow very strongly and the country surpassed the United States as the second largest global consumer of coal in 2016 (EIA, 2017). Coal demand in India is expected to nearly double by 2040 as the country continues to install new coal-fired electricity generating plants, as well as rising demand by industry, in particular, primary metals, cement, chemicals and petrochemicals sectors. Despite the significant increase in coal consumption, coal's share in overall energy consumption in India is projected to decrease from 49% in 2015 to 43% in 2040, due in part to policies promoting renewable and nuclear-based power generation. Coal demand in other non-OECD Asia, the Middle East and Africa will nearly double by 2040 because many countries in the region are taking advantage of the relatively low cost of coal, sourced locally or by importation, in efforts to meet the rapidly growing demand for electricity. However, the use of coal for home and industrial heating is expected to decline in most regions of the world due to increased use of natural gas and biogas. Coal use has been declining in OECD countries, and the trend is expected to continue because of increasing competition from natural gas and renewables, increasing energy efficiency is driving down electricity demand, and stringent environmental regulations are being enforced.

4.4.3.   Future of renewable energy

Renewable energy is an important and growing part of the ongoing global energy transformation, and is now recognized by many countries as the first-choice for providing quick access to electricity, expanding, upgrading and modernizing power systems. Most countries in the world have established renewable energy policies and targets, and potential investors from the private sector are playing increasingly prominent roles. Solar and wind are particularly popular, accounting for most of the investments in recent years, driven largely by falling costs: the cost of wind turbines has fallen by nearly a third in just six years, and solar photovoltaic (PV) modules now cost 20% of the prevailing prices six years ago. Renewables accounted for around 14% of the total global final energy consumption evenly split between traditional biomass and modern renewables, and 24% of global power generation in 2017. Projections based on current active policies and investments around the world, show a growth to around 20-35% contribution to power generation by 2030. Just six countries led by China account for nearly 80% of current renewable energy use, but many other countries have active renewable energy development programs and policies. The International Renewable Energy Agency has developed an aggressive yet feasible roadmap for global energy transition that positions the world on the path to a sustainable environment by keeping the rise in average global temperatures well below 2°C, ideally to 1.5°C in the present century. This would require aggressive decarbonization of energy through major improvements in efficiency, electrification of energy services, and deployment of renewable energy which together can deliver more than 90% of needed reductions to energy-related CO_2 emissions (IRENA, 2019). However, this is feasible only if the share of electricity in final energy use increases from 19% in 2016 to around 50%, and around 70% of all cars, buses, two- and three-wheelers, and trucks would be powered by low-carbon electricity. Much of the projected growth in renewable energy for power generation will be in emerging nations, many under pressure to fast-track 'electricity for all by 2030'. Most of the expected growth in hydro-energy will be in sub-Saharan Africa, while China, United States, and Europe are leading the world in the deployment of solar and wind energy, often referred to as variable renewable energy (VRE). Solar energy is particularly suitable for deployment as utility units serving mini-grids, individual households, or small communities, making them particularly attractive in sub-Saharan Africa which will account for most of the people without access to electricity globally in 2030. One major advantage of PVs is modularity which allows for very small start-up and easy scale-up, from mini-units that fill personal needs to large-scale power plants that serve communities and vital establishments, and feed grids. The United States leads the world in terms of VRE mini-grid capacity deployment, serving sensitive installations: hospitals, universities, industrial plants, military facilities, and small communities, driven largely by the need to provide reliable and resilient energy service to critical units in the face of extreme weather and frequent grid outages. Many states of the US and other countries also have strong policy incentives that promote installation of small units for homes and commercial establishments. The State of California recently enacted a law which requires all new homes to feature solar roof panels. Most utility projects range in size from one megawatt to several hundred megawatts, and deployment lead times vary from days to months depending on project size. Small units have battery backup to resolve the problem of intermittency, while larger establishments use standby diesel generators or pumped storage.

Power generation has been the primary target of most renewable energy policies, and most of the developments in renewable energy use have been in the power sector. However, extensive potentials abound in other end-use areas - heating and cooling in buildings and industry, as well as transportation. These sectors together account for a large portion of global energy use, and about 60% of energy-related CO_2eq emissions (IEA 2017). Already, modern renewable energy technologies provide energy for water and space heating, cooling and cooking in buildings, and heating and steam generation in industries. Solar PV systems range from small-scale heating units to large-scale district heating networks, and industrial facilities supplying medium-temperature (100-250°C) process heat, while solar concentrator systems can supply heat up to 400°C, high enough to raise superheated steam for power generation. Many technologies are already available commercially, including heat pumps that can store surplus renewable energy-generated electricity in form of heat for later use in space heating and a wide range of industrial processes. Modern bioenergy accounts for by far the largest share of renewable energy use for direct modern heating and cooling, but is still only about 10% of the total biomass energy, the balance being traditional biomass. The major challenge for wider deployment of renewable energy, in particular, those that are bioenergy-based, is the low price of alternative fossil fuels in recent years which has slowed down investments. Solar thermal and geothermal heating technologies are also experiencing rapid growth, although from a relatively low starting base. Geothermal energy is becoming increasingly important in countries that have the resource and use for heat raising and power generation is expected to continue to rise. Hydroenergy is still by far the largest renewable power source globally and will remain so over the next two decades, although deployment in developed countries will reduce due to environmental concerns, but will grow in the emerging nations.

Penetration of renewable energy in the transport sector has been relatively slow, but potential opportunities are substantial. Renewable biofuels are already in use commercially: bioethanol is being blended with conventional gasoline in some developed countries, helped by blending policy mandates; biodiesel is playing an increasing role in transportation because of its lower carbon intensity compared with oil-derived diesel; biokerosene is being blended with jet fuel for aircraft although it accounted for less than 0.1% in 2018. Extensive opportunities exist for a greater use of renewable fuels in all forms of heavy transportation - aviation, marine and road freight - which together account for around 35% of transport energy demand. However, in spite of the clear potential in decarbonizing transport, investment in new biofuel production capacity has been declining, due largely to the prevalent low prices of oil in the global market. Linking electricity with other sectors, in particular, heating, cooling, transport (sector coupling) will help to promote more extensive use, solving the problem of variability in solar and wind energy, and integrating a rising share of variable renewable energy (VRE) into existing power systems. It will provide continuous energy for various applications, while also expanding the use of renewables in other end-use sectors. For example, a VRE system can supply grid power when availability coincides with demand, but can also use surplus electricity to power heat pumps that feed thermal grids for district heating and cooling, or individual building thermal systems. Effective control of the production and distribution systems that couple variable renewable electricity with thermal applications will require 'smart' electrical and thermal energy control systems, many of which are already available and in commercial use. Availability of affordable power storage systems will be crucial for the development and wide deployment of variable renewable energy (VRE) technologies, and there are many different energy storage technologies for different applications. These include pumped/hydro storage, compressed air energy storage, thermal storage, fuel cells, supercapacitors, batteries, etc. Pumped hydropower is the most mature which is in use worldwide and accounts for the vast majority of global electricity storage capacity estimated at 145 GW in 2015 (IHA, 2016), and projections show that capacity will more than double by 2030. In a typical installation, surplus electricity generated from any energy source during night time or idle energy from a VRE plant can be used to pump water to a high level for generation of hydropower during high-demand periods, or to cover periods when solar and wind energies are not available, and can provide 24 hours or more of backup. However, hydro-storage is expensive and only suitable for large installations.

Electrical battery storage is the most mature for small-scale applications and the demand is increasing rapidly, driven primarily by the growing market for electric vehicles. Batteries are being deployed to store unused electricity from VREs, for later use when direct solar electricity is not available. These include household and solar PV systems, island wind systems, and off-grid VRE systems for rural electrification. Total global electric battery capacity is low currently, around 0.8 GW, but is projected to increase to around 250 GW by 2030. There are many types of electric batteries but lithium-ion batteries are set to dominate the market because of their high energy density, efficiency, and relatively long life, followed by relatively low-cost advanced deep cycle lead-acid batteries, sodium sulphur and advanced flow batteries. Lithium-ion batteries are used widely in consumer electronics as well as plug-in electric vehicles, airplanes, ships and stationary solar PV systems. Cost of battery storage systems is expected to continue to decline rapidly due to increasing demand driven largely by the auto industry, and could become widely deployed as residential utility-scale storage in a few years. Already, in countries where grid power supply is erratic, many homes and commercial enterprises rely on deep-cycle lead-acid battery-inverter backup systems to store power when available.

4.4.4. Future of nuclear energy

The future of nuclear electricity has been uncertain for the last decade due to a combination of factors: declining price of natural gas; the impact of policy-driven rapid expansion of renewable resources on electricity prices; and national policies in several countries following the Fukushima Daiichi nuclear power plant accident in 2011 are all affecting the growth prospects of nuclear electricity. Many other countries that have historically accounted for the majority of nuclear power development are de-commissioning old nuclear power plants, slowing down on new construction and developing other low-carbon energy sources. The future prospects of nuclear power generation in Western Europe are unclear: some countries in the sub-region have either established firm policy not to build new plants or to de-commission existing plants - Belgium, Spain, Switzerland, Germany, while others, in particular France and the United Kingdom have not made clear policy statements on the future of nuclear electricity. Most of the United State's 99 operable nuclear power reactors are nearing their lifespans and four were decommissioned recently. Only one new reactor has been added in the last 20 years, although the licenses of some reactors have been extended. The United States Energy Information Administration (EIA, 2018) projects a continuing decline in nuclear power generation over the next two decades, largely because of the increasing competitiveness of natural gas.

The future of nuclear electricity in Asian and East European countries is bright. Starting around 2000 and driven by aggressive economic development in Asia, nuclear energy has become an important part of the energy mix in the region. Another motivating factor has been the global pressure on the region to reduce dependence on fossil fuels (mostly coal). Nuclear power offers an attractive low-carbon alternative for rapidly adding large baseload power generation required to power the aggressively increasing economies in the region with minimal greenhouse gas emissions, and most of the inputs can be sourced locally. Around 450 nuclear power plants are operating in different countries of the world, about 60 are under construction, and over 500 are planned or proposed (WNA, 2018). China is accelerating the deployment of nuclear electricity plants and nuclear capacity is growing at around 11% per annum. According to the International Atomic Energy Agency (IAEA, 2016), 40 of the 68 nuclear reactors which were under construction in 30 countries in 2015 were in China, India, Korea and Russia, and China alone accounted for 28 plants.

Demand for electricity in the industrialized nations is stagnating or even declining due to increasing efficiency across the energy supply and use spectrum, and growth has been shifting from OECD to non-OECD countries. Demand is projected to surpass that of OECD within the next few years, rising to around 60% of the total global demand by 2040. This rapid growth will no doubt stimulate the development of locally available and appropriate electricity generating options, including nuclear power. It is not surprising that most of the nuclear electricity plants under construction or planned, are in emerging nations. Nuclear power is a reliable and secure energy source that can be deployed in many countries, considering the small amount of fuel required to generate enormous energy. Furthermore, the industry is largely insensitive to changing dynamics of the global oil market. Nuclear power generation is considered environmentally benign and with zero carbon emissions but this is not true when associated carbon emissions in the production of infrastructure inputs and decommissioning are taken into account. Furthermore, while the direct negative impact of nuclear power generation on climate change may be low, the potential damage of radioactive emissions and waste to the environment and human/ecosystems health could be enormous and just as enduring considering the exceptionally long half-life of most radioactive wastes.

The energy intensity of nuclear fusion, that is, the amount of energy produced per unit weight of fuel is very large, around 10,000 to 20,000 times larger than in fossil fuel combustion. This is one of the major attractions because countries which have no uranium resources can easily import small quantities to generate very large amounts of energy. Furthermore, fuel accounts for only about 5% of the total generating costs and most other inputs can be sourced locally, which makes it the lowest-cost baseload electricity supply option in many countries. A typical, large 1000 MWe nuclear power plant requires about 75 metric tons of low-enriched ore in its core at start-up, with annual top up of only 25-27 metric tons of fresh enriched fuel each year. It means that nuclear power plants can maintain continuous operation for decades without worrying too much about turbulence in the global primary energy market. The recent shift away from nuclear power generation as a result of accidents appears to be in reverse and investment is picking up again, with nuclear energy accounting for 2% of the growth of total primary energy demand in 2017. There are sufficient uranium reserves to power the current level of capacity for over a hundred years and undiscovered/unevaluated resources are believed to be at least ten times the known reserves. Moreover, the design of reactors in terms of safety and efficiency has improved tremendously in recent years and commercialization of fast breeder reactors which have been under development for many years could sustain nuclear power production indefinitely.

Around thirty (newcomer) states are considering, planning or starting nuclear power programs, while twenty more have expressed interest, and, with the support of the International Atomic Energy Agency (IAEA), are working with technical partners in planning infrastructure development. However, it is unclear whether these expected new additions will be sufficient to stall or reverse the downward trend in the global nuclear electricity generation. More than half of the around 450 reactors currently in operation are over 30 years old and, while there may be new constructions (probably not in the same regions), IAEA expects a continued decline over the next decade or two, in spite of the fact that capacity in Asia is expected to increase by around 43% over the same period. There has been a major shift of global energy investments from fossil fuels to low carbon technologies in the last few years, but the shares of energy efficiency and renewables were ten times and fifteen times the share of nuclear power generation. On the other hand, the continued growth in energy demand in the developing world could stimulate the re-emergence of nuclear electricity as a prominent component of the global low-carbon electric power generation mix. Many nuclear power countries in the world are committed to nuclear non-proliferation and are concerned about the possibility that newcomers could move from peaceful use to nuclear weaponry, but some, in particular, Russia and China are involved in virtually all the planned new capacities in the developing world. Another major issue is the capability of these newcomers for coping with safety and waste management issues, considering the seriousness of the three accidents in the last three decades, all of which occurred in the developed world.

If fuel in a nuclear reactor gets too hot it will melt the casing, and the radioactive materials contained will be released. In a very serious fuel melt event the radioactive materials may escape the reactor and containment structures, posing high risk to people and the environment, as happened at both the Fukushima Daiichi and Chernobyl plants. Used nuclear fuel is intensely radioactive and although the radioactivity diminishes quite quickly, it remains at levels which could be dangerous to life for thousands of years. Spent fuel requires shielding, can only be handled remotely by experts and requires special measures for its final disposal. It also continues to generate a significant heat load, both inside and outside the reactor for several years and requires active cooling during this time. There is still no environmentally sustainable method of disposal after around eighty years of active nuclear technology deployment. It is noteworthy that the United States is still debating the potential final safe resting place for the nuclear waste accumulated in the atomic bomb project of the early 1940s, still potent and currently stored on temporary sites. The same is true of waste stored on sites of nuclear plants that are currently operating or de-commissioned all over the world. The Chernobyl nuclear accident in 1986 is the world's most serious and, after nearly forty years, decontamination is still in progress, zones remain excluded and large areas are desolate. Most of the radionuclides (short-lived iodine-131 and long-lived caesium-137) were deposited as dust in the vicinity of the plant and over much of Europe but fine material was carried far and wide, and high into the stratosphere which made possible long distance migration detected in many countries all over the world. Recent studies have shown a significant rise of thyroid cancer in people who were children in the area at the time of the accident, and settlements as far away as 200 kilometers from the nuclear plant are still radioactively contaminated. Milk produced has been found recently to contain five times the acceptable level of radiation for adults, and twelve times for children, and this level of radiation is expected to persist until at least 2040. About 200 metric tons of highly radioactive material is still in the reactor core, encased in concrete. A New Safe Confinement (NSF) structure with a lifetime of 100 years is currently under construction, funded largely through international donation. This will facilitate a safe decommissioning of the plant and management of the radioactive waste. Hopefully, by then, the world would have figured out how to sequestrate nuclear waste permanently. The Fukushima Daicchi nuclear accident of 2011 was caused by a tsunami which followed an earthquake; decommissioning is just starting and is expected to take decades. There is little doubt that a lot of lessons have been learnt from the three nuclear accidents in the last three decades or so and current global efforts and cooperation will likely make nuclear energy safer but the fact that the most advanced nuclear nations of the world still have no clear plans about safe, permanent disposal of nuclear waste should be of major concern. Also, it is unclear whether the around twenty emerging nations/newcomers can safely and effectively manage nuclear safety and waste disposal, or respond appropriately to potential emergencies.

Another major cause for concern on nuclear electricity proliferation is the potential effect of radioactive nuclear waste on the environment, in particular, the oceans. Over the past half-century, radioactive materials were dumped or buried in ocean beds all over the world, including de-commissioned reactors, and large amounts of liquid and solid nuclear waste. Although this has been prohibited by an international agreement, the damage is done already in view of the longevity of potency. Since the waste remains potent and dangerous for hundreds and may be thousands of years, it is unclear how long the storage containers will hold under very hostile environment before failure which will lead to contamination of soil and aquatic life. How this potential contamination will affect marine life or humans, is still unclear, but there is considerable concern in the scientific world that both short-lived radioactive elements, such as iodine-131, and longer-lived elements such as cesium-137 and plutonium can be absorbed by marine bacteria and other marine life, and transmitted up the food chain, to fish, marine mammals and humans (Grossman, 2011). Nuclear waste is currently encased in copper and concrete containers which have been stored on temporary sites in many countries for decades because there is still no acceptable technology for permanent disposal.

4.4.5. Future of global economy, human development and energy demand

Virtually every aspect of human development depends critically on energy: for heating, cooling, transportation, electricity, manufacturing and health technologies, consumer products, etc. While efficiency improvements across all sectors will continue to moderate growth in energy demand, the virile growth in global economy, population and prosperity will remain primary drivers of energy demand. However, recent events have highlighted the unpredictability of future trends: the current Covid-19 global pandemic virtually collapsed the world's economy in just weeks: air and road travel dropped by 85%; demand for oil dropped by the same amount; world fossil energy production, trade and investment collapsed; and millions lost their jobs around the world. Although these events have resulted in an unprecedented decline in energy-related emissions, the relief can only be temporary: the global economy will spring back but many 'new normals' will emerge. The exponential rise in 'work from home', virtual meetings, on-line education, telemedicine, e-commerce; remote socialization will likely be sustained in OECD countries, China and India which account for around three-quarters of global primary energy demand. Oil demand probably will not return to previous levels for a decade or more because of the drastic drop in road and air transportation (many airlines, small energy companies and businesses which are currently comatose around the world may never recover) but demand for the largely fossil electricity that will power the 'new normals' will rise significantly and, unless there is a major shift to low-carbon electricity, the current significant drop in global energy-related emissions will not be sustained. Also, a prolonged global decline in oil demand and prices will impact negatively on new investments in renewable energy and energy efficiency improvement unless strong policy support instruments emerge. However, while the economic and human development ramifications of the severe disruption of the global energy system will be serious in the short term, the situation is unsustainable. The very large investments that have been made in the global fossil energy infrastructure in the last decade or so (oil field, gas pipeline and liquefied natural gas infrastructures, coal mines, export terminals, etc.) will mature; demand for many other uses of fossil fuels (apart from transportation and electricity) which are essential for human development will continue to rise (production of primary metals and materials, chemicals, petrochemicals, fertilizers, polymers, pharmaceuticals, etc.), but many new technologies and business strategies that promote environmental carbon neutrality will likely emerge in most sectors of the global economy.

Chapter 5
Human-related emissions

5.1.   INTRODUCTION

Many natural processes release toxic emissions which end up in the Earth's atmosphere and can cause issues including climate change and human health problems: desert dust, volcanoes, vegetation decay, human and animal respiratory systems, etc. The consequences of environmental pollution (severe weather, heat waves, flooding, draught, weather and climate change) have always been part of human existence but they have always occurred in different places at different times. However, many human activities also pollute the atmosphere and emissions, particularly from energy use and agriculture have intensified since the early part of the last century to the extent that they are now making these consequences more frequent, more widespread, and more devastating.

5.2.   TOTAL ANTHROPOGENIC POLLUTION

Global anthropogenic emissions have increased gradually since the Industrial Revolution of the 1800s and have intensified in the last hundred years or so, driven largely by fast economic and population growth. Although energy production and use account for nearly 70% of the total global anthropogenic pollution, there are many other sources (Figure 5.1). Carbon dioxide is emitted mainly from fossil fuel combustion, in forestry and other land use (FOLU), and many industrial processes; methane comes mainly from fossil fuel production and agricultural activities that involve composting, fermentation, landfilling and animal breeding. Nitrous oxide is produced from use of fertilizers and fossil fuel combustion; and industrial processes, refrigeration/air conditioning and production of many consumer goods are the main sources of anthropogenic F-gases, which include hydrofluorocarbons (HFCs), perfluorocarbons (PFCs), and sulfur hexafluoride (SF_6).

5.3.   ENERGY-RELATED EMISSIONS

Emissions from the energy sector have increased exponentially since global industrial activities intensified from early twentieth century: CO_2eq emissions (carbon dioxide, methane and nitrous oxide) have more than doubled in the last forty years. Carbon dioxide is the main constituent of energy-related pollutants, accounting for around 90% and, although the gas is the least potent in terms of atmospheric damage, it has become the benchmark for the evaluation of the anthropogenic effect of all other gaseous pollutants, expressed in carbon dioxide equivalent (CO_2eq). Over 99% of energy-related emissions come from the production and utilization of fossil fuels (coal, oil and gas), with coal accounting for around 45%. Oil was the greatest pollutant in the 1970s but has been overtaken by coal. The regional distribution of emissions has also changed significantly: in the early part of the 20th century, virtually all emissions originated from the United States and Europe, today together they account for less than 30%. Just two countries (the U.S.A. and China) accounted for around 45% of the global CO_2eq emissions in 2017, estimated at 32.9 billion metric tons (Figure 5.2). Carbon dioxide emissions stagnated for two consecutive years up to 2016 due primarily to strong efficiency improvements and low-carbon energy technology deployment, but then started rising again in 2017, reaching a record 33.1 Gt CO_2eq in 2018, believed to be a result of the unusually hot and cold weather in some parts of the world. Although emissions from all fossil fuel use increased, the power sector accounted for nearly two-thirds of emissions growth and coal use in power generation was the source of around a third of the total global energy-related emissions in 2018, mostly in Asia. China, India and the United States accounted for 85% of the net increase in emissions over the last two years (IEA, 2019d).

Different sectors of the economy generate different types and levels of pollution emissions. The energy supply sector processes (energy extraction, conversion, storage, transmission, and distribution) that deliver final energy to the end-use sectors (industry, transport, building, agriculture, forestry - is the largest contributor to global greenhouse gas emissions and accounted for about 42% of total energy-related anthropogenic GHG emissions in 2016, the largest single source. Industry contributed around 21% which came from fossil fuels used for in-plant power generation, metallurgical, chemical and mineral production, as well as feedstocks for chemicals and petrochemicals production (Figures 5.3-5.7). The building sector was the least polluting, contributing only about 6%, although around four times higher when pollutions from power generation are shared between sectors on the basis of electricity use. Much of the increase in energy-related anthropogenic pollution in the last two decades or so has been driven by large increases in the production of energy-intensive materials such as ferrous, non-ferrous and non-metallic primary materials, chemicals and petrochemicals, and all these processes depend heavily on fossil fuel energy and electricity while oil and gas are also precursors to chemicals and petrochemicals production. Most of these heavy industries are now in emerging and developing countries, many of which have also expanded very rapidly fossil-fueled power generation capacities.

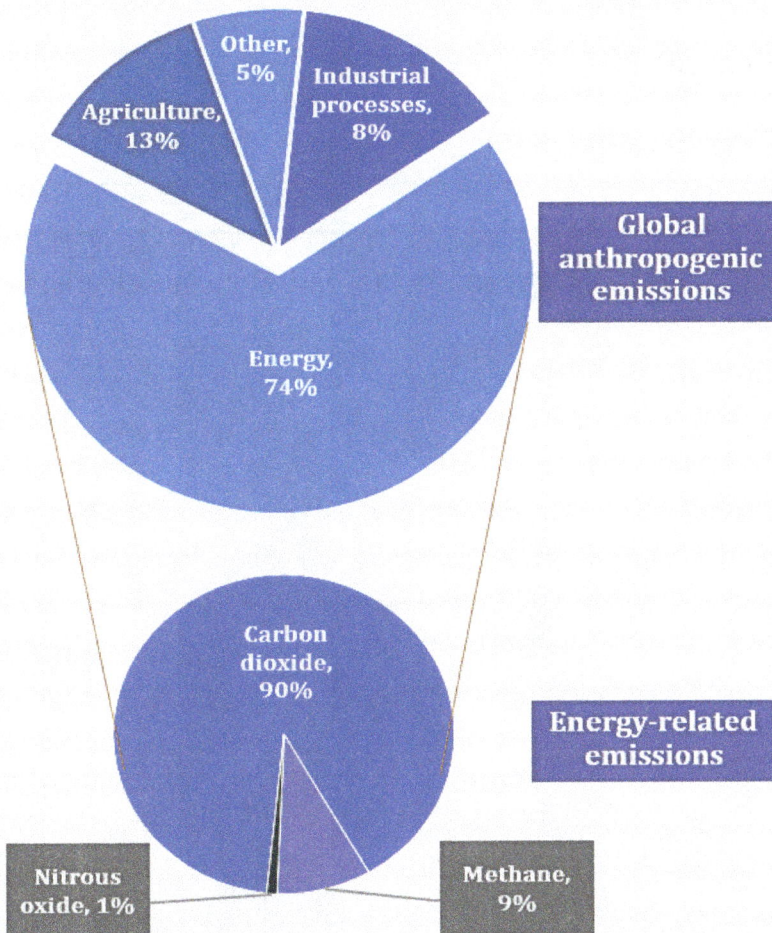

Figure 5.1 Growth trend and estimated shares of global anthropogenic greenhouse gas (GHG) emissions *(Data from IEA, 2017e. 2018; IPCC, 2014).*

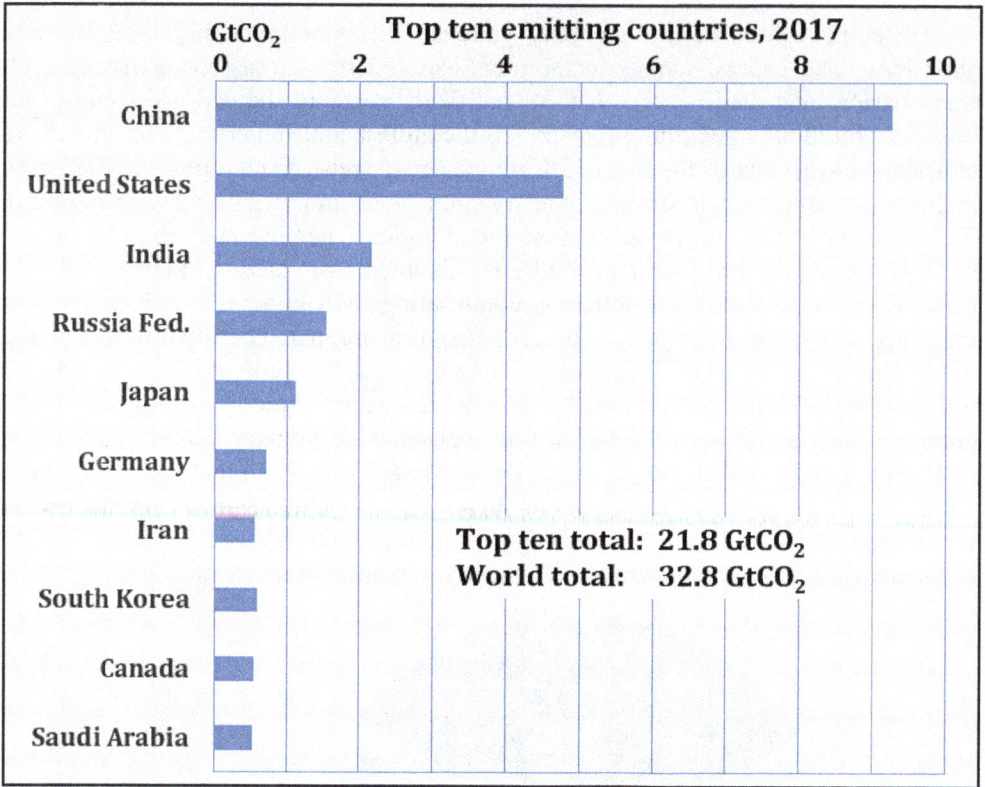

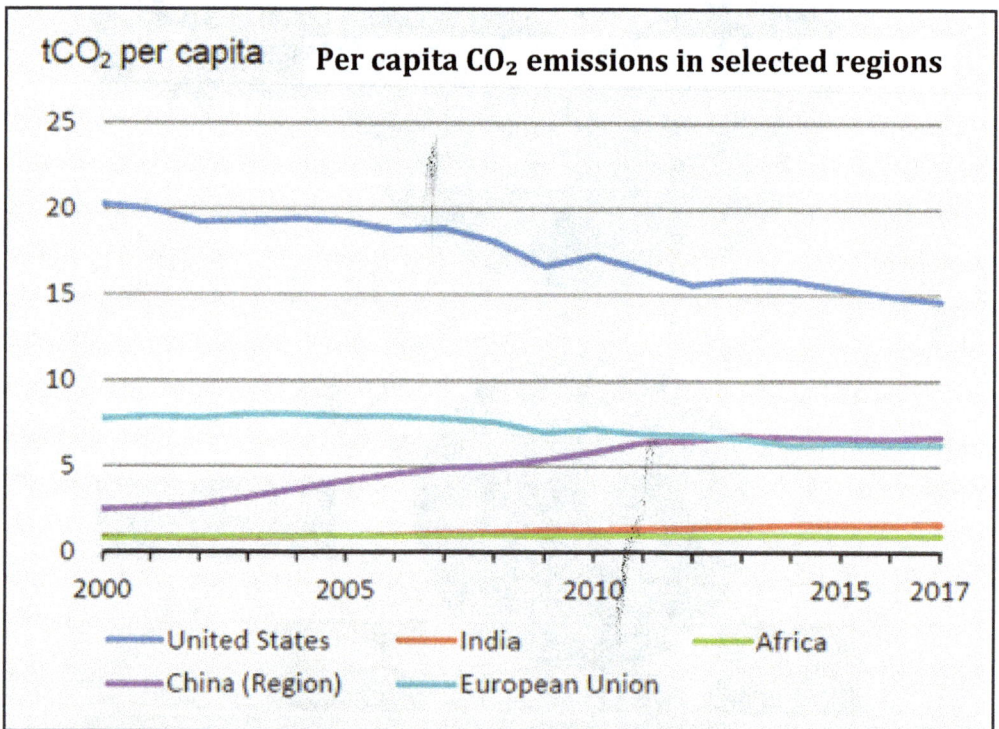

Figure 5.2 Top ten emitting countries and per capita emissions in 2017
(Data from IEA, 2019d).

(a)

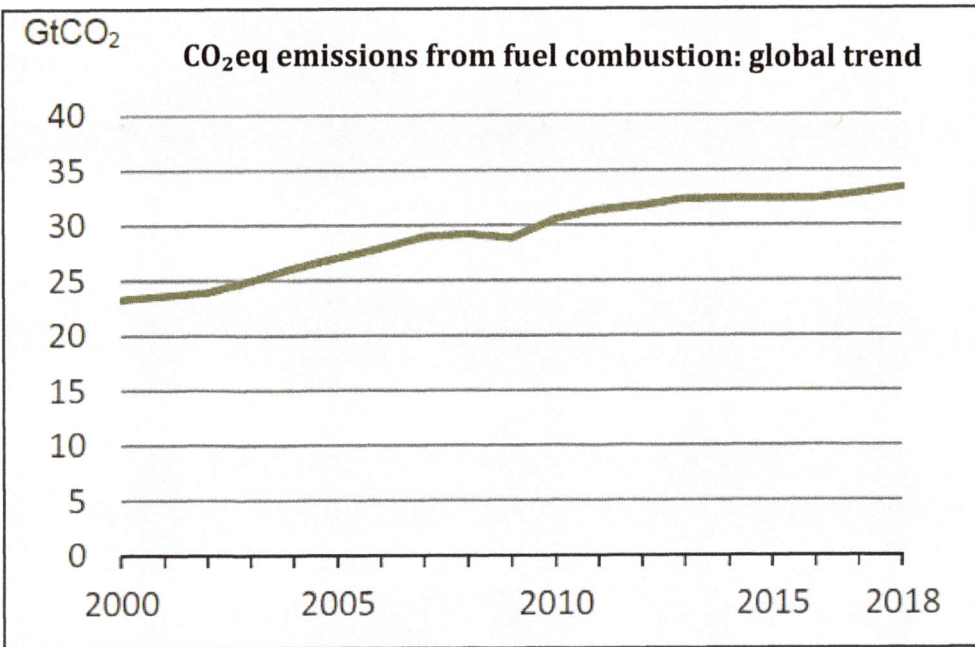

(b)

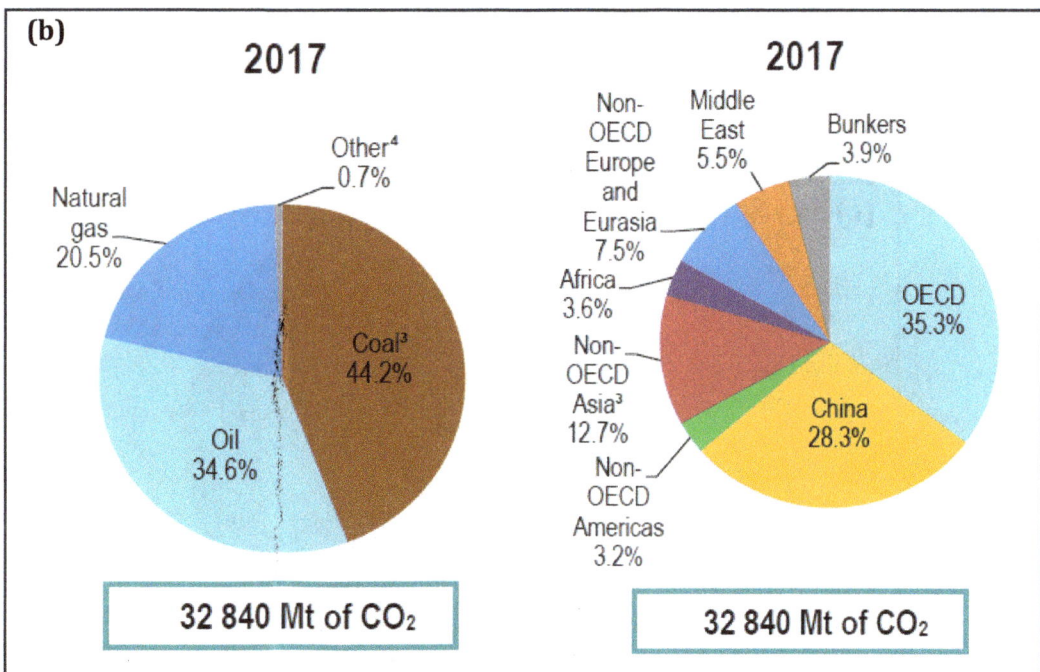

Figure 5.3 Global CO_2eq emissions (a) by fuel (b) by region 2017 *(IEA, 2019a).*

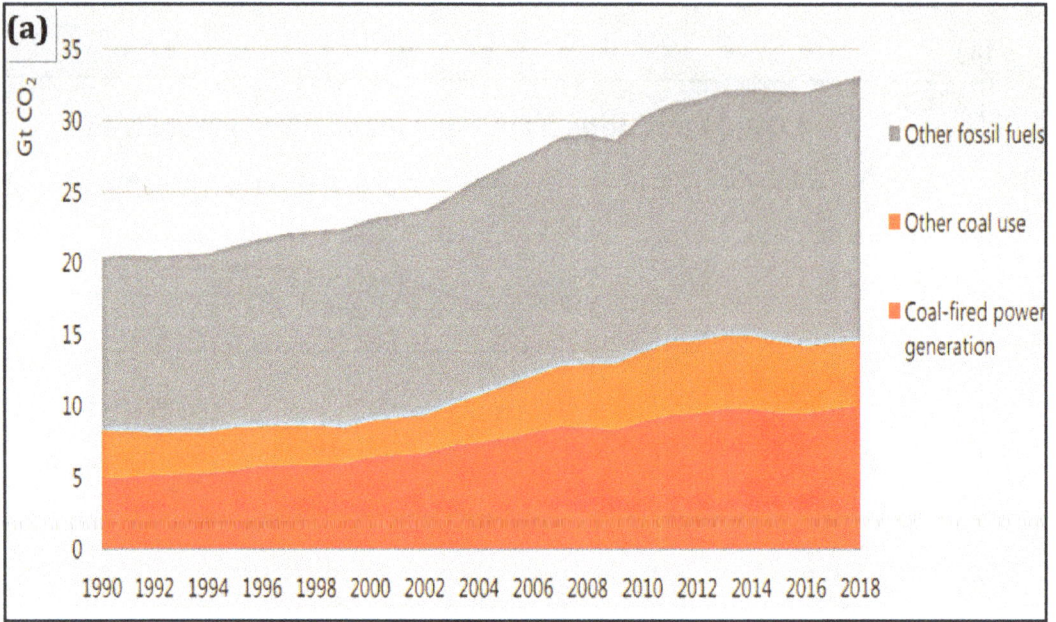

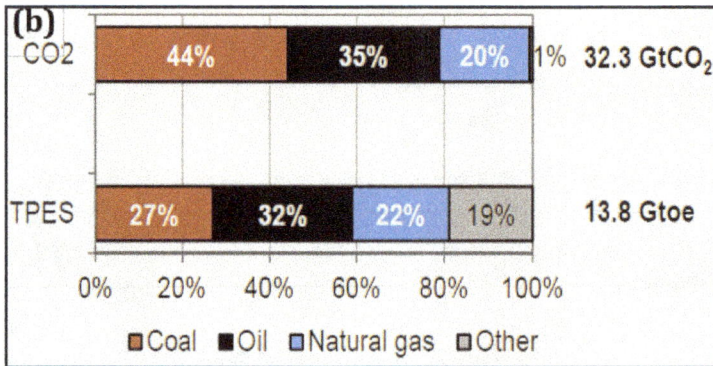

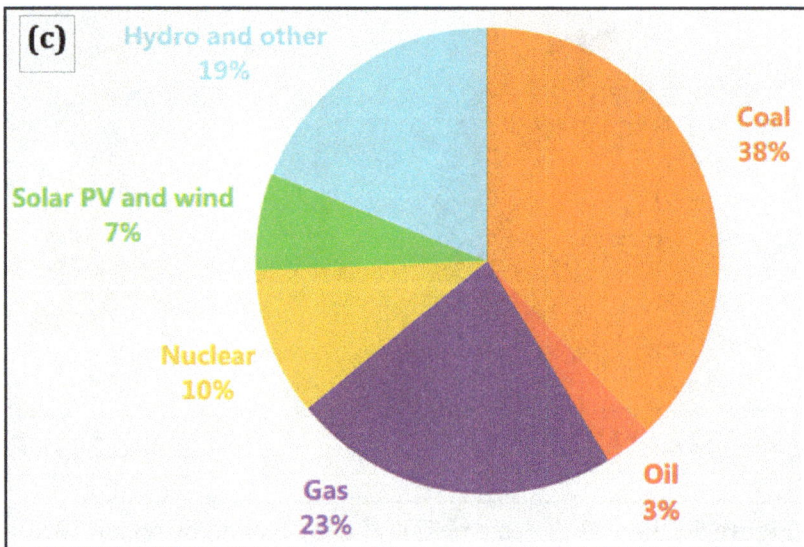

Figure 5.4 (a) Global energy-related carbon dioxide emissions by source, 1990-2018   (b) Total primary energy supply and CO_2 emissions, 2016 (c) Electricity generation mix, 2018 *(IEA, 2018a, 2018b).*

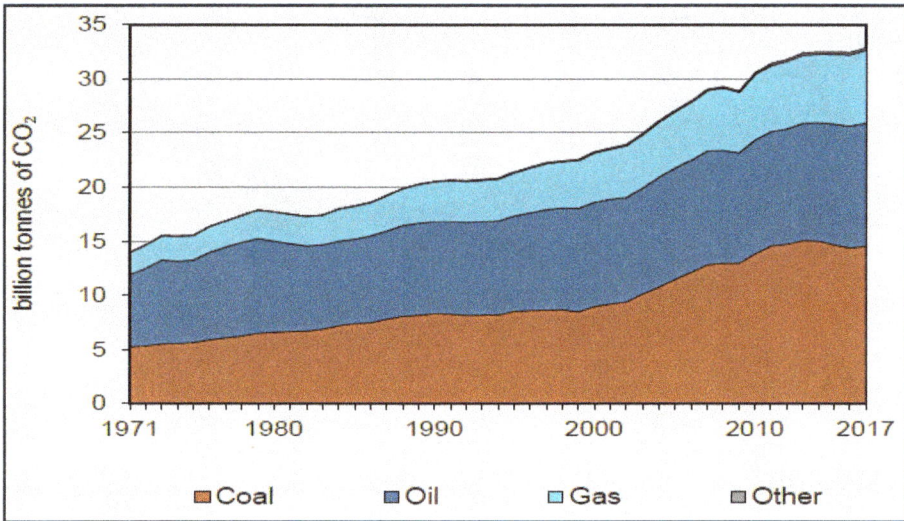

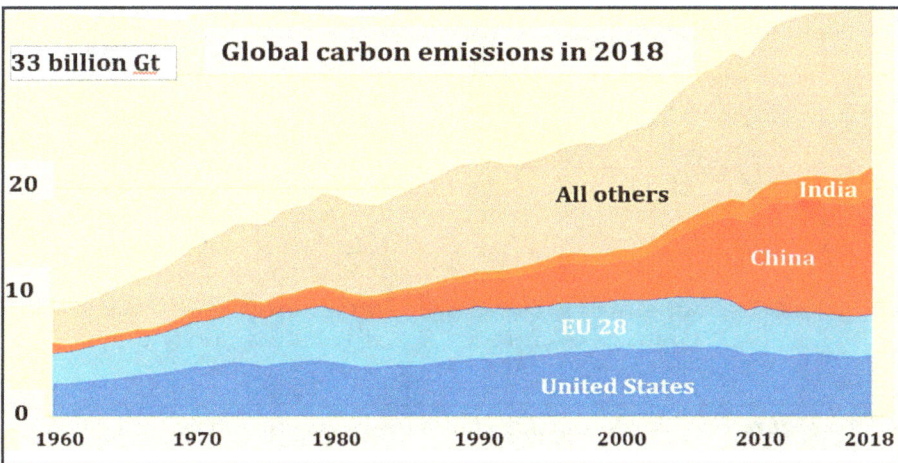

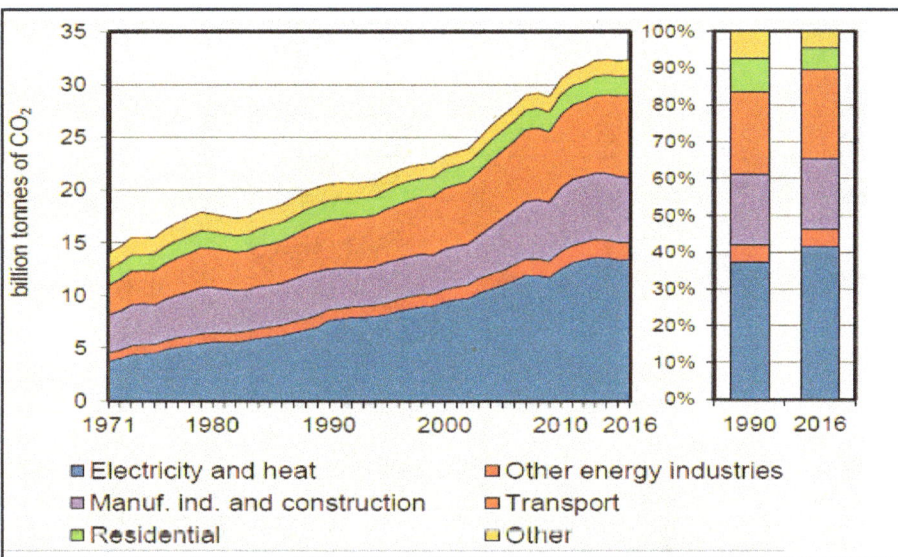

Figure 5.5 Global energy-related carbon dioxide emissions by source, region and economic sector *(IEA, 2019d)*.

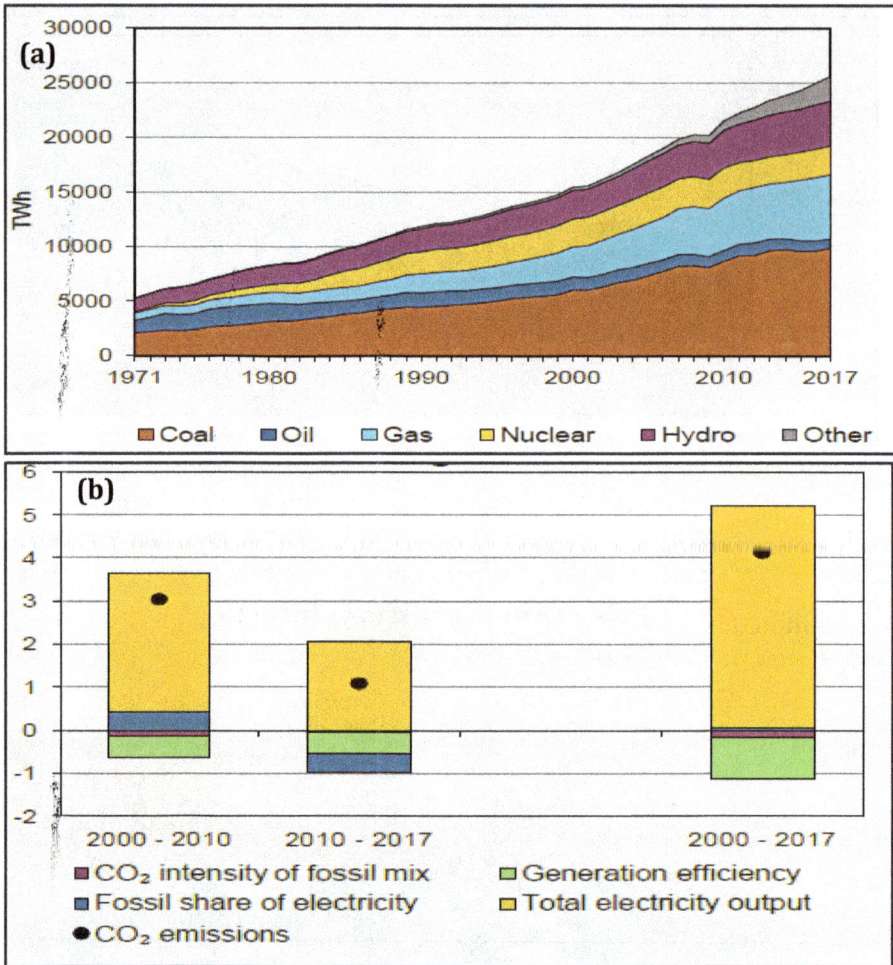

Figure 5.6 Carbon dioxide emissions in electric power generation (a) by source (b) driving factors *(IEA, 2019d).*

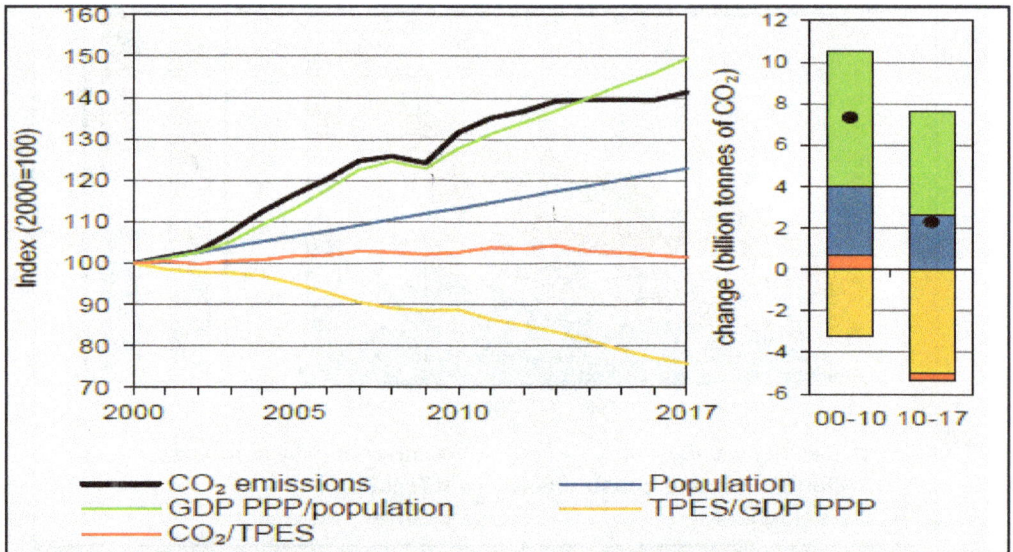

Figure 5.7 Total carbon dioxide emissions and driving factors *(IEA, 2019d).*

All recent projections indicate that emerging countries will account for nearly all of the increase in primary energy consumption (and related carbon dioxide emissions) over the next two decades. Production is often inefficient and energy intensity is high. Emissions by region and economic sector are presented in Figure 5.5. However, even within the emerging regions there are significant differences: while Asia accounted for 53% in 2017, Africa contributed only 4%.

Energy-related emissions come from the three major sectors of the global economy: Industry, building and transportation. Power generation, the most polluting energy enterprise is classified with industry but related emissions are distributed between the three sectors on the basis of electricity use. The energy supply sector, from extraction to end-use is the largest consumer of primary energy and, because many of the processes are inefficient, the sector is also the largest contributor to energy-related anthropogenic greenhouse gas emissions which account for over two-thirds of the global total. Virtually all energy production processes (drilling, mining, conversion, transportation, distribution and captive energy use) produce substantial anthropogenic gaseous and particulate emissions. Production processes degrade landscapes, contaminate groundwater, oceans, rivers and aquifers. Coal mining, preparation and transportation (in particular, opencast mining) damage large areas of land and emit particulate dust and methane; many countries flare enormous amounts of methane gas during oil and gas production; and there are significant losses during transportation of all fossil fuels. Gas flaring emits black carbon and about 400 million metric tons of CO_2eq per year; significant amounts of oil and gas escape into the environment during production and transportation (fugitive losses). Fossil and nuclear power plants contaminate land masses, rivers and oceans with heat and radioactive solid waste; acid rain formed by reaction between anthropogenic acidic gases and moisture causes environmental problems which damage infrastructure, plant and vegetation, and threaten ecosystems of rivers, lakes and oceans; solid waste products of energy (in particular, polymer and electronic waste) and traditional use of biomass are causing increasingly serious environmental problems worldwide. Most of these pollutants are hazardous and have been associated with human illnesses.

5.4. EMISSIONS FROM FOSSIL FUEL USE

Fossil fuels power global economic growth and human development, and currently account for about 80% of the total world's primary energy use (oil 31%, coal 26%, gas 23% in 2018). In spite of global efforts to develop less carbon-intensive energy resources, the share of fossil fuels within the world energy supply has remained relatively unchanged over the past four decades and all recent projections show that fossil fuels will continue to dominate the primary energy mix in the foreseeable future, still accounting for around 78-80% in 2040 and beyond. Fossil fuels power global electricity generation (around two-thirds), and, together with electricity, supply nearly all the energy required by the three exponentially growing sectors of the global economy: industry, transportation and building. Efficiency gains across all sectors of the economy and all regions have been moderating growth of energy demand that would have been commensurate with the fast growth of the global economy and population, and projections indicate a modest growth of around 35% over the next two decades, compared with over 100% and 25-30% growth in the economy and population respectively. The positive impacts of efficiency improvements on emissions have also been substantial.

Over 40% of the total global primary energy consumption is used to produce electricity, all forms of generation have negative and significant environmental impact on air, water and land, and nearly half of the total global energy-related anthropogenic emissions come from electricity and heat generation. Electricity is central to all aspects of economic and human development: manufacturing, food production, healthcare, education, conducive household environment, etc., and is an important component of every person's environmental footprint. Electricity is the world's fastest-growing form of end-use energy consumption and demand is expected to grow by about 70% over the next two decades, accounting for around two-thirds of the global primary energy increase over the period, with developing and emerging nations accounting for most of the growth. These countries depend on local deposits of mostly low-grade coals, the most polluting of all fossil fuels (lignite coal has around twice the carbon footprint of natural gas), and all recent projections predict that they will also account for most of the future growth in GHG emissions. Energy-related emissions rose to a historic high of 33.1 Gigatonnes CO_2eq in 2018, a nearly 2% (569 million metric tons) rise from 2017, equivalent to the total emissions from international aviation, and China, India and the United States accounted for 85% of the net increase. Just two countries: the United States and China accounted for around 45% of the total global energy-related emissions and fossil fuels filled around 80% and 85% respectively of the primary energy requirements in both countries. However, coal accounted for 60% of the primary energy consumption in China compared with around 13% in the United States. Great differences exist in per capita emissions from all sectors of the global economy: total emissions of carbon dioxide per capita amounted to 4.4 metric tons in 2017, but more than one-half of the global population is below 2 metric tons (Figure 5.2). Per capita emissions in China almost tripled since 2000, reaching values similar to those of the European Union (EU) in the early 2010s; while population grew by less than 10%, total carbon dioxide emissions almost tripled. Between 2000 and 2017, India doubled emissions, but its per capita value is still one quarter of that of the European Union. Africa has the lowest per capita emissions among all regions - around one tenth of the US - and did not show any increase over the period. Should India and Africa reach similar levels of per capita emissions as those of the EU, an additional 13 $GtCO_2eq$ (more than one third of current levels) would be released in the atmosphere.

5.4.1.   Emissions from power generation

While emissions from all sectors of the economy and from all fossil fuels increased in 2018, the power sector accounted for nearly two-thirds of emissions growth and coal used for power generation accounted for nearly a third of the emissions, mostly in Asia. Emissions from electricity and heat generation in emerging and developing countries have doubled in the last two decades, compared with a significant fall in industrialized countries. China alone accounted for around two-thirds of the emissions' increase in 2018, coming mainly from large increases in coal-fired power generation and energy-intensive production of primary metals, chemicals, chemicals, petrochemicals and cement. Electricity/heat generation and transport account for two thirds of total CO_2eq emissions (Figure 5.5) and have been equally responsible of almost the entire global growth in emissions since 2010; the remaining third is split between industry and buildings. Sources vary across countries: while transport is predominant in many American countries, in Asia one-half of emissions derives from power generation and less than one sixth from transport. The picture changes after reallocating emissions from power generation to the final sectors: industry accounts for slightly less than one half of total emissions, buildings and transport for one quarter each. The buildings sector uses one half of the electricity that is consumed globally (21,000 TWh in 2017), industry the other half since transport is not yet electrified to any significant extent. Most of the buildings consumption takes place in the OECD countries while most of industrial demand is in Asia.

Around one half of the global increase in emissions between 2000 and 2017 came from power generation in Asia: India and China alone pushed emissions from electricity generation up by 200 MtCO$_2$eq annually. The huge increase in demand estimated to be over 600 TWh annually since 2000 and the fact that there has been no notable decrease in the carbon intensity of generation, have been the main drivers of the increase in global emissions from electricity generation. The increasing role of Asia, heavily reliant on coal-fired plants, meant that despite falling carbon intensities across most major producers in the industrial regions of the world in the last two decades, the world average remained relatively flat. However, improvements in renewables penetration and efficiency of power plants over the past decade or so have contributed significantly to the decline in emissions per unit of electricity generated: the annual growth rates of emissions between 2010 and 2017 decreased to half the values of the previous decade.

Fossil fuels (mostly coal, 39% and natural gas, 22%) supplied 64% of global power generation in 2018, and the most dramatic growth has been in China. Expected shifts in energy mix to less carbon-intensive sources (wind, solar, nuclear, natural gas) will help reduce primary energy use per unit of power output and the CO$_2$eq intensity of delivered electricity by more than 30% over the next two decades or so, but most of the reduction will be accounted for by the industrial nations who will consume the lowest electricity. Although emissions from power generating plants account for over 40% of the total energy-related emissions, outputs from specific plants depend on the technology adopted and fuel, and are unique to the individual facility because efficiency counts, but fossil fuels (coal, oil and natural gas) are by far the highest pollutants. Coal-fired power plants fueled by young coals (lignite and sub-bituminous) are the most polluting, producing around fifty times the emissions from renewable and nuclear power plants for the same power output, and these are the cheapest and most widely available grades in the developing world. Coal-fired power plants are the largest contributors to global energy-related emissions, accounting for about a third (over 10 Gt) in 2018. In fact, there has been a rise in the demand for coal in the last two years after a significant fall in the previous two years and coal-fired power plants were the single largest contributors to the growth in emissions observed in 2018. The International Energy Agency (IEA, 2019d) assessed the impact of fossil fuel use on global temperature increases, and found that emissions from coal combustion were responsible for over 0.3°C of the 1°C increase in global average annual surface temperatures above per-industrial levels, making coal the single largest source of global temperature increase. Nevertheless, the contribution of coal to global energy-related emissions has reduced to 30% from around 44% in 2016, due partly to closure of inefficient plants in the developed world and China but also to commissioning of more technologically advanced coal power plants.

Natural gas is less polluting at the generating plant but more polluting than coal at the production and transportation stages. As discussed earlier, when assessed on a lifetime cycle basis, natural gas could be as highly polluting as mature coal depending on efficiency along the production and transportation chain. The relative advantage is the potential for capture of natural gas pollution at the production stage. Natural gas is rapidly replacing coal as the fuel of choice for power generation, particularly in Europe and North America driven by strong environmental policies and, in the case of United States, abundant supplies of cheap natural gas. However, because coal is relatively abundant and affordable in the emerging regions, in particular Asia, the fuel outcompetes all other power generation energy sources and the impact of natural gas use on energy decarbonization efforts has not been as significant as desirable. Furthermore, although many aging coal-fired plants in the developed world are being retired and very few new plants are being commissioned, most of the plants in the emerging world are relatively new, 12 years old on the average, meaning they have decades to go before reaching their planned end of production in about 30 to 50 years. Also, India plans to increase coal-fired power generation by around 25% over the next two to three years and China accounts for about 55% of the total global coal power capacity. While China is shutting down small highly-polluting coal power plants, new, more efficient coal power plants are being built or planned, with a total capacity of around 300 GW, more than the combined capacity of the United States' 350 coal power plants. China is also building many more coal-to-fuel and coal-to-chemical plants and around two thousand coal power plants are currently under construction or planned in different parts of the world, mostly in the emerging economies but also in Poland and Australia. Most new plants are expected to come on stream in the next few years and will increase the total global coal power capacity by around 30%. Nevertheless, it is estimated that switching from coal to gas reduced coal consumption by about 60 million metric tons in 2018, helping to avert 95 million metric tons of carbon emissions, decreasing total energy-related emissions by about 15%. The switch was most significant in China and the United States, each reducing emissions by 40-45 million metric tons. Furthermore, most new coal-fired power projects all over the world are adopting high-efficiency clean-coal technologies which include high-technology combustion systems, coal gasification, and emissions' capture and processing.

5.4.2. Emissions from industry

Energy and industry have been closely interdependent since the Industrial Revolution, and their future remains intertwined. Energy powers all kinds of industries - manufacturing, non-manufacturing/services, construction, agriculture that produce food and goods for mankind. On the other hand, consumer demand for end-use energy and the wide range of varied products of industries drives energy demand, provides the impetus for massive investments in energy infrastructure, and propels continuous development of new energy technologies. Energy cost is always a major component of virtually any industrial production process, and this has provided a strong incentive for a persistent pursuit of higher efficiencies across the energy production and end-use spectra. Industrial energy demand growth would be at par with economic growth but for the extensive gains in manufacturing and other energy end-use efficiencies over the last two decades or so. Energy-intensive manufacturing (iron and steel, copper, cement, paper, food, glass, chemicals, primary metals) accounted for nearly 70% of the energy use by the sector in 2018, and the chemicals industry was the fastest growing user of fossil fuels in the industrial sector, both for energy and as feedstock for the production of chemicals and petrochemicals. Much of the increase in energy-related anthropogenic pollution in the last two decades or so has been driven by large increases in the production of these energy-intensive materials, in particular, steel, cement, chemicals and petrochemicals which depend primarily on fossil fuels. Most of these heavy industries are now in emerging and developing countries, many of which have also expanded very rapidly their coal production and coal-fueled power generation capacities. All recent projections indicate that these countries will account for nearly all of the increase in industrial primary energy consumption and carbon dioxide emissions over the next two decades. Production is often inefficient and energy intensity is high, lower-carbon energy options are few, more energy is required for a unit output compared with developed countries, and, inevitably, energy-related emissions remain high. However, industrial emissions are closely related to GDP output, hence there are significant variations even within the emerging regions: contribution of industrial-related emissions from Africa in 2018 was less than 10% that of Asia.

Over half of the primary energy consumption by industry is not used as fuel but as feedstocks for the production of polymers, chemicals, petrochemicals, fertilizers, lubricants, etc. In spite of projections of over 100% growth in the global economy over the next two decades or so, mostly in emerging regions, efficiency gains and growing use of less carbon intensive energy and manufacturing technologies will help reduce emissions from the sector relative to GDP by 35-50%, and contribute to a nearly 45% decline in the carbon intensity of global GDP. The industrial sector energy demand is higher than any other end-use sector, and accounts for 50-55% of total global final energy use. However, in spite of the leadership in end-use energy consumption, the sector accounts for only 19% of emissions from fuel combustion (25% when fossil fuel use as feedstock is taken into account and 36% when emissions from heat and power generation are allocated on the basis of end use). The developed world is transiting to low energy-intensive economy (light manufacturing and services), leaving the emerging world to host energy-intensive industries such as primary metals' extraction and processing, cement, chemicals, petrochemicals, fertilizer production powered mainly by fossil energy sources. Furthermore, manufacturing is 2 to 3 times more carbon and energy intensive than services, and, within manufacturing, energy intensity of production of primary materials could be two orders of magnitude higher than light manufacturing. This explains why most of the future energy-related emissions will come from the emerging regions. It should be noted also that the developed world is becoming increasingly dependent on the emerging countries for the supply of primary materials, thereby driving rapid expansion of production facilities and infrastructure in these regions.

Industry relies heavily on electricity for most process operations. Mostly driven by electricity and coal, industry's total energy consumption increased globally by 1 Gtoeq in the last two decades and reached almost 3 Gtoeq, 30% of the global final consumption. While many countries experienced decreases, China and India tripled their consumption estimated at 1 Gtoeq and 0.2 Gtoeq respectively. Emissions from industry in China (including indirect emissions from electricity) exceeded the total emissions of the United States in 2018. Much of the increase in energy demand for the iron and steel and non-ferrous metals industries was filled by coal; the chemical sector experienced a big increase in natural gas and coal consumption (almost exclusively in China), while oil demand remained flat; electricity increased at comparable rates in all industrial sub-sectors. Globally, coal combustion in the iron and steel industry alone was responsible of 2 billion metric tons of CO_2eq of emissions. There is as yet no viable alternative to coal use in the blast furnace which accounts for over 70% of the world's iron and steel supply.

5.4.3. Emissions from transport

Transportation currently accounts for about 25% of the global energy-related CO_2eq emissions. The world's fleet of vehicles will at least double by 2040, to around 2 billion, and most of the growth will be in emerging nations, with fossil fuels (mainly oil and other liquid fuels) currently supplying about 95% of the sector's fuel demand. However, growth in demand for transportation fuels will be modest, 30% or so, responding to increasing efficiency of conventional vehicles across all regions of the world, and the share of fossil fuels is projected to decline to about 88% as the use of alternative fuels (natural gas, biofuels and electricity) slowly increases. Electric vehicles have a strong potential for decarbonizing the transportation sector, depending on the sources of energy used in the manufacture and operation. Currently fossil fuels account for most of the manufacturing energy of electric vehicles and routine battery charging and unless there is a fast and significant transition to lower carbon power sources, the lifecycle carbon footprint could be twice as high as for comparable conventional internal combustion engine vehicles. Penetration of electric vehicles into the global auto market is projected to rise from 5 million in 2018 to 150-350 million (currently 5%, increasing to 6-18% of global vehicle population) by 2040 (different levels of optimism and assumptions), mostly in the light-duty vehicle sector assuming current problems of high cost, support infrastructure, low and unstable global oil market prices, strong competition from increasingly fuel-efficient, more affordable and environment-friendly conventional cars, and consumer preferences can be surmounted.

Emissions from transport increased by 2% annually at the global level in the period 2000-2017, reaching 8 $GtCO_2eq$ and increase was fairly uniform across all regions. Road transport, mostly for passenger travel, accounted for three quarters of total transport emissions and it is the mode that increased the most in absolute terms (2%), second only to international aviation which grew at 3%. Such a big increase was mostly propelled by Asian countries: China and India increased their emissions from road transport by a factor of four and three respectively since 2000 and similar growth rates were experienced by other countries in the region. However, despite such increases, levels of road emissions per capita in many Asian countries still remain lower by an order of magnitude than in the United States where transport accounts for more than one third of total emissions. In view of the projected rise in the vehicle population and air travel over the next two decades, emissions will continue to increase, though moderated by higher efficiencies across all sub-sectors of transportation. The impact of the growth of electric vehicle market penetration will largely depend on the source of electricity for manufacture and operation.

5.4.4.   Emissions from Buildings

The building sector comprising residential and commercial users accounts for about 20% of the total delivered energy consumed worldwide, but only 6-9% of the total global energy-related emissions come directly from the sector, although the emissions level more than doubles when emissions associated with electricity used by the building sector are factored in. The projected world average growth rate in energy demand by the sector over the next two decades is 1.5% but there will be significant variations by country and region, not only in growth rates but in type and amount of energy consumed. For example, growth rate in the United States will be about 0.1%. OECD, 0.6%, Mexico, 1.9%. The lower growth rate in the industrialized countries is a result of the expected relatively slow growth rate in the economies and population, as well as improvements in building insulation shells and efficiency of appliances. On the other hand, the economies, population, urbanization and prosperity will grow in non-OECD countries where over 80% of the world's population resides. The region's projected population as well as economic growth rates are more than twice that of OECD. Delivered energy consumption in the residential sector is expected to grow by 48% and will account for around 13% of total world delivered energy consumption in 2040 but the bulk of the growth (around 80%) will be in non-OECD countries. Growth rate in non-OECD countries is around 2.1%/year compared with 0.6%/year in OECD countries. Considering the expected faster growth rate in the building sector, particularly in the emerging world, growth in energy demand should be much higher than the projected values. However, buildings and appliances are becoming increasingly energy-efficient and new technologies coming on stream will improve efficiencies at even faster rates.

Natural gas is the dominant residential and commercial source of energy in OECD countries but electricity is more important in the emerging economies of the non-OECD countries. Even in OECD countries, demand for electricity is growing at a fast rate of

1.0%/year compared with 0.6%/year for natural gas, and is expected to surpass natural gas as the leading source of residential delivered energy within the next ten years. Electricity share of world residential energy consumption is around 40%, increasing to about 45% in 2040. In spite of the vibrant current and projected growth in global housing demand over the next two decades, energy efficiency is playing a big role in driving down end-use energy demand by the building sector as modern appliances and advanced building technologies and materials proliferate across all regions of the world.

The building sector is the least polluting of all the end-use sectors, and accounts for less than 10% of GHG emissions from energy end-use. It could be significantly higher when considered on a lifecycle basis since the sector is a major consumer of electricity, much of which comes from fossil fuels. The dynamic growth of infusion of energy-efficient technologies into all aspects of the sector, - from energy-efficient building to fittings, appliances and smart remote control systems - should help to further drive down emissions from the sector. However this will depend on the extent of decarbonization of power generation, since around 90% of the energy demand growth in the sector will be met by electricity. Current policies in many developed countries promoting self-power generation with modular solar PV systems should also help in reducing pollution from the sector to a very small proportion of the total end-use emissions. Some international organizations and private enterprises are also helping to spread self-power solar PV systems across the developing world in order to fast track access to electricity, from small personal units that charge phones and power lamps, to rooftop PV systems that can power homes and small communities.

5.4.5. Non-energy Emissions

Around a third of the global anthropogenic emissions come from non-energy related sources: forestry and land use, agriculture and livestock production, and non-energy related industrial processes. Agriculture, forestry and other land use (AFOLU) activities account for 12-13% of carbon dioxide, 44% of methane and 82% of nitrous oxide emissions from human activities globally. Agricultural activities (biogenic sources) such as wet rice cultivation, soil fertilization, and vegetarian livestock production release significant quantities of methane into the atmosphere. For example, the digestive process of a cow emits about a quarter of a kilogram of methane into the atmosphere daily. Significant methane emissions also emanate from decaying vegetation, especially in landfills, hydro dams, artificial lakes and reservoirs. Agriculture accounts for over a third of non-energy emissions, non-energy industrial processes about 20%, with the balance coming from other sources. Non-energy anthropogenic emissions have received relatively little attention even though they account for nearly a third of the global total emissions. Furthermore, while energy-related emissions are predominantly carbon dioxide, agriculture emits mainly methane and nitrous oxide which have 20 and 300 times respectively of carbon dioxide's global warming potential. Clearly, there is substantial scope for reduction in non-energy emissions, notably in agriculture, use of landfills for waste disposal, uncontrolled incineration of waste, bush burning, etc.

5.5. OUTLOOK ON ENERGY-RELATED EMISSIONS

In spite of their poor environmental credentials, fossil fuels continue to meet more than 80% of total primary energy demand and over 90% of energy-related emissions are from combustion. Carbon dioxide comes mainly from power generation and transportation, methane originates mainly from oil and gas extraction, transformation and distribution, and accounts for about 10% of energy sector emissions. Much of the balance is nitrous oxide from energy transformation, industry, transport and buildings. Although methane and nitrous oxide gases account for very small proportions of the total anthropogenic greenhouse gases, their global warming potentials (GWPs) are much higher than for carbon dioxide. The level of polluting emissions from energy use is closely related to the amount and mix of both primary and end-use energy consumption, and the main focus of global mitigation effort is to decarbonize the entire energy value chain.

The global economy is rising, population growth is dynamic, and the quest for human development is strong, all powered by energy. The world is faced with a major challenge of providing adequate energy to meet exponentially increasing global demand in an environmentally sustainable manner (more energy with lower carbon emissions). The International Energy Agency (IEA) has stated that energy must be fully decoupled from global economy and the rise in energy-related emissions must be reversed within the next two decades if the world is to embark on the track towards the achievement of the set goal of limiting the rise of global temperature by the end of the century to 2°C, preferably 1.5°C. As discussed earlier, even this small temperature rise could have a catastrophic impact on the global climate. Also, the International Panel on Climate Change (IPCC, 2018) recently assessed all available scientific, technical and socio-economic literature relevant to global warming and concluded (with a high degree of confidence) that, if pollution continues at the current rate, global warming will likely reach 1.5°C above pre-industrial levels between 2030 and 2052. The maximum temperature reached will depend on the point at which net zero global anthropogenic emissions is achieved and sustained. Even then, warming from the cumulative net global emissions up to the time of net zero emissions will persist for centuries to millennia and will continue to cause further long-term changes in the climate system such as ocean warming, sea level rise and associated impacts. Also, the International Energy Agency (IEA, 2019g) projects that achievement of the 1.5°C target will require the deployment of much stronger and integrated and coordinated global action and policy instruments to further reduce the projected level of emissions for 2040 (based on current and expected actions) by around 40%.

World energy-related emissions currently account for around 65% of the total global emissions, and will continue to grow over the next two decades, be it at half the pace of the previous two decades. A growth of only about 10% is projected by 2040, despite the fact that population will have risen by about 25% and global GDP will have more than doubled. However, growth in emerging countries will be significantly faster compared with the OECD countries. Emissions are projected to peak in OECD countries by 2030, and then start to decline, falling by around 15% by 2040 compared with 2016, but the contribution of non-OECD countries (excluding China) is expected to rise by around one-third, accounting for about 50% of the total global emissions by 2040. China alone contributed about 60% of the growth in emissions from 2000 to 2016, and will account for about 35% of the total global energy-related emissions in 2030, after which energy decarbonization policies announced recently by the country should cause a gradual decline to about 25% in 2040, still more than double that of the United States.

Energy-related emissions growth will be mitigated by increasing energy efficiency and gradual shift from coal to lower carbon energy sources, in particular, natural gas and renewables for power generation. By 2040, the carbon-intensity of the global economy is likely to fall by half, with substantial contributions across all regions. Energy efficiency gains are expected to be a major contributor to this achievement because they translate to lower energy input and associated emissions for a unit of economic output. The greatest impact of energy decarbonization will be in the power generation sector where the use of coal is expected to decline in favor of natural gas and renewables. However, most of this shift in fuel mix will be in the developed nations. Energy-intensive manufacturing currently accounts for around 27% of total global delivered energy to the industrial sector and the proportion is unlikely to increase over the next two decades despite an overall increase in the demand by the industrial sector. The ongoing shift by developed countries to less energy-intensive manufacturing will lower emissions from these regions and the trend is expected to intensify. However, these industries are shifting to the emerging nations, in particular, Asia which rely heavily on coal, and often cannot afford to deploy more efficient manufacturing technologies, hence the achievement of the developed nations in lowering emissions from industry will be largely offset by significant increases in the emerging nations. Emissions from liquid fuels, mostly from transportation will flatten in the OECD countries, again due mainly to the positive effects of improved energy efficiency, but will grow substantially in the developing nations, in particular, India and China due to the expected rapid growth in both personal and commercial transportation. The building sector contributes less than 10% to global energy-related emissions (although significantly higher when emissions from electricity are shared) and further reduction is projected. Although energy demand in the sector, in particular, electricity is projected to rise sharply over the next two decades, it will be mitigated by substantial improvements in building energy efficiencies (building envelope, fittings and appliances).

The fossil fuel share of total global primary energy supply (TPES) was 81% in 2016, roughly the same level for nearly thirty years (EIA, 2018). Over the period, coal and oil jointly represented about 60% of TPES and almost 80% of CO_2 emissions. The contribution of gas has also been stable at around 20% while the balance of 19% accounted for by lower carbon sources (renewables and nuclear). Coal has the highest carbon intensity and accounted for about 44% of global energy-related emissions in 2016. Details of emissions in terms of trends over the last three decades, fossil fuel source, economic sector and region are presented in Figures 5.4-5.6. While the use of coal has declined significantly in Europe and North America over the last few years, it is still the predominant fuel in Asia, accounting for around 60% of the emissions from that region in 2017.

Two regions contributed three-quarters of the total energy-related emissions in 2017: Asia (53%) and Americas (22%) and accounted for 85% of the net increase in emissions in 2018. China alone accounted for 28% while the United States contributed about 15%. Events in these regions will largely determine the global outlook on anthropogenic emissions over the next two to three decades. China's outlook to 2050 shows that a combination of efficiency improvements, technology implementation, fuel substitution, and emission control policies will hold emissions level flat in the next thirty years, overall, and in all sectors of the economy, with the exception of industry where an annual growth rate of around 0.6% is projected. However, the country is under severe pressure to provide more primary energy and electricity for its extensive energy-intensive industrial sector as well as the increasingly middle-class population, and coal is the only major primary energy available locally (natural gas resources are also being developed rapidly and the country is linked to the Russian gas pipeline). China currently operates more than half of the world's coal power plants and many more are under construction or planned. All recent projections show that the country will remain the world's largest emitter of energy-related pollution in 2040. The United States has been shutting down coal-fired and nuclear power plants in favor of natural gas-fired plants largely because the unit cost of coal-powered electricity can no longer compete in view of the increasingly stringent environmental regulations and declining costs of natural gas. Not even the recent policy reversals in favor of coal production and use have succeeded in stemming the rapid downturn in the industry. Primary energy consumption is expected to remain flat over the next twenty years, and so will energy-related emissions, although there will be significant sector variations. While emissions in the transportation and building sectors will continue to fall, contribution by the power sector will remain flat, but there will be significant rise in the industrial sector emissions.

5.5.1.   Outlook on global primary energy demand

Energy supply, mix and use largely determine the nature and extent of energy-related emissions. All recently published outlook reports on future global primary energy demand predict a rise of around 35% over the next two decades or so, in response to the growth in the world economy which is expected to double over the same period, and growing prosperity across all regions. Fossil fuels which currently account for over 80% of the total primary energy demand will remain dominant throughout the period, still accounting for nearly 80% of the demand. Much of the projected growth in primary energy demand is attributed to non-OECD nations, many of which continue to rely heavily on fossil fuels to meet the fast-paced growth of energy demand. Emissions depend primarily on energy demand and use and reduction would depend largely on the extent to which energy demand can be decoupled from global economy growth and decarbonized. Developing an outlook for energy even over the period of a few years can be very complex and inaccurate because of the many unpredictable variables and necessary assumptions. Nevertheless, tools including extensive database and software packages have greatly improved accuracy over the last two decades or so. Most projections consider several scenarios: Emerging transition (ETS) which takes into account current, expected and anticipated national and global policy instruments and technology developments over the next two decades or so, and faster, even faster transition scenarios (FTS, EFTS) which identify potential options for repositioning the world on a feasible pathway towards achieving sustainable environmental targets. Primary energy consumption grew at a rate of nearly 3% in 2018, the fastest rate over the last decade, almost double its 10-year average, and fossil fuels filled nearly 85% of the demand. Just three countries: China, United States and India accounted for more than two-thirds of the global increase in energy demand. Inevitably, carbon emissions grew by 2%, the fastest growth rate in nearly a decade. While the growing economy was a significant trigger, much of the energy demand increase may be weather-related: the unusual severity of cold weather and heat waves (both potential consequences of climate change) and a significant rise in demand for heating and air conditioning across the world's major centers of energy demand. Most of the rise in energy demand was for natural gas and electricity consumption but there was also a significant rise in the demand for coal for electricity generation and heating even in the developed world.

The expected fast growth rate of the global economy (GDP) will lead to a significant rise in energy demand, although the continuing decline in energy intensity (energy used per unit of GDP) driven by increasing efficiency in energy production and utilization will likely moderate energy demand and limit rise to no more than about 30-35%. All recent published outlooks and projections by reputable independent international organizations show that fossil fuels will still account for around 80% of global primary fuel consumption for the next two to three decades. This demonstrates clearly the complexity of mitigating climate change in a world that requires provision of more energy for less carbon, and shows that the world is on an unsustainable path towards the declared goals for a sustainable environment. Despite the continuing rapid growth in renewable energy in 2018, it provided only a third of the increase in power generation and accounted for only a quarter of the total primary energy use for power generation in 2018. Decarbonizing energy is the surest path towards a sustainable environment but there are formidable challenges: while the industrial world is set to continue and elevate its achievements over the last decade in lowering energy carbon footprint, most of the growth in energy demand over the next two decades or so will be in the emerging economies and fossil fuels, in particular coal, will account for most of the consumption. These regions will host most of the world's energy-intensive production capacity for the primary building blocks of the global economy: primary metals, cement, chemicals, petrochemicals, polymers, fertilizers which depend primarily on coal as energy sources (heat and electricity) as well as feedstocks. Furthermore, most of these countries cannot afford to adopt currently available technologies for decarbonizing energy. In effect, expected significant drop in energy-related emissions in the developed world will be largely neutralized by major rises in the emerging nations. China is already adopting cleaner coal technologies which compare favorably with natural gas in terms of carbon intensity but other countries in the region will need international assistance.

Extensive scientific data indicate (with a high degree of confidence) that human-related emissions have been the dominant cause of the observed global warming since the mid-20th century. The Earth's surface temperature is projected to rise over the 21st century under all possible emission scenarios, and there is substantial scientific evidence in support of the theory that a warming global environment will make extreme weather more frequent and intense, occurring in many more places than usual. It is very likely that heat waves will occur more often and last longer, flooding will be more frequent and severe, oceans will continue to warm and acidify, ocean levels will continue to rise, extreme weather storms and tornadoes will become more frequent and devastating because they are powered by heat derived from warming oceans. Extreme precipitation events will become more intense and frequent in many regions of the world, impacting on local weather. All these events will continue to fuel demand for more energy for adaptation including heating and cooling and, without major progress in decarbonization, result in higher anthropogenic emissions.

5.5.2. Outlook on emissions from power generation

Electricity generation accounts for about 42% of energy-related CO_2eq emissions, but the expected shifts in energy mix to less carbon-intensive sources (wind, solar, nuclear, natural gas) will help reduce primary energy use per unit of power output and the CO_2eq intensity of delivered electricity by more than 30%. Natural gas which has the lowest carbon footprint of all fossil fuels, will likely play a critical role in helping to decarbonize the power sector in future. The fuel is reliable and efficient for power generation when available at competitive prices. Also the ease of handling, transportation, and flexibility make it well suited to meet peak demand and backup intermittent renewables. However, the extent of its role depends on several uncertainties: stronger public sentiments against nuclear, hydro and coal power could enhance the role of natural gas as a fuel of choice, while a strong increase in the deployment of renewables could reduce its growing role in power generation. Also, the role of natural gas in the electricity generation mix will vary across countries and regions, prominent in natural gas-rich countries and regions, but relatively low in countries that have limited access to inexpensive natural gas. Renewable energy, in particular, solar and wind will play an increasingly significant role in decarbonizing power generation, with growth in deployment across all regions, although at different rates.

Investments in fossil fuel power plants (coal and natural gas-powered) remain strong and are projected to continue to rise over the next decade. Since these plants typically have long lifetimes of 30-40 years, they will account for large amounts of future emissions and will inevitably move the world further from the path to a sustainable environment. Around 40% of global electricity and a third of the annual global energy-related emissions come from about ten thousand coal-fired power plants operating in 80 countries across all regions of the world. Many countries in the industrial world are closing coal power plants for economic and environmental reasons and nineteen including the United Kingdom and Germany plan to phase out coal power plants completely, but many more mostly in the emerging world are either building or planning new plants in response to exponentially increasing demand for electricity by the growing economies and population. Several recent studies have proposed potential pathways to the carbon budget to keep global temperature rise to 2°C or even lower (1.5°C) by the end of this century, in accordance with the Paris-2015 Agreement. A study by Pfeiffer *et al.* (2018) estimates that current and planned fossil power plants will emit around 600 $GtCO_2eq$ above the levels compatible with the average 1.5°C - 2°C scenario. The entire planned capacity needs to be canceled and around 20% of current operating capacity have to be stranded to meet the climate goals set out in the Paris-2015 Agreement. The International Energy Agency (2017) projects that nearly all fossil fuel power plants need to close, replaced by low-carbon electricity which accounts for 96% of global power demand, in order to place the world on the pathway to net-zero emissions by 2060. However, considering the situation assessments and projections discussed earlier, the chances of either option are very remote.

5.5.3.   Outlook on emissions from transportation

Transportation currently represents about 25% of the total global CO_2eq emissions from combustion, and growth over the next two decades or so is expected to be modest, responding to increasing efficiency of conventional vehicles across regions, in spite of the expected rapid growth in global transportation. The expected growth of electric vehicle (EV) population over the period, representing 6–35% (depending on level of optimism) of the global transport population by 2040 will likely have the greatest impact in the light-duty vehicle sub-sector. However, growth of EV market penetration is one of the greatest uncertainties in energy use projection analyses, largely because of lack of critical infrastructure, policy uncertainties, and entrenched consumer preference which may be difficult to switch in view of increasing competitiveness of conventional vehicles on virtually all fronts. It should be noted also that China and the United States currently account for about three quarters of the total global population of electric vehicles but the two countries sourced 61% and 75% respectively of their electricity from fossil fuels (coal and natural gas) in 2019. In effect the net environmental benefit of electric vehicles on climate change is currently minimal, although the positive effect on urban environment would be substantial. However, recent projections show that fossil fuel power generation in both countries will drop to 50-55% by 2040-2050. The high potential of electric vehicles in decarbonizing global transportation is not in doubt but power generation would need to be decarbonized at a much faster rate than currently projected and the various issues with EVs discussed earlier must be resolved. Furthermore, shifting support policies, increasing sophistication of conventional vehicles, and volatile global oil market are potentially significant constraints to the growth of electric vehicle market growth.

In a recent projection by ExxonMobil (2018) several different scenarios of energy demand, end-use and associated emissions were tested, based on various assumptions and uncertainties, a hypothetical scenario was developed in which all light vehicles on the roads are electricity-powered by 2040, totally eliminating the demand for oil in the light-duty vehicle sector of transportation. In order to achieve this, global sales of light-duty vehicles would need to be 100% all electric (50 times from existing levels), starting in 2025 requiring sales of around 110-140 million electric vehicles annually henceforth (more than a hundred times the number of electric vehicles sold in 2016). Even with this unrealistic scenario, total energy-related emissions would be reduced by no more than about 5% and, although emissions from light-duty vehicles would reduce to zero, emissions from power generation would rise with the increase in electricity demand, and CO_2eq emissions from the power sector could increase by about 15%, with coal accounting for 60% of the increase. Even though this analysis may seem biased, coming from a petroleum company, several independent organizations have come to similar conclusions, notably the International Energy Agency (IEA, 2017b), and Energy Information Administration (EIA, 2017) as exemplified by the following statement from IEA:

> *"Electric vehicles (EVs) are in the fast lane as a result of government support and declining battery costs but it is far too early to write the obituary of oil, as growth for trucks, petrochemicals, shipping and aviation keep pushing demand higher.....rising oil demand slows down but is not reversed before 2040 even as electric car sales rise steeply".*

Perhaps the most optimistic scenario on electric vehicle market penetration was presented in a recent IEA publication (2019e), based on the EV30@30 initiative launched at the Clean Energy Ministerial meeting in 2017 which aims to promote market penetration of EVs to a 30% share of the global new vehicle stock in 2030. This translates to about 250 million EVs available for sale in 2030, with passenger light duty vehicles (PLDV) and battery vehicles accounting for around 50%.

In the recent past several countries in Europe have announced policies to stop the sale of diesel cars by 2040. This was clearly based on the assumption that diesel engines are more polluting than petrol equivalents. This is in fact not true of modern diesel vehicles which feature advanced catalytic converters that reduce emissions. Furthermore, diesel vehicles deliver more kilometers per liter of fuel, hence pollution rate could be actually lower. In order to check this, manufacturer's specifications of five 2018 models of a popular car were analyzed in the course of this study, all with 1.6 liter engines, four petrol and one diesel. Emissions for the four petrol models range from 110 to 140g/km CO_2eq, compared with 93g/km CO_2eq for the diesel model: the petrol engines deliver around 20 km/liter compared with around 30 km/liter for the diesel car. It should be noted however that the quoted figures are from auto manufacturers, and independent verification is needed. Also, even if the announced policies of banning sales of ICE cars by 2040 is unrealistic, any potential reduction in the nitrogen oxides emitted by diesel engines would be enormously beneficial since most of the effect is local and the negative impact on human health can be substantial. Furthermore, it has accentuated consumer awareness, interest and reactions. Sales of diesel cars have plummeted, but there has been no major rise in the sale of EV, largely because people are confused and prefer to delay investments in new vehicles. Furthermore, the market share of electric vehicle sales in Europe in 2019 was only 3.5% and around one-third were hybrids which use both electricity and gasoline.

Interest in hydrogen-powered fuel cell electric vehicles (HEV/FCEV) has been strong for some time but the major issue was the development of a capable hydrogen fuel cell. This problem has now been largely resolved, and the market for HEVs is emerging and several big auto manufacturers are now offering a variety of models. The difference between BEVs and HEVs is that BEVs run on batteries that have to be recharged, while HEVs have fuel cells which draw hydrogen from the tank and produce electricity on board. This means that they can stop at a hydrogen refueling station to buy liquid hydrogen, just like a conventional vehicle, and a full tank can deliver 400-450 kilometers, more than most electric cars currently on the market. Liquid nitrogen is currently being produced mainly from fossil fuels, but can also be produced from biomass, and water, and, like EVs, the only emissions are heat and water. Electric and hydrogen vehicles hold very high promise as a very effective mitigation of transport pollution, particularly in cities currently plagued by incessant and toxic smog. One potential problem is the fact that production of hydrogen by any method requires a lot of electric power most of which will likely come from fossil-fueled power stations. However, most power stations are located on the outskirts of, or well away from cities, and emissions can be captured and processed on site. There is little doubt that the electric vehicle market will continue to grow and moderate emissions from the transportation sector. However, most of the projected growth will occur in the developed countries while emerging economies will account for most of the future growth in vehicle population. Furthermore, intra-country policies that are promoting the replacement of ICEs with electric vehicles are also stimulating (be it inadvertently) the export of inefficient used vehicles to the emerging countries, most of which have no strong emissions' policy control policies. Since there are no stratospheric country boundaries, pollution from any source anywhere in the world will likely have the same effect on global climate change.

5.5.4.   Outlook on emissions from industry

Economic growth is the primary driver of energy use and associated CO_2eq emissions. The industrial sector is the highest polluting in the energy end-use sector, contributing around 25% of CO_2eq emissions in 2016 and not much had changed in 2018. Projections show that, in spite of the fact that GDP is expected to double by 2040, growth in emissions will be modest due to efficiency gains and growing use of less carbon intensive energy and manufacturing technologies which will help reduce emissions from the sector relative to GDP by about 50%, and contribute to a nearly 45% decline in the carbon intensity of global GDP. However, there will be significant regional variations: most of the emerging countries that will largely host future energy-intensive industries rely heavily on coal, the most carbon-intensive fossil fuel for power generation as well as the production of primary metals, cement, glass, chemicals, petrochemicals, all of which are indispensable globally for economic growth. It is important to note that high proportions of these products are exported to the developed economies.

5.5.5.   Outlook on missions from buildings

The building sector (residential and commercial) is the least polluting of all energy end-use sectors, accounting for only 6-10% of the total global emissions despite the fact that the sector currently accounts for around 20% of the total global delivered energy consumption. However, considering that buildings rely strongly on electricity, this contribution rises to around 27% when emissions from electricity are allocated to the consuming sectors. Rapid population growth, rising urbanization and incomes are all fueling demand for modern homes, which in turn stimulates growth of commercial services. Residential energy demand is projected to grow by about 20%, consistent with the projected population growth over the next two decades or so, with non-OECD nations accounting for around 40% of the total growth, and China and Africa alone will account for about 30% of the growth in demand (ExxonMobil, 2018). Around 90% of the demand growth will be met by electricity, hence the average worldwide household electricity will rise by about 30% across all regions over the period, moderated by rising efficiency in the sector - energy efficient building construction, fittings, appliances, and consumer products, and increasing awareness among consumers of the multiple benefits of optimized energy use. Residential electricity use is projected to rise by about 75% by 2040, driven largely by around 150% increase in non-OECD countries (about 250% rise in India and Africa), and coal will likely be the main power generation fuel.

5.5.6   Outlook on emissions in the emerging world

Energy is an indispensable propellant of human development and, with the sector accounting for around three-quarters of global pollution emissions, decarbonizing energy will be critical in achieving global efforts to slow down the negative impact on the environment. Also, even if the announced policies of banning sales of ICE cars by 2040 is unrealistic, any potential reduction in the nitrogen oxides and black carbon emitted by diesel engines would be enormously beneficial since most of the effect is local, with potentially devastating effect on human health and other components of the ecosystem. The International Energy Agency (IEA) currently supports countries through the provision of energy emissions statistics and training of country officials in policy, modeling and energy statistics. IEA also has a Clean Energy Transition Program which targets emerging countries. As laudable as these programs are, the effect will be minimal unless the countries have financial and technological support to improve efficiencies across the energy supply and consumption chain. Electricity which accounts for around 40% of energy-related emissions is increasingly becoming the fuel of choice across the world, filling around 20% of global final energy consumption, and is projected to rise even faster in future, with the emerging world accounting for the bulk of the demand. Decarbonizing the sector particularly in the developing world will be critical to the achievement of global effort to mitigate the impact of anthropogenic emissions on the climate. Around a third of current global investments in energy is in electricity generation in developing economies but low, regulated consumer prices and the need for cost recovery are disincentives for the much-needed strong investments and compromising the adoption of high-efficiency or low-carbon energy technologies.

Projections on the feasibility of the United Nations' target of access to clean domestic energy all by 2030 are not very optimistic. Currently, around 3 billion people - more than 40% of the global population and about 50% of the population in the developing countries - lack access to clean domestic energy. Most of them rely on traditional wood, charcoal, and biomass for cooking and heating, often in poorly-ventilated environment. This has been linked by the World Health Organization (WHO, 2018) to around 3 million premature deaths in 2016 in addition to the 4.2 million who died as a result of outdoor (ambient) pollution. More than 90% of air pollution-related deaths occurred in low- and middle-income countries, mainly in Asia and Africa. While progress on access to clean domestic energy has been gathering momentum in parts of Asia, backed by targeted policies focused mainly on proliferation of the use of liquefied petroleum gas (LPG), sub-Saharan Africa is far from being on track. Projections show that the number of people globally without access to clean household energy will decline by only 18% by 2030, with around 820 million of them (56% of the population) in sub-Saharan Africa. In effect, increased access to electricity, mostly in emerging countries will likely continue to depend heavily on fossil fuel-powered generation although renewables will also play a significant role, and the continued dominance of fossil fuels and traditional biomass use in emerging nations will account for most of the projected rise in anthropogenic emissions across the regions. Also, any dramatic improvement in household ambient pollution over the period, particularly in the sub-Saharan region is unlikely.

The global dynamics of primary energy consumption have changed significantly in recent years and an even more dramatic change is expected. Just two decades ago, Europe and North America accounted for more than 40% of global primary energy demand compared with 20% for developing economies in Asia. This situation is projected to completely reverse by 2040: Asia will account for half of the growth of natural gas, 60% of the rise in wind and solar PV, more than 80% of the increase in oil and nearly all of the growth in wind, coal and nuclear energy. Asia will also dominate power generation: already, six of the world's top ten power companies in terms of installed capacities are Chinese utilities. Asia's share of oil and gas trade will rise from around half of the current world total to more than two-thirds by 2040. The share size of China in terms of economy and population means that any changes in energy use make a significant impact on the global scenario. China's economy is slowing down and also transiting from high-intensity manufacturing to less energy-intensive production. Furthermore, growth of the renewable and nuclear energy sectors has been rising, hence a gradual decrease in energy demand rate is expected, and energy intensity will also start to decrease. Overall, there should be a significant fall on CO_2eq emissions of the country by 2040 but the country will remain the world's largest source of energy-related pollution. Significant shifts from coal to natural gas, renewable energy and nuclear power are projected for the country, and the consequent effects on CO_2eq emissions are expected to be substantial because of the size of the country's economy. China's choices will play a major role in determining global energy-related emission trends, in particular, the country's transition towards a more services-based economy and deployment of low-carbon energy, accounting for around 40% of new global installed solar PV and wind capacity over the next twenty years or so. This would be a significant positive development in the global emissions mitigation effort since the country is likely to account for around 40% of the expected growth in electricity demand, the highest source of energy-related pollution over the period. The country is also projected to overtake the United States by 2030 as the world's largest producer of low-carbon nuclear electricity.

China currently leads the world in the global EV market, accounting for about 25% of the total global sales in 2017, and will account for about 40% of the projected global investment in electric vehicles by 2040. It is unlikely that the country can do much to reduce coal use and emissions from industry considering the extensive use of coal for power generation and fossil fuels both as energy sources and as feedstock for primary metals, non-metals and chemicals/petrochemicals production, and the major global demand for its products. However, decarbonizing electricity could have a significant impact in its transportation sector, particularly the electric vehicles sub-sector which currently depends heavily on coal electricity for manufacture and operation. On the contrary, energy demand by India continues to grow at more than 2.5 times that of China, representing more than a third of the global increase, and most of the growth will be filled by coal and nuclear power. However, India's contribution to the global total emissions in 2016 was only a quarter of China's emissions, and recent projections show that China will still account for two times India's emissions in 2050.

Around 80% of the global population consuming less than 100 Gigajoules of energy per capita per year (the minimum that stimulates healthy human development growth) live in the emerging world which will also account for most of the world's population growth over the next two decades or so. Rapid growth of the economies and urbanization, expected massive growth of the middle class population and the desire for higher living standards will drive energy demand across the regions, although the demand will be partly offset by efficiency gains and declining energy intensity. The declared UN policy goal of modern energy for all by 2030, which may be overoptimistic, will require around 25% more energy than projected in the emerging transition scenario of the International Energy Agency in order to reduce energy poverty to about a third by 2040. A rapid rise in the deployment of power generation plants in the developing world is expected over the next two decades and most will be powered by fossil fuels. Asia Pacific accounted for 75% of global use of coal in 2018, and most of it was produced within the region. The coal was used for the production of electricity, heat, steel, copper, aluminium, cement, chemicals and petrochemicals, fertilizers, much of which ended up in the developed world.

Various recent projections expect an exponential growth of electric vehicle from the current 5 million to 150-250 million by 2030-40. Lithium-ion battery and electric motors are the main components of EVs both of which require input of some special metals, notably copper, lithium, nickel, cobalt, platinum and palladium. Copper is already an important metal for the electrical sector of the global economy and electrical systems of internal combustion vehicles, and an additional 3 million metric tons a year will be required to fill the future needs of the EV industry. Nickel is an important steel alloying element and is also used in many other industries but an additional 1.3-1.5 million metric tons a year will be required by the EV industry by 2040. The other metals are critical stockfeeds of many industries but current requirements are relatively small. However demand is expected to rise 3-5 fold by 2040 because of the needs of the EV industry. The expected steep rise in demand for these metals raises several environmental issues: The world currently sources most batteries (EV, auto, solar power storage) from Asia; nearly all the metals required for manufacture are currently being sourced mainly from the emerging world, mining is relatively crude, inefficient and highly polluting; extraction and refining are energy-intensive, powered mainly by coal-fueled electricity. The battery manufacturing process accounts for around 40% of the energy required for producing an EV, again filled mostly by coal electricity since Asia is currently the primary source. There are plans for some European manufacturing facilities but most of the critical metals will still come from the developing world where there are relatively few environmental regulations. Mining of these metals which is often open-cast is already creating major environmental issues including toxic mineral dust, contamination of rivers and water supplies and destruction of vegetation and landscapes. Mineral experts doubt that current and projected production of these metals can meet the expected demand in 2040 but, more importantly, production of EV batteries and required materials will account for a significant proportion of the expected rise in global energy demand-related anthropogenic emissions over the next two decades, most of which will come from the emerging regions.

Another potential environmental issue with electric vehicles is the non-availability of appropriate technologies for the disposal of discarded lithium-ion batteries. Lithium-ion batteries have memory issues and energy retention declines with increasing recharge cycles. When an EV battery loses a significant part of its electricity storage capacity, typically 20%, it can no longer power a vehicle effectively and has to be changed. Estimates vary but current batteries which cost as much as a medium-sized ICE vehicle ($15,000-20,000 to replace) will need to be changed after around 150,000-250,000 kilometers or ten years, which means that there could be up to 30 million discarded used batteries by 2040, assuming around 10% of the expected 250-350 million EV vehicles on the roads would have needed battery replacement. However, the battery still has high power storage capacity which makes second-life use possible, for example in stationary electricity storage applications, notably for surplus power in utility-scale intermittent renewable energy installations such as in buildings, thereby increasing the lifetime use of the battery by around 70% before reaching end-of-life. While lead-acid batteries used in ICE vehicles are a hundred percent recyclable, there is currently no viable technology for lithium-ion batteries. Lithium-ion chemistry is very complex, dangerous, and very sensitive to heat, often leading to fires and explosions, and there have been many fire incidents in airplanes, electric vehicles, laptops, cell phones, etc. Furthermore, there is currently no standardization of either the chemistry or the technology, which makes it difficult to develop appropriate recycling technologies. Apart from the fact that li-ion batteries contain very expensive and carbon-intensive materials such as lithium, cobalt, nickel, copper, graphite which could be recovered by recycling batteries after second-life use, unrecycled, discarded batteries will create new environmental issues in the next two decades unless effective recycling technologies emerge. Furthermore, recycling discarded batteries will reduce the battery greenhouse emissions attributable to the vehicle on a per-kilometer basis by about 42% (ICCT, 2018) while also reducing the demand for virgin raw materials.

5.5.7.   Outlook on effect of global economic, social and exigencial forces on fossil energy use

Another major environmental mitigation challenge is the prime role of fossil energy in the global economy. The United States has recently emerged as the world's largest producer of oil and gas and more than doubled its oil output over the last decade to over 15 million barrels a day in 2019 (the highest in the world), driven by the wide deployment of fracking technology. The country is also poised to becoming one of the world's largest exporters of fossil fuels (oil, coal, liquefied natural gas). Crude oil is the world's most valuable export product, followed by refined petroleum oils (from crude oil); automobile exports take a third place. The global oil and gas export market is dominated by the United States, Saudi Arabia and Russia Federation which together accounted nearly 50% of the total global output in 2019, while coal production is dominated by Asia Pacific countries, Russia and the United States which produced 90% of the global coal output in the same year. The United States holds the world's largest reserves of coal and even though the country is making effort to reduce internal coal use, the export drive has picked up in recent years and new liquefied natural gas processing and export facilities are being commissioned. The economies of Russia, many other European and emerging countries depend primarily on the production and export of oil and gas, while many countries (Indonesia, China, India, Australia, Russia, Poland, South Africa) rely heavily on the production and export of coal. Russia is expanding its very large natural gas pipeline networks across Europe and Asia; Africa already has intra-country and regional natural gas pipeline networks which are currently being extended to Europe; and the global market for liquefied natural gas has been growing very rapidly in the last decade, dominated by Australia, Qatar, the United States, Russia, Iran, Canada. On the other hand, many countries, notably Japan and most countries of Europe have virtually no fossil energy resources and import most requirements. The economies of many emerging countries depend critically on exportation of fossil fuels, accounting for nearly all the foreign exchange earnings and very large up-stream down-stream employment. Also, every major oil and gas company in the world and many Asian coal producers are still committing to projects that do not align with the Paris Climate Change Agreement which targets limiting the increase in global average temperatures to 1.5 degrees Celsius by the end of the century. This is because fossil energy production and export are major driving forces of the global economy and across all regions of the world in terms of revenue (taxes and foreign exchange earnings) and employment generation across the whole spectrum of production, servicing, conversion and distribution. It is difficult to imagine that, for environmental reasons, any country (developed or emerging) can contemplate a reduction in fossil energy production, local use and export to save the environment, because of the obvious and potentially severe negative impact on its economy. Furthermore, the fluctuating dynamics of the global oil market which is closely tied to the global economy impact negatively on global efforts to decarbonize energy: every time oil prices decline, investments in renewable energy use, energy efficiency improvements and energy decarbonization also decline. Fossil energy use has also become a major political pawn in some developed countries, causing significant instabilities in the deployment and sustenance of effective energy-related environmental mitigation policies that survive political cycles. Policies are developed and reversed depending on the local exigencies and political climate. It is interesting to note that the United States, the world's largest consumer of primary energy, the largest cumulative emitter, currently the world's second largest source of energy-related emissions and one of the leading architects of the Paris-2015 Protocol has pulled out of the Agreement and is in the process of reversing most of the country's existing environmental mitigation policies because of the effects on its economy.

Another potentially significant challenge to global efforts to decarbonize energy is the effect of projected slower growth of power demand in the developed world due largely to improving efficiencies. This is expected to slow the speed with which renewables can penetrate power generation since it is hard for a new renewable power plant to compete commercially against existing fossil fuel facilities. Therefore penetration of renewables in the power fuel mix will depend on the pace at which existing power stations are retired. Furthermore, the intermittency of most renewable energy resources and therefore the need for a fossil fuel-powered base for most grids are problematic: although solar and wind units can power small communities and installations, the main contribution to the global energy demand and consumption is still through integration with fossil-fueled power grids. While solar and wind power plants are already competing with gas-fired peaking plants to manage short-run fluctuations in supply and demand even across country borders, conventional power plants remain the main sources of power grid stability and system flexibility. Another issue with renewable energy, in particular, bioenergy is the enormous amount of resources required, notably land and irrigation resources which are also in critical demand for the production of food considered more urgent in view of the fast global population growth rate. However, in spite of all these potential obstacles, there will be significant decarbonization of energy in the coming decades but this will be due mainly to shifts in energy mix and improved efficiencies across all sectors of the global economy.

There is growing doubt on the commitment of the over a hundred countries that signed the Paris-2015 Agreement and submitted country plans, considering the recent monumental and consequential policy reversal by the United States which will inevitably impact negatively on the commitment of most other countries. While Europe-28 is very much on course because of the central policy institution and enforcement strategies, there is little to show that other countries are making much progress in meeting their commitments which in any case are insufficient to place the world on the path to a sustainable environment. China, the world's largest polluter which accounted for 28% of the global energy-related emissions in 2018 has released laudable and comprehensive energy decarbonization plans but the country appears to be moving in the opposite direction, driven by local exigencies of providing power to the teaming and increasingly middle-class population, while also producing more power to the increasingly global Chinese manufacturing industry. Various projections show that China needs to reduce its coal power capacity by over 40% from current levels in order to meet the reductions required to hold global warming at the target limit. On the contrary, the country has been expanding its coal use for energy and as chemicals/petrochemicals feedstock very rapidly. China is pursuing a policy that is clearly out of alignment with the Paris-2015 Agreement. The country has been responding to strong pressures from organized industries for construction of many more coal power plants.

In summary, as long as the global economy remains strong, the population keeps rising and prosperity grows and spreads across all regions of the world, growths in primary energy demand and related emissions are inevitable and countries will continue to prioritize local exigencies of energy equity and energy security over global climate change. Furthermore, the economies of many countries across all regions of the world and economic stratifications depend critically on the upstream/downstream fossil fuel industry for employment generation and export earnings. Crude oil and refined products are the world's two leading export products and investments in exploration and infrastructure are very strong. If the world were to stop mining fossil fuels today, the global economy would collapse: transportation would freeze; production of critical products that drive the global economy would halt: primary metals; cement; chemicals and petrochemicals; plastics; fertilizers; drugs and pharmaceuticals, cosmetics and bodycare products; synthetic fibers for clothing, tire manufacture, upholstery, and many other products; carbon fiber composites for sports equipment, airplane/automotive components; lubricants; bitumen for road surfacing and roofing; and many more. Renewable energy development would freeze without vital inputs such as iron and steel and cement; the electric vehicle market will collapse because of shortage of batteries the manufacture and charging of which currently depend heavily on fossil fuels. On the other hand, a dramatic drop in demand for energy would collapse the global economy, as is happening in real time due to the current global Covid-19 pandemic which virtually closed down the global economy.

Moving the world towards a sustainable environment would require strong and resilient mitigation interventions: resilient policy instruments that survive political cycles that promote aggressive decarbonization of power generation and transportation; improved efficiencies across the whole spectrum of primary energy production, conversion and use; and moderation of lifestyle habits of consumers which is driven largely by want and impulse rather than need (see Chapters 7 and 8). Also, the new normals that will emerge from the current world health, economic and energy crises could have positive effects on environmental pollution and climate change mitigation (see Chapter 4). However, it is unlikely that there will be any major change in projections of primary energy demand and the continued dominance of fossil fuels. Energy demand will rebound once the global economy picks up. Carbon dioxide emissions which declined by an average of 17% in April 2020 compared with 2019 due to collapse of the global economy and the near-total halt in travel and commuting has since surged again to within 5% of the 2019 level once the global economy started to reopen.

The world can achieve carbon neutrality without closing down the fossil energy industry, and various options are discussed in Chapter 7. Furthermore, the common law of supply and demand is very effective in moderating the fossil energy industry. For several years now the industry has faced strong headwinds due largely to the unstable global oil prices which have been consistently low, caused by oversupply. The current global oil price is sharply below breakeven levels of many producers especially most in the United States who use fracking technology. Even before the global Covid-19 pandemic, many of the country's producers were already filing for bankruptcy and perhaps no more than 15% can survive for long even if oil prices rise to the pre-pandemic level of around $40 per barrel. The current global campaign to defund fossil energy is also impacting negatively on the industry. A growing number of investors are moving away from fossil energy stocks and targeting socially responsible investing. A huge amount of oil, natural gas and coal supplies have been taken off line by producers and the industry could have major trouble attracting future capital. Oil and gas ompanies must spend heavily to maintain production from existing facilities even without commissioning new ones. Sustained starving of the fossil fuel industry with funds could swing the world in the opposite direction of supply deficit in a couple of years and this could be equally disastrous for the global economy.

Chapter 6

Consequences of Environmental pollution

6.1. INTRODUCTION

The environment sustains life on earth in many ways, including regulation of global temperature, weather, climate, and there are many natural control mechanisms in place. Although many of the events that are experienced on earth - weather and climate change, severe weather, draught - are in fact natural phenomena, there is ample scientific evidence that human activities particularly in the last century are disrupting many of the natural control systems and making the negative impacts on the Earth more widespread, intense and severe. Human activities especially since the beginning of the Industrial Revolution have consistently undermined the natural balance in the carbon cycle as designed by nature, with various activities that produce anthropogenic gases, dust and other chemical compounds which end up in the atmosphere. Release of carbon dioxide into the atmosphere from fossil fuel and land use perturbs the natural carbon cycle. Over half of the anthropogenic emissions is removed from the atmosphere by carbon sinks in terrestrial ecosystems through enhanced photosynthesis, but also in the oceans, with potentially negative consequences. Increases in the natural concentration of the greenhouse gases in the stratosphere reflect more heat back to earth and greatly disturb nature's control system which enables the sustenance of life on the planet, and could have a profound effect on climate and ecosystems. Global average temperature is rising, causing melting of ice caps and warming of oceans; ocean levels are rising and causing coastal flooding in many regions of the world; warming oceans are energizing storms causing landfalls to be more widespread, severe and devastating; increased concentrations of dust and smog alter the natural response of clouds to the Sun's radiation, interfering with the natural processes of moisture aspiration and transpiration and altering local weather; and the consequences of anthropogenic pollution are impacting severely on human and ecosystem health.

The natural atmospheric ozone layer also known as 'good ozone' is an effective shield which prevents most of the harmful ultraviolet rays from the Sun (especially the potentially lethal UV-B radiation) from reaching the Earth and causing human health problems including skin cancer, eye diseases, immune deficiency, and respiratory diseases. UV-B radiation affects the physiological and developmental processes of plants and could cause fundamental changes in the biochemical cycles of plants, and plant diseases. UV-B radiation causes damage to the natural developmental processes of marine life and reduces survival rates of a very important group of marine organisms known as phytoplankton. These organisms play a vital role in sustaining aquatic food pool, as well as the formation of oil and gas under the seabeds. Ozone is also released into the lower atmosphere from fossil fuel combustion and many industrial applications. This type is often called 'bad or dirty ozone' because it is a potential source of smog which is a health and safety hazard, particularly in urban areas.

Human activities - fossil energy production, conversion and transportation, power generation, road, air and ocean transportation, forest fires, biomass burning, use of basic biofuels, and others release potent black carbon, volatile organic compounds, and other aerosols, mostly into the lower atmosphere (troposphere). Although most of these substances have a short life in the atmosphere, the potentially negative effects on humans and the ecosystem are extensive. Apart from causing smog over cities (which reduces visibility and interferes with the natural cloud activities), they have been linked to various diseases, premature deaths, and observed issues with agriculture and aquatic life. Synthetic as well as naturally occurring biopolymers degrade and become brittle when exposed to UV-B radiation, and special additives used to stabilize and protect these materials from radiation damage (such as bisphenols) are also potentially injurious to human health.

6.2. SOURCES OF ENVIRONMENTAL POLLUTIONS

The four major spheres that make up the ecological system (see Chapter 1) are characterized by intricate intra and interactions, all interdependent and interconnected. The interactions between these spheres to a large extent sustain life on earth and determine the local weather and global climate. The functions of the spheres are controlled by natural regulatory processes which are often unpredictable and imperfect in many ways. For example, the Earth's rotation around the Sun plays a critical regulatory role but, for known and unknown reasons there can be slight changes in its orbital characteristics, enough to cause significant climate changes which have caused extreme weather; transformed fertile lands into deserts and caused interchanges between oceans and land masses for millennia. Events in Earth's structure and core are dynamic, leading to natural occurrences such as earthquakes, volcanoes and warm springs. However, discussions in earlier chapters have shown that human-related emissions are interfering with the natural environment and compromising its regulatory processes. The negative impacts are multi-dimensional and are becoming increasingly severe. Anthropogenic emissions end up in the troposphere, the lowest part of the atmosphere that is in direct contact with life on earth. Within a few weeks coarse particulates drop off while the gases and ultrafine particles rise to the next layer of the atmosphere (stratosphere) that controls climate on earth, but the residence time in the troposphere is long enough to cause significant changes in local weather and severe human health and other ecosystem problems.

6.2.1. Tropospheric (ambient, lower) layer pollution

The first fifteen to twenty kilometers of the atmosphere, known as the troposphere or lower atmosphere is the part of the atmosphere that is in direct contact with life on Earth. It is also the zone that largely determines the weather; it is the source of oxygen that sustains human and life and the depository for exhaled carbon dioxide that is crucial to the survival of plant and vegetation on Earth. The layer controls the direction of the winds at any given time, and the water cycle which involves evaporation from the Earth's surface, seeding, precipitation and condensation in the troposphere resulting in clouds, rain, snow, slits etc. It also controls the movement of winds which form thunderstorms, hurricanes and tornadoes, heat waves and season patterns around the world. While the relative proportions of the constituent gases of the troposphere remain fairly constant, the water vapor is variable, depending on evaporation and transpiration processes on the Earth's surface which is the source. The evaporation process plays a prominent role in controlling temperatures on the Earth's surface because the energy that sustains the endothermic processes is sourced from the solar energy and thermal radiation re-emitted or reflected from the Earth's surface. The natural activities controlled by the troposphere determine the weather pattern on the Earth and, ultimately the regional and global climate systems (weather pattern over many decades). In effect, any events that alter the natural balance of the tropospheric activities will have an immediate and direct impact on life on earth, as well as well as the weather and ultimately, the climate. Human activities produce a wide range of chemical compounds that can accumulate in the troposphere and   affect global weather, human and animal health, aquatic life, vegetation, etc. see Chapter 5).

A significant proportion of natural and anthropogenic emissions of minute particles, also known as aerosols remains suspended in the lower atmosphere 10-15 km above the Earth's surface for a few days to a few weeks, the main sources and physico-chemical nature of aerosols have been discussed in some depth in Chapter 5. During their short lifetime, they can cause severe problems for people and the global ecosystem: when these particles are sufficiently large, they are visible, dense, scattering and absorbing light and obscuring visibility. Perhaps the most visible effect of tropospheric pollution is the photochemical smog that hangs over cities all over the world. The World Health Organization (WHO) estimates that around 92% of the world's population live in places where air quality levels fall far below WHO limits and, although ambient air pollution affects both developed and developing countries alike, low- and medium-income emerging countries (particularly in Western Pacific and South-East Asia regions) experience the most severe effects. In a recent study by WHO (2018) covering around 3,000 cities and towns in 103 countries, only four OECD countries (Italy, Poland, Turkey and Mexico) featured in the first 500 most heavily polluted cities, while fourteen Indian cities were among the world's twenty most polluted, with Kanpur topping the list. The impact of tropospheric ambient) pollution has a much more direct and severe impact on life on earth because it is the part of the atmosphere that is in direct contact with the Earth's surface. However, because it is mostly locally sourced and there is significant physico-chemical variability in content, the effect on life is much more localized. Mining operations, oil and gas drilling, fossil fuel transportation, conversion, and use, some industrial processes all release toxic gases into the troposphere. Other potential sources include volcanoes, desert dust, power generation, agriculture and land use, forest fires, transport, oil spills, and household basic energy use. The potential impact of aerosols on health depends on the chemical composition, concentration and proximity, all of which depend on the sources.

6.2.2.   Stratospheric layer pollution

The Earth's climate system is powered by solar radiation, around 65-70% of which reaches the Earth's surface in the visible part of the electromagnetic spectrum (insolation). The Earth absorbs around 50%, water vapour and dust account for about 15-16%, and the balance is reflected back to space. Global temperature is determined by the balance between incoming solar energy and the radiation reflected back into space by the Earth's surface (albedo), and has been relatively constant over many centuries. As discussed in Chapter 1, activities of greenhouse gases in the stratosphere regulate the global average temperatures which have been around 14°C for over a century up to 1980 or so. Without this natural control, the Earth would be around minus 18°C, too cold to sustain life. However, there are wide temperature variations across the globe, from around zero in the Antarctica to around 50-60°C in the Libyan desert. Anthropogenic greenhouse gases that end up in the stratosphere enhance the natural concentrations and heat-trapping capabilities. It means that more of the Earth's reflected heat is trapped and reflected back, thereby warming the Earth.

6.2.3.   Ozone layer pollution

A thin layer of ozone is present in the lower layer of the stratosphere and natural chemical reactions in the layer which involve formation of ozone from natural atmospheric oxygen, decomposition and recombination, use up most of the dangerous rays from the Sun thereby preventing them from reaching the Earth and causing health problems. Although thickness of the layer varies significantly across regions, the average global concentration has remained stable for centuries. Fluorocarbons and other similar gases released largely from refrigeration and air conditioning may end up in the stratosphere where they react with the natural ozone, thereby reducing concentration and the capability of the layer to shield the world from the harmful rays of the Sun.

6.3. INDICATORS OF ENVIRONMENTAL POLLUTION

Gaseous and fine particulate compounds released into the atmosphere by human activities: energy production and use, agriculture and land use, and many others (anthropogenic pollutions) alter the natural mechanisms that control weather and climate; they cause severe problems for life on earth. Many of the negative impacts of environmental pollution depend on the location of the emissions in the atmosphere: emissions that are light enough to be air-borne stay in the lower atmosphere (troposphere) for a few weeks, most fall back to earth but, depending on wind dynamics, the gaseous and light particulate components may rise to higher levels, ending up in the middle atmosphere (stratosphere). The negative impacts of ambient pollution are largely instant and local but pollution of the middle atmosphere is gradual, fairly uniform across the global environment, and the most urgent negative impact is global warming which is the main cause of climate change. The problem of human-related emissions has been discussed in some depth in Chapter 5 but the key points are summarized below.

- Global anthropogenic emissions have increased gradually since the Industrial Revolution of the 1800s and have intensified in the last hundred years or so, driven largely by fast economic, population and prosperity growth. Although energy production and use account for nearly 70% of the total global anthropogenic pollution, there are many other sources: carbon dioxide is emitted in forestry and other land use (FOLU) and many industrial processes; methane comes from animal and rice production; and most of the fluorocarbons come from refrigeration/air conditioning and household goods industries. Forestry and land use produce about 11% of carbon dioxide emissions while methane comes partly from oil, gas and coal production but also from agricultural activities that involve composting, fermentation, landfilling and animal breeding. Nitrous oxide is produced from use of fertilizers and fossil fuel combustion. Industrial processes, refrigeration and many consumer products are the main sources of anthropogenic F-gases, which include hydrofluorocarbons (HFCs), perfluorocarbons (PFCs), and sulfur hexafluoride (SF_6).

- Emissions from the energy sector have increased exponentially - CO_2eq emissions (carbon dioxide, methane and nitrous oxide) have more than doubled in the last forty years. Carbon dioxide is the main constituent of energy-related pollutants, accounting for around 90% and, although the gas is the least potent in terms of atmospheric damage, it has become the benchmark for the evaluation of the anthropogenic effect of all other gaseous pollutants, expressed in carbon dioxide equivalent (CO_2eq). Over 99% of energy-related emissions come from the production and utilization of fossil fuels (coal, oil and gas), with coal accounting for around 45%. Oil was the greatest pollutant in the 1970s but has been overtaken by coal. The regional distribution of emissions has also changed significantly. In the early part of the 20th century, virtually all emissions originated from the United States and Europe, today together they account for less than 30% of energy-relate emissions. Just two countries (the U.S.A. and China) accounted for around 45% of the global CO_2eq emissions in 2016, estimated at 32.6 billion metric tons. Carbon dioxide emissions stagnated for two consecutive years up to 2016 due primarily to strong efficiency improvements and low-carbon technology deployment, but then started rising again in 2017, reaching a record 33.1 Gt CO_2 in 2018, believed to be a result of the unusually hot and cold weather in some parts of the world but also because of the perennially low price of oil in the global market which tends to depress investments in alternative energy. In effect, increasing energy use causes global warming which energizes severe weather, the consequences of which stimulate more energy use.

- Although emissions from all fossil fuels increased, the power sector accounted for nearly two-thirds of emissions growth and coal use in power generation accounted for around a third of the total global energy-related emissions in 2018, mostly in Asia. China, India and the United States accounted for 85% of the net increase in emissions over the last two years. Emissions from other key sectors of the global economy have also increased steadily across all regions of the world, although moderated by efficiency gains. Various projections show that emerging countries will account for most of the future growth in the global economy, population, demand for primary and end-use energy, and energy-related emissions in the coming decades.

6.4.   WEATHER AND CLIMATE

Weather describes the conditions of the atmosphere at a certain place and time with reference to temperature, pressure, humidity, wind, and other key parameters, collectively known as meteorological elements. Weather is also described by the presence and movement of clouds, precipitation, and the occurrence of spatial phenomena, such as rainfall, snowfall, heat waves, thunderstorms, dust storms, tornados, etc. Climate takes into account all the above variables over a period of time ranging from months to years, projected in either direction to thousands and millions of years in order to arrive at a global climate pattern. Weather is largely local and can change from hour to hour, day to day, month to month, or year to year. On the other hand, climate generally refers to an aggregate of weather statistics over much wider areas for a decade or more. Natural fluctuations in solar energy output due to the Earth's rotation, and consequent fluctuations in the amount of incoming shortwave radiation (SWR) can cause changes in the Sun's energy that hits the Earth's surface and the amount that is reflected, thereby altering the balance either way, so can human activities that release emissions of gases and aerosols (anthropogenic emissions) into the atmosphere which enhance the natural concentrations of greenhouse gases or deplete ozone concentration. However, while the natural greenhouse effect either heats up or cools the Earth as necessary, anthropogenic greenhouse gases almost always heat up the Earth, while depletion of ozone caused by pollution enables hazardous rays of the Sun to reach the Earth. The most familiar features of weather are the day-to-day and day-to-night variations in temperature, heat waves, cold spells, humidity, precipitation, rainfall, windiness, atmospheric pressure, and cloud structure, while increasing global average temperature is the most important cause and indicator of climate change. While weather can change from day-to-day, even within the day, climate change happens slowly over decades, hundreds, maybe even thousands of years. Global climate is taken as the average across the world, and could be in terms of average temperature or precipitation patterns.

Any or combinations of the Earth-Environment natural phenomena can cause weather or climate change, resulting in extreme weather, desertification, negative health impacts, etc., and there are many natural processes which can cause global warming or cooling, with significant impact on global climate. For example, the movement of tectonic plates, volcanic activities, and changes in the Earth's orbital axis can cause major climatic changes, resulting in the appearance and disappearance of long periods of warm or cold global weather, and this explains why the Earth has experienced many periods of ice ages over the last one million years, some lasting several hundred years. El Niño (warm phase) and La Niña (cold phase) are opposite phases of a natural climate cycle known as the El Niño-Southern Oscillation (ENSO) cycle. ENSO swings between warmer and cooler seawater in the tropical pacific and frequently causes climate pattern across the region to swing back and forth every three to five years on the average, each swing lasting around one year or more, leading to significant differences in the average ocean temperatures, winds and surface pressure, and rainfall across parts of the tropical Pacific. During El Nino, rainfall is below average in the western tropical Pacific around Indonesia and above average in the central and eastern areas where ocean temperatures are higher than average. The effects of La Nina on temperatures and rainfall across the region are reversed. The cause of ENSO is unclear and the frequency can be quite irregular. Although the main effects are localized, the side effects can impact on weather in far away countries that border the Pacific Ocean, causing severe flooding in places like North American states and draughts in nations that border the western Pacific Ocean. In fact, El Nino is regarded as the Earth's most influential climate pattern. There is also ample geological and paleo-climatological evidence that oceans and land mass have interchanged over time as a result of earthquakes and other massive land movements. For example, the Sahara desert is believed to have been covered by ocean thousands of years ago during the warm, wet age and many areas were fertile land.

6.5. GLOBAL WARMING

Natural fluctuations in solar energy output due to the Earth's rotation, and consequent fluctuations in the amount of incoming short wave radiation (SWR) can cause changes in the energy balance, so can human activities that release emissions into the atmosphere: gases which modify natural concentrations of greenhouse and ozone gases, and aerosols which cause severe human health problems, but there is ample scientific evidence that the average global decadal temperature has increased fairly steadily by nearly around 0.6°C in the last four decades. This may seem insignificant but small changes in the Earth's average temperature can have very big impacts on the global climate. Perhaps the most important indicator of climate change, is global warming, caused largely by anthropogenic GHG emissions which end up in the stratosphere (middle atmosphere), increasing the concentrations and heat-trapping capabilities of the natural GHGs.

As discussed in Chapter 1, many gases known as greenhouse gases (GHGs) present in the Earth's stratosphere (middle atmosphere, between about 20 km and 40 km or so above the Earth's surface) are capable of trapping, holding and releasing the Sun's thermal energy. Although they are present in very minor quantities relative to nitrogen and oxygen, they play a crucial role in the control of the amount of heat that is retained by the Earth's surface, and therefore the Earth's average surface temperature. The main natural GHGs are water vapor, carbon dioxide, methane, and nitrous oxide, but several others [ozone, chlorofluorocarbons (CFCs) hydrofluorocarbons (including HCFCs and HFCs)] are also active. Natural carbon dioxide is a product of respiration and methane is produced primarily by anaerobic (oxygen-deficient) decomposition of organic matter in biological systems. The compound is also produced by termites which harbor bacteria that are capable of breaking down organic matter in wetlands and swamps. Agricultural activities (biogenic sources) such as wet rice cultivation, soil fertilization, and vegetarian livestock production release significant quantities of methane into the atmosphere. For example, the digestive systems of ten cows emit about one metric ton of methane into the atmosphere annually. Significant methane emissions also emanate from decaying vegetation, especially in landfills, hydro dams, artificial lakes and reservoirs. Water vapor is the largest contributor to the natural greenhouse effect and plays an essential role in the Earth's climate. However, the amount of water vapor in the atmosphere at any given time is controlled mostly by air temperature, rather than emissions, and therefore the maximum varies across the globe. The atmosphere can retain around 7% more moisture for every one degree centigrade rise in temperature, hence, a column of air in the tropics may hold ten to twenty times more moisture than a similar column in the polar regions. For these reasons, water vapor is considered as a *feedback* agent rather than a *forcing* to climate change, and therefore usually not included in lists of anthropogenic greenhouse gases.

The average global temperature has risen steadily since the beginning of the industrial revolution around 1750 and there is little doubt that there has been an uptake of energy by the global climate system. In effect, total radiative forcing is positive, and the largest contribution has been the steady increase in the atmospheric concentration of carbon dioxide. As discussed in Chapter 1, different types of anthropogenic pollutants have different effects on the environment, quantified in terms of radiative forcing (RF) which is the difference between the heat energy absorbed by the Earth from solar radiation and energy radiated back to space, resulting from a change in concentration of a particular substance, or the properties of the Earth's system. Each compound has a specific RF value associated with it; the more positive the value, the greater the contribution of warming influence, while negative values indicate cooling influence. Stratospheric pollutants contain compounds which have positive RF (carbon dioxide, methane, nitrogen oxides), and the effect on the natural mechanisms by which greenhouse gases control the Earth's temperature has been discussed in some depth in Chapter 1. The effect of anthropogenic fluorocarbons on the natural ozone layer was also discussed. Volcanic eruptions often release aerosols into the atmosphere and, while much of it stays in the troposphere, a substantial proportion of chemical compounds can rise above 10 km or so into the stratosphere and contribute to radiative forcing, although the effect usually fades off within about two years. The total anthropogenic RF for the main gases (CO_2, CH_4, N_2O and halocarbons) in 2011 relative to 1750 was 3.00 W/m^2, with CO_2 alone causing over half (1.68W/m^2). The most important halocarbon is dichlorodifluoromethane (CFC-12) which contributed 0.337 W/m^2 to the total. The Montreal Protocol has been very effective in reducing the use of fluorocarbons (F-gases) in the developed countries, but the replacement gases (hydrofluorocarbons, HFCs) also belong to the F-gases family and are potent greenhouse gases but with lower global warming potentials than fluorogases and they do not destroy natural ozone. Ammonia is also used in industrial refrigeration. However, as discussed earlier, F-gases are still in common use in the emerging world and the negative impact on climate will resonate beyond their borders.

There is ample scientific evidence to support the belief that the RF effect of energy-related emissions is positive, and the average global temperature is rising, exemplified by warming oceans, rising water levels, increased flooding, increasingly intensive extreme weather, melting arctic ice, etc. Furthermore, human activities (land clearing, construction, agriculture) have changed and continue to change the Earth's surface topography and atmospheric composition. Some of these changes have direct or indirect effects on the energy balance of the Earth and wind movement, and are thus potential drivers of weather and climate change. A consistent rise of the global average temperature by even less than 1°C can have potentially severe and diverse consequences on earth - moisture aspiration and transpiration, extreme weather; warming, rising and acidifying oceans; desertification, poor crop yields, etc.; and extensive meteorological records over the last two hundred years or so, confirm significant enhancement of these variables.

6.6.   WEATHER AND CLIMATE CHANGE

The clear distinction between weather and climate has been discussed in section 6.4. Weather is characterized by the presence and movement of clouds, precipitation, and the occurrence of spatial phenomena, such as thunderstorms, dust storms, tornados, etc.; climate takes into account all the above variables over a period of time ranging from months to years, projected in either direction to thousand and millions of years in order to arrive at a global climate pattern. The World Meteorological Organization (WMO) defines climate as the average of variables (mainly temperature, precipitation and wind) over 30 years, but more detailed analysis also considers frequency, magnitude, persistence, trends, and some other variables in the determination of global climate. Although weather and climate are interdependent, the clear distinction is the fact that weather defines the state of the atmosphere at a given time and place, and can change from hour to hour, day to day, month to month, or year to year. Climate on the other hand, generally refers to an aggregate of weather statistics over wider areas for a decade or more. It is therefore easier to predict climate than weather. Short-term changes in climate are known as weather changes but climate change refers to changes in the statistical properties of the weather pattern over of long period of time, covering several decades. In effect, climate is the long-term pattern of weather across regions.

6.6.1.   Weather change

The local weather is largely controlled by heat exchange between the Earth and the atmosphere, and the local wind dynamics, both of which determine evapo-transpiration, cloud systems, rainfall, etc. The winds carry moisture from the oceans and, through complex reactions in the cloud system, dump rain or snow on the Earth's surface. This explains why local weather could be very unstable, often changing from hour to hour, with rain or snow falling on one area while adjacent areas remain dry. Global warming increases the ocean temperatures and enhances the moisture-holding capacity of winds, hence both rain and snow can intensify and winds can become more energized and destructive. Fine carbon aerosols emitted from fuel combustion can alter the natural seeding processes in the clouds and alter the pattern of rainfall, while deforestation alters the natural heat exchange (albedo) between the Earth and the atmosphere causing draught. Civil construction can alter local land surface topography to the extent that wind dynamics change significantly, affecting the local weather.

6.6.2.   Climate change

Climate is determined by a complex mix of natural processes, such as the rotation of the Earth around the Sun; changes in the equilibrium of energy transfer between the Sun and the Earth; the pattern of distribution of surface moisture by winds, ocean currents and other mechanisms; and activities in the five different 'spheres' that make up the Earth's system: the atmosphere, lithosphere, hydrosphere, cryosphere, biosphere. The atmosphere surrounds the Earth and is filled with gases. This is the most variable part of the climate system where the composition and movement of gases can change radically within short periods of time. The lithosphere (land) contains all the world land surface (and the semi-solid and liquid land deep beneath the surface). The lithosphere is very uneven, made up of mountain ranges, hills, deep valleys, plains, vegetation, buildings, etc. All these surfaces have different solar absorption and reflection characteristics (albedo) hence contribution to global warming is variable. Furthermore, uneven terrain acts as wind barriers or speed enhancers, and therefore can significantly affect wind movements in terms of intensity, variability and direction.

Changes in the hydrosphere (a collection of oceans, rivers and lakes), include variations in temperature, salinity, acidity, but the pace is very slow compared with the atmosphere. The cryosphere (really a sub-division of the hydrosphere), comprising frozen water, ice sheets and glaciers is also relatively consistent. The main climate-related activity in this sub-system is the continuous process of reflection or absorption of the Sun's rays (albedo) which determines the melting rate of the ice. Any changes in the natural balance are slow and become evident after long periods of time. The biosphere which comprises all life on Earth profoundly influences climate through the complex control of natural activities, especially the natural balance of CO_2eq in the atmosphere. Living humans and animals inhale oxygen and exhale carbon dioxide which plants need for photosynthesis, returning oxygen to the atmosphere; forests and oceans absorb substantial proportions of carbon dioxide, serving as 'carbon sinks'. Climate varies across regions because of variations in the determinant interactive factors: variations in the intensity and distribution of the Sun's energy that reaches the Earth and the amount that is reflected; elevation and topography, land use pattern, the latitudinal location which largely determines how much of the Sun's energy reaches the region and for what duration during the Earth's annual rotation cycle around the Sun. For example, regions located near the Equator (zero latitude) receive sunlight year-round, whereas those near the poles hardly receive any and temperatures rarely rise above freezing. Other variables are vegetation on the Earth's surface, proximity to oceans, and, not the least, the biosphere which is the sum total of all life on Earth.

The radiative balance between incoming solar shortwave radiation (SWR) and outgoing longwave radiation (OLR) is influenced by several global climate variables. Natural fluctuations in solar energy output due to the Earth's rotation, and consequent fluctuations in the amount of incoming SWR can cause changes in the energy balance, so can human activities that release emissions of gases and aerosols into the atmosphere which modify natural concentrations of greenhouse and ozone gases. Other human activities such as deforestation, construction of buildings and infrastructure can also alter the Earth's surface albedo significantly. It is estimated that about half of the incoming solar shortwave radiation is absorbed by the Earth's surface. The fraction of SWR reflected back to space by gases and aerosols, clouds, and by the Earth's surface (albedo) is about 30%, and about 20% is absorbed in the atmosphere by clouds and greenhouse gases. The Sun provides its energy to the Earth primarily in the tropics and the subtropics, from where it is partially redistributed to the middle and high latitudes by atmospheric and oceanic transport processes. The global climate system has been monitored over the last hundred years or so, especially since the 1950s when sophisticated instrumentation and satellite technology became available, and the changes observed have led to the conclusion that global average temperature, the main indicator of climate change is rising, with potentially serious consequences for life on Earth. Apart from global warming, there are many other negative effects of anthropogenic pollution, notably acidification of the oceans, acid rains, tropospheric smog, etc.

Observations in the climate system are based on direct physical and biochemical measurements, and remote sensing from ground stations and satellites, which provide a comprehensive view of the variability and long-term changes in the atmosphere, the oceans, the cryosphere and at the land surface, and records date back to the mid-19th century. The database is extended back hundreds to millions of years, based on paleoclimate reconstruction and future directions are determined by modeling. It is relatively easy to predict climate response to anthropogenic emissions over the next two decades or so (near-term/decadal climate prediction) because there is a strong body of knowledge and baseline data on most of the important variables that will determine climate change, and also because there is considerable inertia in the response of the environment to climate forcing, in particular, the oceans. In effect, current data on anthropogenic emissions will largely determine the direction of climate change over the next two decades or so. There is extensive data dating back to the 19th century on multiple independent climate indicators, from high up in the atmosphere to the depths of the oceans. They include changes in surface, atmospheric and oceanic temperatures, glaciers, snow cover, sea ice, sea level and atmospheric water vapor. Many independent research groups have analyzed these extensive database and all have come to the same conclusion that the world has warmed since the 19th century. Records of global anthropogenic emissions from 1850 show an exponential rate of increase from the 1950s, believed to have been responsible for a steady rise in the global average temperature, although each year and even decade is not always warmer than the last. Even though the average rise is below 1°C over the last hundred years or so, the negative consequences are diverse and potentially serious. More than 90% of the excess energy absorbed by the climate system since at least the 1970s has been stored in the oceans. As the oceans warm, the water expands, accounting for much of the observed rise in sea levels over the past century. Melting of glaciers and ice sheets due to rising temperatures also contribute to the rising levels of the oceans. A warmer world is also a moister one, because warmer air can hold more water vapor, and global analyses show that specific humidity, which measures the amount of water vapor in the atmosphere, has increased over both land and the oceans. Regional climates - monsoon systems, tropical phenomena, cyclones - are the complex results of processes that vary strongly with location and so respond differently to changes in global-scale influences. Monsoons are the most important modes of seasonal climate variation in the tropics, and are responsible for a large fraction of the annual rainfall in many regions. Their strength and timing are related to atmospheric moisture content, land-sea temperature contrast, land cover and use, atmospheric aerosol loadings, and other factors. Increasing global temperature will intensify monsoons in the future, affecting larger areas, because atmospheric moisture content increases with temperature.

Anthropogenic greenhouse gases that reach the stratosphere enhance the natural concentration of the gases and their ability to absorb reflected heat from the Earth and reflect it back, thereby warming the Earth. Fluorocarbons deplete ozone and reduce its concentration in the lower part of the stratosphere, causing more of the Sun's harmful rays to reach the Earth's surface. Because aerosols are distributed unevenly in the atmosphere, they can heat and cool the climate system in patterns that can drive changes in weather, in particular, precipitation. Aerosols from large volcanic eruptions that enter the stratosphere sediment out within a year or two but, in that short period, can cause stratospheric cooling. Aerosols also come from power generation, internal combustion engines, bush burning and use of traditional bioenergy. There are uncertainties about the net effect of natural and anthropogenic aerosols on climate since aerosols can affect climate in many ways. First, they scatter and absorb sunlight, which modifies the Earth's radiative balance. Aerosol scattering generally makes the planet more reflective, and tends to cool the climate, while aerosol absorption has the opposite effect, and tends to warm the climate system. The balance between cooling and warming depends on aerosol physico-chemical properties and environmental conditions. The presence of sulphate aerosols in clouds is believed to modify the morphology of the cloud droplets, making them more reflective. However, most recent studies agree that the overall radiative effect from anthropogenic aerosols is to cool the planet.

Black carbon (BC) is one of the most potent radiative forcers in atmospheric pollutants, and its relative proportion and composition will largely determine the overall effect on the climate. Black carbon absorbs ultraviolet radiation from the Sun and therefore disturbs the planetary radiation balance; the radiation is converted to heat energy which has the potential for global warming; and because atmospheric residence time of BC is usually very short, contribution to global warming is substantial. The global warming potential over a 20-year time frame (GWP20) is 2,421, that is, one metric ton of BC would have the same integrated radiative effect over 20 years as 2,421 metric tons of carbon dioxide (Sims *et. al.*, 2015). On the other hand, black carbon reduces sunlight that reaches the surface and the amount that is reflected back to space, causing some cooling. Many studies have also suggested that atmospheric black carbon (BC), the most prominent component of energy-related particulate matter (PM) emissions can alter the atmospheric temperature above, within and below clouds, and consequently alter cloud distribution, reduce cloud albedo and change the density of small particles in clouds which act as seed for water vapor condensation. This could change the lifetime, reflectivity and stability of clouds and alter changes in the hydrological pattern of rainfall (U.S.E.P.A, 2012; Bond *et al.*, 2013; IPCC, 2014). Although the interactions between atmospheric pollution and clouds, and the effect of atmospheric pollution on weather are largely localized and variable because of the short atmospheric lifetime, the cumulative effect over long periods could result in global warming and climate change. Black carbon, brown carbon, (a relatively absorbing and hence warming portion of organic carbon), as well as others, especially variable organic compounds (VOC) and sulphur dioxide (SO_2) (which forms sulphates) have forcing negative values, indicating cooling effect. Therefore, variability of the concentration and physico-chemical properties of emissions must be taken into account to understand the net climate and health impacts of any emission source or mitigation (UNEP/WMO, 2011).

Black carbon settles on snow and ice, thereby reducing the surface reflectivity (because it is black) and the amount of diffusive reflection of incoming solar radiation (land surface albedo) is reduced, causing the surface to absorb more heat, accelerating its melting. This is believed to be partly responsible for the gradual reduction and disappearance of ice in the Arctic, the Himalayas and other glacial and snow-covered regions of the world. The Arctic region is particularly vulnerable because the atmosphere is drier and temperature inversions that inhibit atmospheric vertical mixing are frequent, and particulate matter tends to remain in the atmosphere longer than in other regions of the world (U.S.E.P.A., 2012). The melting of snow and ice causes warmer oceans and increased ocean levels, both of which impact on storm patterns and strength, resulting in severe weather (snow albedo feedback) (Flanner *et al.,* 2007; Quinn *et al.*, 2008). It has been estimated that the effect of black carbon on snow albedo may be responsible for as much as a quarter of the observed global warming. Clearly, the effect of black carbon on the climate is complex and multi-dimensional, and the net effect on climate is not yet well understood. However, there is substantial scientific evidence that BC can have direct and indirect effects that can result in global warming or cooling depending on the season and the composition. There is also evidence that the global net mean human-induced RF due to the change in BC emitted from fossil fuel and biomass combustion between 1750 and 2011 was high, estimated to have been 0.64 W/m^2 (Myhre *et al.*, 2014). This makes BC the third largest contributor to anthropogenic RF, after carbon dioxide (1.68 W/m^2) and methane (0.97 W/m^2). Recent studies indicate that BC has contributed much higher globally averaged RF than previously estimated, perhaps as high as 1.1 W/m^2 (Carmichael, 2008; Quinn *et al*, 2011; Bond *et al.*, 2013).

Climate science has developed very rapidly in the last five decades or so and a lot is now known about the variables which determine global and regional climate, weather patterns, etc. This has made it possible to forecast weather patterns with considerable degree of accuracy. However, there are still many unknowns and uncertainties, and this explains why forecasts are often dead wrong. For this reason, predictions and projections about the climate and weather are always qualified with levels of confidence, from very low to very high (from exceptionally unlikely to virtually certain). As discussed earlier, there are many natural phenomena that can cause climate change, for example, changes in solar irradiance, atmospheric trace gases and natural aerosol concentrations. Paleoclimatic reconstructions have generated extensive data on probable climate changes over the previous thousands of years, prior to the instrumental period. It is therefore possible to place currently observed changes in the perspective of natural climate variability.

Evidence abounds showing that the Earth system has always responded and will continue to respond to natural forcings - solar, volcanic, orbital, natural variations in atmospheric composition. For example there are known variations in the Earth's orbital parameters as well as changes in its axial tilt, both of which affect the seasonal and latitudinal distribution and magnitude of solar energy received at the top of the atmosphere, which in turn determines the Sun-Earth energy balance and intensities, and durations of local seasons. Solar irradiance (TSI) is the nearly periodic 11-year cycle of changes in solar radiation which shifts from maximum solar conditions (with many sunspots) to solar minimum (with, basically, none). Over the period, the radiation ejected from the Sun varies in quantum, spatial distribution, and composition. During the solar maximum conditions, there is a slight increase in the total solar irradiance but the ultraviolet (UV) content is increased significantly, leading to generation of stratospheric ozone in the mid-to-upper stratosphere, which ultimately results in greater ozone concentration in the tropical lower stratosphere. The combined effect of increased UV and long-wave radiation helps warm that region. How this impacts on climate is still a subject of intensive research but there is ample evidence that solar cycles influence tropospheric rainfall patterns in different ways across the globe (Rind *et al.*, 2008). Also, total solar irradiance, though small in magnitude, does appear to affect sea surface temperatures, especially at latitudes (such as Northern Hemisphere sub-tropics during summer) where cloud cover is small and irradiance is abundant. However, satellite observations of total irradiance (TSI) changes from 1978 to 2011 show a small positive radiative forcing (RF) (0.05 W/m^2) compared with values for anthropogenic GHGs (3.0 W/m^2) and will become increasingly insignificant in the coming years. The recognition of the fact that natural phenomena can cause significant climate forcing has facilitated current conclusions about climate change which are determined based on comparison of current phenomena with paleoclimatic database to determine whether or not they are unusual.

Perhaps the most convincing evidence of climate change over the last 200 years or so is the result of in-depth analysis of observational records of the global atmosphere, land, ocean and cryosphere systems over the period. There is incontrovertible evidence that the atmospheric concentrations of GHG gases, in particular, CO_2, CH_4 and N_2O have increased substantially; land and sea surface temperatures have increased over the last 100 years; observations from satellites and *in situ* measurements indicate reductions in glaciers, Arctic sea ice and ice sheets; data on radiative budget and ocean heat content suggest a small imbalance; and satellite data on atmospheric temperatures show increases in tropospheric temperatures and decreases in stratospheric temperatures. The fact that the stratosphere and troposphere exhibit opposing temperature changes is a strong indication that the Sun is not the main driver (otherwise the response would both either positive or negative). The opposing response is consistent with known GHG effects. Studies and publications on climate change are extensive and most of the presentations are too technical for environmental activists and policymakers. Recent reports of some of the world's leading authorities on climate change have been evaluated and the important conclusions are summarized below.

* It is virtually certain that human influence has warmed the global climate system, resulting in changes in temperatures near the surface of the Earth, in the atmosphere

and in the oceans as well as changes in the cryosphere, the water cycle, and some extremes. While some of these observations may also have been caused by natural phenomena, there is ample evidence that the contribution is insignificant.

* More than half of the observed increase in global mean surface temperature (GMST) from 1951 to 2010 is very likely due to the observed anthropogenic increase in greenhouse gas (GHG) concentrations.

* The atmosphere and oceans have warmed, the amounts of snow and ice have diminished, sea levels have risen, and the concentrations of greenhouse gases have increased. Each of the last three decades has been successively warmer at the Earth's surface than any preceding decade since 1850.

* Ocean warming dominates the increase in energy stored in the climate system, accounting for more than 90% of the energy accumulated over the last fifty years or so. The frozen parts of the planet, known collectively as the cryosphere, affect and are affected by local changes in temperature. Over the last two decades, the amount of ice contained in glaciers globally has been declining every year for more than 20 years, the Greenland and Antarctic have been losing mass, glaciers have continued to shrink almost worldwide, and Arctic sea ice and Northern Hemisphere spring snow cover have continued to decrease in extent.

* The rate of sea level rise since the mid-19th century has been larger than the mean rate during the previous two millennia. Over the period 1901 to 2010, global mean sea level has risen by about 0.2 meter and projections indicate the rise could be as high as 2 meters by the end of the century, submerging coastal areas in many countries if pollutions continue unabated.

* The atmospheric concentrations of carbon dioxide, methane, and nitrous oxide have increased to levels unprecedented, carbon dioxide concentrations have increased by 40% since pre-industrial times, primarily from fossil fuel emissions, but also from net land use emissions. The oceans have absorbed around 30% of the emitted anthropogenic carbon dioxide, causing ocean acidification. This is one of the strongest evidences of human influence on the climate system. The consequences for aquatic life can be devastating.

* A large majority of global land areas have experienced decreased frequency of cold extremes and increased frequency of warm extremes since the middle of the 20th century, consistent with warming of the climate. Warm days and nights have become more frequent while frequency of cold days and nights has decreased. There is an increasing trend in the frequency and intensity of heat waves across most regions. Increases in heavy precipitation have probably also occurred over this time, but vary by region.

* Climate change, whether driven by natural or human forcing, can lead to changes in the likelihood of the occurrence or strength (or both) of extreme weather and climate events. While there is no sufficient evidence that the annual numbers of tropical storms, hurricanes and major hurricanes count in the North Atlantic basin have increased over the past 100 years, there has been a substantial increase in storm

activity, and the frequency and intensity of the strongest tropical cyclones in that region since the 1970s consistent with warming oceans. Storms need energy derived largely from the oceans to form, move and intensify; a warmer Earth's surface will also intensify heat waves and make them more frequent. One major implication is the associated rise in the demand for cooling energy which is likely to come mostly from fossil power. This has happened in the last two years and caused fossil energy use and energy-related emissions to increase at higher rates compared with several previous years.

* The total radiative forcing (RF) between the Earth and the Sun is positive, and has led to uptake of energy by the climate system, the largest contribution caused by the increase in the atmospheric concentration of CO_2eq since 1750. It is very likely that anthropogenic forcings, dominated by GHGs have contributed to the warming of the troposphere, while anthropogenic ozone-depleting substances, have contributed to the cooling of the lower stratosphere.

* Continued emission of greenhouse gases will cause further warming and long-lasting changes in all components of the climate system, increasing the likelihood of severe, pervasive and irreversible impacts for people and ecosystems.

* A rise in global average surface temperatures is the best-known indicator of climate change, and there has been a steady rise over the last century or so. It is estimated that human activities have caused between 0.8°C and 1.2°C global warming above pre-industrial levels, likely to reach 1.5°C between 2030 and 2052 if it continues to increase at the current and projected rates. Some projections indicate that the rise could be as high as 2.5°C by the end of the century. Furthermore, recent scientific evidence shows that the rise in global temperatures over the past 150 years has been far more rapid and widespread than any warming period in the past 2,000 years and none of the pre-industrial extreme weather events that are often quoted by global warming sceptics is remotely equivalent either in degree or extent to the warming over the last few decades. The pre-industrial warming events affected only limited regions of the planet at a time unlike the warming of the last five decades or so which has been planetwide. The unusual heat waves and cold spells also affected limited regions at different times often separated by centuries.

* Future climate will depend on committed warming by past as well as future anthropogenic emissions, and natural climate variability. It is very likely that heat waves will occur more often and last longer, and that extreme precipitation events will become more intense and frequent in many regions. The oceans will continue to warm and acidify, and global mean sea level will continue to rise.

* Global greenhouse gas emissions show no signs of peaking: global emissions from energy and industry have continued to increase, reaching 55 $GtCO_2$eq in 2019 (33 $GtCO_2$eq from energy use). Clearly the world is not on any feasible pathway to sustainable environment which requires that emissions in 2030 are 25% and 50% lower than current levels by 2030 to put the world on a least-pathway to limiting global warming to 2°C and 1.5°C respectively (Figure 1).

*   Even if human-related emissions were stopped today the negative impact on some components of the climate system could persist for centuries, due to the long lifetime of some of the gases and the inertia of the main sinks, in particular, the oceans. Aerosols have a lifetime of weeks, methane (CH_4) about 10 years, nitrous oxide about 100 years, and hexafluoroethane (C_2F_6) about 10,000 years. Carbon dioxide is more complicated as it is removed from the atmosphere through multiple physical and bio-geo-chemical processes in the ocean and on land, all operating at different time scales. It is estimated that about 10-40% of CO_2eq emission will remain in the atmosphere for as long as 1000 years.

*   Climate change will amplify existing risks and create new risks for natural and human systems. Risks are unevenly distributed and are generally greater for the disadvantaged people and communities in countries at all levels of development.

*   While climate change is currently the most topical of all environmental issues, it is in fact not the most urgent since the negative effects are slow and some may not manifest for hundreds of years. Furthermore, human-related emissions are mainly enhancers, not the only cause of climate change. However, ambient pollutions caused mostly by urban transportation, power generation, use of traditional biomass, land use, etc., have been associated with around 7.5 million global deaths annually, around half arising from domestic use of traditional energy by around 80% of the global population, mostly in emerging countries who have poor access to modern energy. Furthermore, while most of the negative impact is local and in real time, some of the emissions will eventually end up in the stratosphere and enhance climate change, and the impact will be global. Clearly provision of clean energy to a world that is likely to increase in population by around 25% in the next two decades is a more urgent problem. For many countries in the emerging regions which will host most of the global primary industries and also account for most of the future growth in the global economy, population and prosperity, coal is the only readily available and affordable primary energy source. Most recent projections expect the regions to account for most of the future growth in global energy-related emissions.

Evidence abounds showing that the Earth system has always responded and will continue to respond to natural forcings - solar, volcanic, orbital, natural variations in atmospheric composition. For these reasons, many phenomena associated with climate change today have always been part of natural occurrences which have caused drastic changes in various parts of global environment over time. For example, there is ample evidence that some of the current deserts were once occupied by water, many drylands and coastlands today were once under the oceans, and massive land movements, volcanoes and tsunamis are all natural processes which impact on the environment. However, these occurred at different times and the impacts were restricted to relatively small sections of the world. There is strong scientific evidence that human activities particularly since the Industrial Revolution, are making the impacts on mankind of some of these natural phenomena increasingly more widespread, more common, more intense and more severe. As discussed earlier, events in three layers of the atmosphere have critical but different impacts on life on earth: the middle atmosphere (stratosphere), the lower-middle atmosphere (lower stratosphere), and the lower atmosphere (troposphere or ambient atmosphere). Some of these risks may be existential to civilization, posing permanent, large, negative, and irreversible consequences to humanity. While the negative impacts of stratospheric pollution are gradual, worldwide and manifest largely as global warming, the effects of tropospheric pollution on human health and well-being of the ecosystem are largely local, instant, and potentially devastating for life on earth.

Annual global greenhouse gas emissions (GtCO₂e)

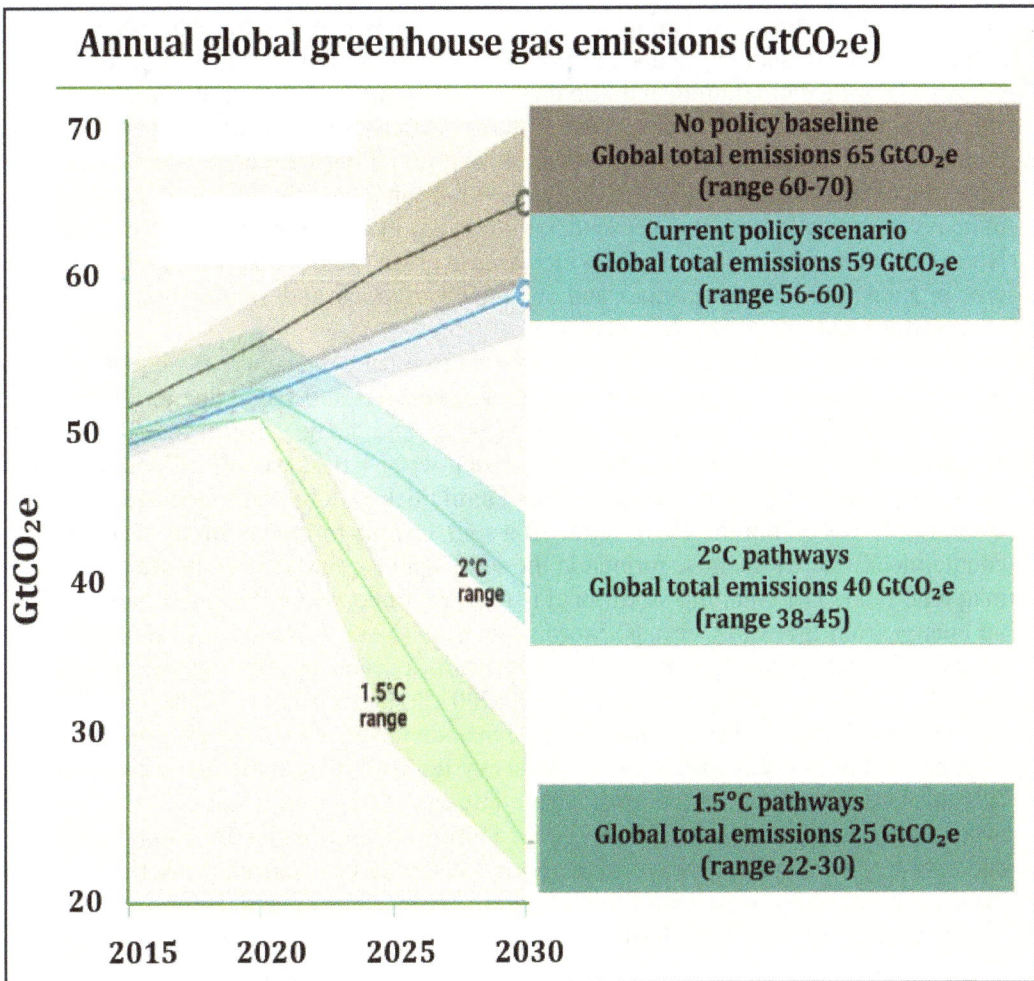

No policy baseline
Global total emissions 65 GtCO₂e
(range 60-70)

Current policy scenario
Global total emissions 59 GtCO₂e
(range 56-60)

2°C
range

2°C pathways
Global total emissions 40 GtCO₂e
(range 38-45)

1.5°C
range

1.5°C pathways
Global total emissions 25 GtCO₂e
(range 22-30)

Figure 6.1 Annual global greenhouse gas emissions and potential pathways to a sustainable environment *(IPCC 2018).*

6.6.3. Extreme weather

Extreme weather is one that is rare at a particular place and/or time of the year. Climate change, whether driven by natural or human forcings, can lead to changes in the frequency and intensity of extreme weather events, such as extreme precipitation events or warm spells, draught, heavy rainfall, flooding, heat waves, etc. The effects are much more diversified and widespread than weather changes. An increase in the average global temperature by just 1°C can influence weather and climate in many ways, in particular, spread, intensity and frequency of extreme weather because the oceans which occupy three quarters of the Earth's surface are the largest heat sinks as well as the originators and powerful energizers of storms.

Higher temperatures boost evaporation which dries out soil and increases drought and desertification; evaporation intensifies as temperatures rise and more moisture in the atmosphere will intensify rainfall and flooding. At present, single extreme events cannot generally be directly attributed to anthropogenic influence, but it has been established that human activities are contributing to global warming which in turn can exacerbate such extreme events as thunderstorms and flooding. Winds take more moisture from the warm oceans and bring more rainfall and flooding when they hit land. As temperatures rise, oceans expand, land and ocean ice melts, ocean waters rise above their normal level and are pushed inland by storms, causing severe flooding. A warmer atmosphere holds more moisture and when the temperatures are below freezing, snowfall can intensify. It is likely that the number of heavy precipitation events over land has increased in more regions than it has decreased. North America and Europe are seeing increases in either the frequency or intensity of heavy precipitation with some seasonal and regional variations and similar trends have been observed in Central and North America.

Hurricanes start as strong winds over warm tropical oceans near the equator, with air picking up moisture from the warm waters of the oceans and rising due to fall in density. The depression that is formed between the rising air and the ocean surface is filled by fresh air which picks up moisture and rises. The rising wet, warm air cools off and the water in the air forms clouds. This process is repeated until thick clouds are formed and the complex system of forces within the cloud starts off a spin. Storms formed north of the equator spin counterclockwise while those formed in the south spin clockwise in response to the Earth's magnetic field. The spinning storm begins to move under the influence of winds, picking up energy from the warm oceans. Storms can travel over thousands of kilometers across oceans, picking up strength and speed from warm oceans and, by the time they make a landfall, speeds could range from around 100-300 kilometers an hour. Such powerful storms can turn into hurricanes and tornadoes and when they land, can cause catastrophic damage to structures, dump heavy rains and cause heavy flooding, especially in the coastline areas. It is interesting to note that the most severe impacts of hurricanes and tornadoes occur far away from the origin, just like the El Nino. Storms and hurricanes are natural phenomena and catastrophic damages caused by extreme weather have occurred many times, in many places over centuries, well before fossil fuel use became widespread. However, the fact that storms need warm waters to form and strengthen means that they will be stronger, more frequent, and more devastating if ocean waters become warmer as a result of human activities. Also, the combined effect of expansion of warmer ocean waters and melting ice caps and glaciers could raise ocean levels by up to two meters by the end of the century, to the point that many coastal towns and cities in parts of the world become submerged, as is currently being predicted for some coastal areas of Europe and the United States.

6.6.4.   Draught and desertification

Draught is an extended period of deficient rainfall relative to the statistical multi-year average (long-term mean) for a region - a season, a year or several years. Desertification on the other hand is defined as a process of land degradation in arid, semi-arid and dry sub-humid regions, resulting from various factors, including climatic variations and human activities. The underlying cause of most draughts can be related to changing weather patterns manifested through excessive build up of heat on the Earth's surface, meteorological changes which result in reduction of rainfall and reduced cloud cover, all of which result in greater evaporation rates. Extended draught could lead to desertification and there are many natural environmental phenomena that can cause draught. However, the spread, persistence and resultant effects of draught are aggravated by human activities such as deforestation and civil construction which alter the soil's natural albedo (difference between the Sun's energy that is absorbed or reflected); overgrazing and poor cropping methods, all of which reduce water retention of the soil; and improper soil use strategies and conservation techniques which lead to soil degradation.

6.7.   OZONE LAYER DEPLETION

As discussed briefly earlier, natural ozone is concentrated in a thin layer mostly in the lower part of the stratosphere, from approximately 15 to 30 km above Earth's surface. The concentration is highest (2-8 parts per million) in the 20-40 km altitude range, although the thickness of the layer varies seasonally and geographically. Ozone in the Earth's stratosphere is created by the Sun's ultraviolet rays (with wavelengths shorter than 240 nm) which supply energy for the dissociation of some of the natural atmospheric oxygen molecules constantly being introduced into the atmosphere by photosynthesis. The molecules split into two highly unstable oxygen atoms which react with oxygen molecules to form ozone, (an exothermic reaction) with the release of heat energy. The net effect of the above two reactions is the formation of two molecules of ozone from three molecules of oxygen, accompanied by the conversion of light energy to heat energy. The ozone formed is unstable and absorbs more of the Sun's ultraviolet rays which causes it to decompose into oxygen molecules and atoms, with the release of heat energy. This continuous set of reversible reactions involving the breakdown and formation of ozone with the accompanying absorption and dissipation of energy is known as the *ozone-oxygen cycle or* Chapman cycle. There is no net ozone depletion because the process produces atomic oxygen that reacts with molecular oxygen to form another ozone molecule. A thin, natural ozone layer is thus created in the lower part of the stratosphere. These reactions play a critical role in utilizing potentially harmful ultraviolet (UV-A & B) radiations from the Sun and converting to less harmful heat energy, thereby preventing all of UV-A and most of UV-B radiation from reaching the Earth's surface and causing human ailments. In effect, stratospheric ozone, in spite of the low concentrations acts as an effective shield, protecting the Earth from radiation damage. Although exposure to UV radiation can be beneficial, notably the production of vitamin D which is essential to human health, over-exposure can lead to serious health issues, including cancer, blinding eye diseases, and premature aging. Human activities in the last few decades, notably in refrigeration and air conditioning have been releasing some chemical gases often grouped as fluorinated gases (or F-gases) into the atmosphere. Apart from the fact that some of these gases are potent greenhouse gases and heat trappers, they also react with the natural ozone, thereby lowering its concentration and creating pathways (ozone holes) for dangerous UV rays to reach the Earth's surface. One CFC molecule can destroy about a hundred thousand ozone molecules and reduce the concentration in the atmosphere, causing ozone holes. Concerted global efforts in the last two decades coordinated by the United Nations have led to a significant reduction in the use of fluorocarbons mostly in the developed world and there are signs that the ozone layer is healing gradually but use of CFC gases is still prevalent in the developing world and effective action needs to spread across all regions of the world.

6.8.   CONSEQUENCES OF CLIMATE CHANGE

The natural greenhouse gases (GHGs) reside in the middle atmosphere and the concentrations regulate the Earth's temperature. Human activities have been shown to increase the concentrations of most of the gases, notably carbon dioxide, methane and nitrous oxide, thereby enhancing their heat-trapping ability and the amount of heat reflected back to earth, and increasing the global average temperature. The global climate system has been monitored over the last hundred years or so, especially since the 1950s when sophisticated instrumentation and satellite technology became available, and the changes observed have led to the conclusion that global average temperature, the main indicator of climate change, is rising, with potentially serious consequences for life on earth. If the trend continues, the average global temperature could increase by up to 2°C or even higher by the end of the century, with dare consequences for current and future generations.

Apart from global warming, there are many other negative effects of anthropogenic pollution, notably warming and acidification of the oceans, acid rains, tropospheric smog, etc., all of which could have potentially devastating effects on life on earth. The International Panel on Climate Change (IPCC, 2013, 2018) has analyzed the very extensive databases from numerous sources on environmental pollution and climate change, many dating back several hundred years, and has concluded with a high degree of certainty that the world has warmed significantly since the 19th century when industrial and human development activities intensified. Strong data-based evidence shows that global surface temperatures have warmed substantially since 1900, and much of the excess energy has been absorbed by the oceans. In summary, anthropogenic pollution is not the only cause of climate change or extreme weather but is a significant contributor notably by making the Earth warmer. The effects of global warming such as warmer oceans and rising sea levels can make bad storms, rainfalls, flooding, heat waves and cold periods worse and much more extensive and destructive. Droughts and wildfires will become more intensive and widespread as a result of changes in evapo-transpiration and stronger winds, crop yields will be lower and the impact on human health would be profound. Many of these negative consequences are already playing out in real time.

There are no atmospheric physical boundaries between countries and regions, the global environment is a single, unified, continuous entity. While the major effects of ambient pollution are localized and often short-lived, the climate-negative consequences of stratospheric pollution are global and enduring. Gases in the stratosphere are relatively well mixed, hence the group is often referred to as well-mixed greenhouse gases (WMGHG). Their life span is in decades or centuries and they are relatively evenly mixed and evenly spread, with changes in concentrations occurring slowly over decades. One important point about stratospheric pollution which is often missed in climate change discourse but needs to be emphasized is the fact that, because the stratospheric environment that determines climate change is unified globally, forcings per unit emission and emission metrics for greenhouse gases do not depend on the geographic location of the source of emission, and the impact of events in one part can resonate in the farthest corner of the world. In effect, any part of the world could suffer the consequences of stratospheric pollution arising from the farthest regions. For example, even though Europe and North America have made significant advances in decarbonizing energy which may have resulted in cleaner local environment, the negative impacts of stratospheric pollution from exponentially rising coal power generation and energy-intensive manufacturing in Asia will be global: both Europe and North America have had unusually severe weather for most of the last decade.

Furthermore, even if all anthropogenic pollution were to stop today, most of the benefits will not become evident for decades, maybe centuries, and future generations will be the greatest beneficiaries.

Global warming leads to several consequences which can impact directly or indirectly on life on Earth. Increased and intensified heat waves can cause death, especially of old people and people with pre-conditions; strengthened tornadoes cause severe damage when they land; ocean temperatures and rise causing flooding in global coastal areas; increasing ocean acidity will impact negatively on global aquatic life; global weather variability could intensify. Several projections indicate that many coastal land areas in several regions could become submerged in the next few decades if the current rise in sea level continues unabated. Increasing temperatures can enhance increased production and migration of some disease-carrying vectors such as mosquitoes which spread malaria and zika virus. It is difficult to quantify the effects of human anthropogenic activities on human health because most diseases are multi-factorial and could be caused by other risk factors. However, there is fairly strong scientific evidence linking respiratory diseases and exposure to anthropogenic pollution. Even in such cases, the contribution of other factors that may cause or aggravate respiratory diseases is unclear.

6.8.1.   Human health and ecosystems' damage

Climate change has been at the center of global discussion on environmental pollution and, although there are many natural phenomena which can cause climate change or generate pollution, it is a well established fact that human activities are interfering with the natural control processes and aggravating the consequences. Global warming which is the most prominent consequence of climate change and emanates from the stratosphere has many health implications, notably illnesses, injuries and fatalities caused by extreme weather (heat waves, tornadoes, flooding, etc). The negative impacts of climate change are global, gradual and enduring, maybe for hundreds of years, but the impacts of ambient pollutions which are largely local can be instant and more devastating because they occur mostly in lowest part of the atmosphere (troposphere) that is in direct contact with life on earth. There are many natural processes which cause tropospheric/ambient/aerosol pollution, notably volcanoes, sandstorms, earthquakes, lightening-induced bush fires. However, many outdoor/indoor human activities are also significant contributors, releasing black carbon, carbon dioxide, nitrous oxide, ozone, and many other potentially harmful chemical compounds into the atmosphere. Aerosols originating from large volcanic eruptions and human activities enter the troposphere but, depending on prevailing wind dynamics, can also reach the stratosphere. Those that reach this level sediment out of the stratosphere within a year or two but, in that short period, can cause stratospheric cooling. However, most aerosol emissions remain in the lower atmosphere for a short time ranging from a few days to a few weeks, causing smog and health issues. Because aerosols, notably black carbon are distributed unevenly in the atmosphere, they can heat and cool the climate system in patterns that can drive changes in weather, in particular, precipitation. Ambient emissions - black carbon; organic carbon; carbon, nitrogen and sulphur oxides; ozone, and many other toxic compounds - impact more directly and instantly on life on earth, causing smog over urban areas, acid rain, toxic household environment, multiple health issues, and having negative impacts on animal, aquatic and plant life. Exposure to ambient pollution can trigger new cases of diseases or exacerbate pre-existing human health conditions and the effects on wildlife, aquatic life, and vegetation can also be devastating. Because tropospheric pollution is mostly locally sourced, there is significant physico-chemical variability in content and intensity in different parts of the locality and the effects on life are much more localized. Much of the pollution in this atmospheric layer comes from local activities: mining operations; oil drilling; fossil fuel transportation, conversion, and use; some industrial processes; household use of traditional biomass. Other potential sources include volcanoes, desert dust, power generation, forest fires, all forms of transportation, and oil spills. The potential impact of aerosols on health depends on the chemical composition, concentration and proximity, all of which depend on the sources.

6.8.1.1. *Impact of pollutions on human health*

The effect of outdoor/indoor ambient pollution on human health varies, depending on the type of pollution, intensity/concentration and duration of exposure, as well as the age and state of health of the individual. Particulate matter is considered the most dangerous component of atmospheric pollution and the fine particles below 2.5 microns in size ($PM_{2.5}$) pose the greatest problems because they can bypass the body's natural defenses, penetrating deeply into the lungs and bloodstream. Children are particularly sensitive to poor air quality because their lungs are proportionally larger than those of adults in relation to their body weight, hence they breathe more. Their immune system is still developing and they are less able to fight health issues that arise from indoor polluted air. Furthermore, babies and other young ones tend to stay around their mothers during cooking, thereby getting exposed to the highest concentrations of aerosols. People with pre-conditions such as heart and lung issues - asthma, chronic obstructive pulmonary disease (COPD), and lung cancer - are also vulnerable, not only because polluted air can make their conditions worse, but also because they tend to spend more time indoors and are therefore exposed to the hazards for relatively long periods. Even healthy people can suffer temporary respiratory problems such as coughing, breathing difficulties and eye irritation. Exposure to ambient pollution can trigger new cases of diseases, exacerbate pre-existing conditions of asthma, bronchitis, emphysema, and other respiratory ailments, or provoke the development and progression of chronic illnesses such as lung cancer and COPD. Ozone is a strong irritant that can cause constriction of the airways, forcing the respiratory system to work harder in order to provide oxygen. Exposure to the gas for extensive periods can cause lung damage and other severe respiratory diseases.

The World Health Organization (WHO, 2018) estimates that around 4 million premature deaths a year globally are linked to outdoor air pollution, mainly from heart disease, stroke, chronic obstructive pulmonary disease, lung cancer and acute respiratory infections, particularly in children. Indoor pollution arising from inefficient use of traditional and fossil fuels (wood, coal, biomass, kerosene) can be just as deadly, accounting for around 3.5 million deaths, the combined effect of both accounting for around 12% of all global deaths (Figures 6.2 and 6.3). Over 90% of air-pollution-related deaths occur in low- and middle-income countries, with nearly 2 out of 3 occurring in WHO's South-East Asia and Western Pacific regions. Ninety-four percent are due to non-communicable diseases - notably cardiovascular diseases, stroke, chronic obstructive pulmonary disease and lung cancer. Close to half of deaths due to pneumonia among children under 5 years of age (about one million deaths a year) are caused by particulate matter (soot) from household pollution. The most vulnerable are women, children and older adults, and those with pre-conditions all of who tend to spend more time indoors where the concentration of the most harmful particulate matter is highest. In fact, air pollution had been identified as the world's largest single environmental risk.

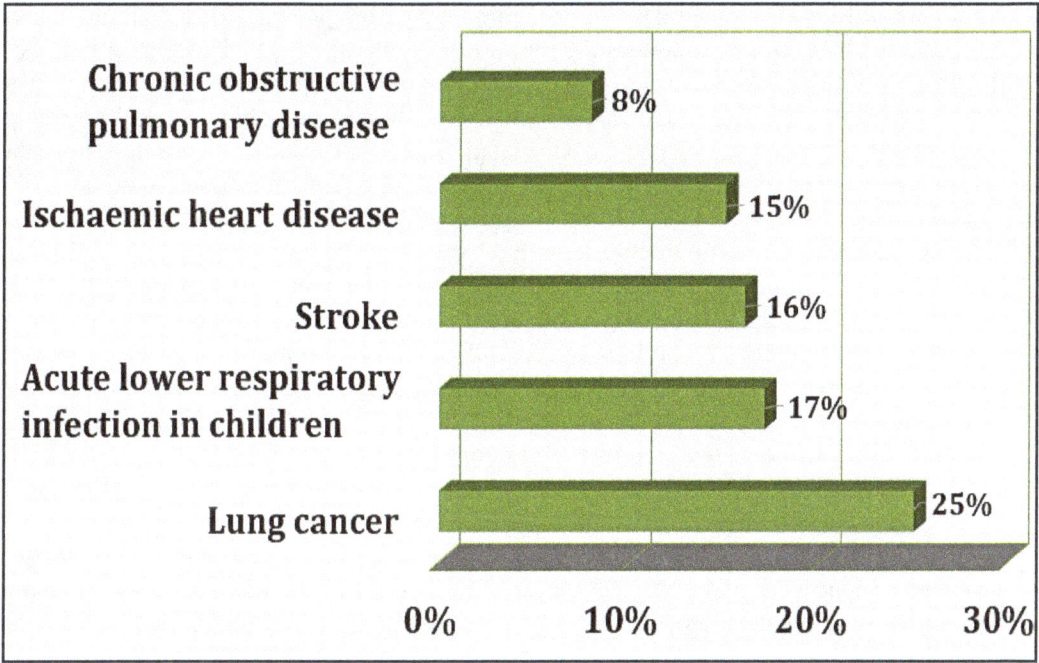

Figure 6.2 Annual global deaths from environmental pollution: Ambient: 4.2 million; Household: 3.8 million *(WHO, 2018).*

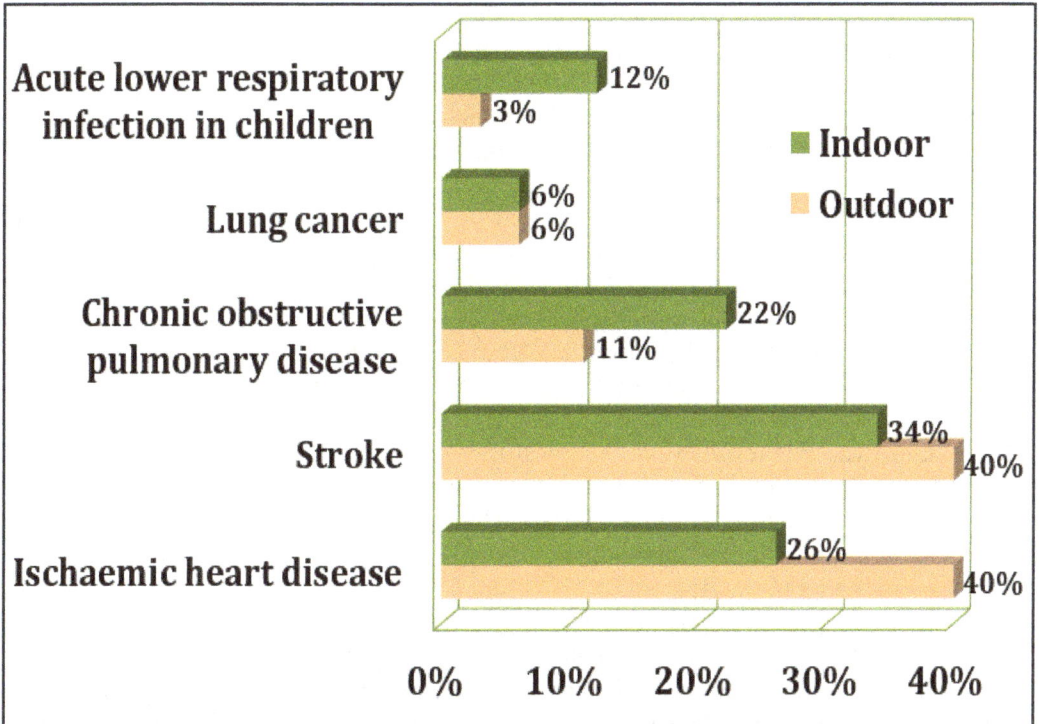

Figure 6.3 Annual global deaths from environmental pollution: Ambient: 4.2 million; Household: 3.8 million *(WHO, 2018).*

About one million children under five years die yearly from pneumonia, with more than half being caused by exposure to household air pollution (WHO, 2018). In addition to toxicity, the proximity of emission sources to potentially exposed populations, and altitude where emissions occur can influence the magnitude of their impact on public health. For example, emissions (black carbon, organic carbon) from automobiles, biomass burning, household energy use occur at ground level compared with high power plant smoke stacks. This likely results in greater exposure and therefore larger health impacts compared with nitrogen and sulphur oxides which come mainly from high smoke stacks, although diesel engines also produce significant nitrogen oxides. Dirty ozone (generated at tropospheric/ambient level from industrial processes), nitrogen dioxide and sulphur dioxide are linked to asthma, bronchial disease, reduced lung function and lung disease. Sulphur trioxide causes irritation of the mucous membranes of the respiratory tract, causing coughing and choking. Inhalation of droplets of acid rain formed by the reaction between sulphur trioxide and atmospheric moisture can also harm the respiratory system, and high levels of carbon monoxide are believed to cause problems in the nervous system. Sulphur and nitrogen oxides in gaseous products of coal combustion often react with moisture, soot and fly ash particles to form a dense smog which not only obscures sunlight but can cause eye irritation and breathing problems. Long-term exposure can cause chronic pulmonary diseases such as asthma, bronchitis, lung cancer, and cardiac problems, including arrhythmias and heart attacks (USEPA, 2012).

Depletion of the Earth's protective heat shield formed by the stratospheric ozone layer enables potentially dangerous ultraviolet radiation to reach the Earth's surface, causing skin burns and skin cancer. Incidences of fluorosis, arsenicosis, selenium intoxication, pneumoconiosis and silicosis are high among rural communities that live near coal mines or burn coals with high fluorine, arsenic, selenium, iron and silicon contents in poorly ventilated environment (Zhang *et al.*, 2004; Finkelman, 2017). Global warming and consequent climate change may have wide ranging and mostly negative direct and indirect impacts on human health, including increasing heat waves that can cause heat strokes and deaths. Atmospheric lead content has been increasing particularly in urban areas, and a significant proportion comes from traffic emissions, due to the lead content of gasoline. This has been reduced significantly in some developed countries in recent years since unleaded gasoline became widely available (though more expensive). However unleaded petrol is not available in most emerging countries and aerial concentrations in urban areas can be very high. Lead accumulates slowly in the blood and when levels reach concentrations of around 800 mg/liter, damage to the central nervous system, kidneys and the brain can occur. Children are particularly vulnerable because the tolerance level is lower and a strong link has been established between high lead exposures and impaired intelligence (Air-Quality, 2017).

Acid deposition is a general term that includes more than simply acid rain. Acid deposition primarily results from the transformation of sulphur dioxide (SO_2) and nitrogen oxides into dry or moist secondary pollutants such as sulphuric acid (H_2SO_4), ammonium nitrate (NH_4NO_3) and nitric acid (HNO_3). The transformation of SO_2 and NOx to acidic particles and vapors occurs as these pollutants are transported in the atmosphere over distances of hundreds to thousands of kilometers. Acidic particles and vapors are deposited via two processes - wet and dry deposition. Wet deposition is *acid rain*, the process by which acids with a pH normally below 5.6 are removed from the atmosphere in rain, snow, slit or hail. Dry deposition takes place when particles such as fly ash, sulphates, nitrates, and gases (such as SO_2 and NOx), are deposited on, or absorbed onto surfaces. The gases can then be converted into acids when they contact water. Sulphur dioxide emissions come from a variety of sources including coal-fired power generation, transportation, industrial processes, fossil fuel combustion, ore smelting and natural gas processing. The main source of NOx emissions is the combustion of fuels in motor vehicles, residential and commercial furnaces, industrial and electrical utility boilers, engines, and other equipment. Nitrogen oxides also come from air used in fuel combustion chambers of fossil power plants. Acid rain is a problem in many parts of the world because many of the water and soil systems lack natural alkalinity such as a lime base and therefore cannot neutralize acid naturally. Increased acidity of rivers, lakes and streams destroys aquatic life and acidified soils harm crops and forests. Acid rain also causes corrosion of road and building metal fixtures and automotive metal components.

Sulphur dioxide and nitrogen oxides in particulate matter (PM) react with moisture in the atmosphere to form reactive acids which fall on the Earth's surface as acid precipitation, rain or snow (wet precipitation), causing damage to metallic structures, soil, vegetation, etc. On the other hand, the acids may react with other chemical compounds in the atmosphere and form gases and salts which are also deposited on surfaces. These reactions occur naturally since sulphur dioxide is produced from many other sources not related to human activities, for example, volcanic eruptions and decaying vegetation. However, burning of fossil fuels (in particular, power generating plants) has greatly increased the concentration of these gases in the atmosphere over decades and acid rain has become a major hazard to human life and the ecosystems. Once in the air, pollutant gases are carried by the wind, and hence deposition can take place a long distance from the source, and if large quantities of acid are deposited in one place then it may have detrimental consequence for human health, buildings, structures, soils vegetation, crops, freshwater, and aquatic life. Acidic sulphur and nitrogen dioxides are powerful irritants, particularly in the respiratory system, especially when they are mixed with PM (synergistic or cocktail effect). Asthmatics are particularly susceptible because these pollutants can cause serious breathing difficulties.

Most materials used in building and other structural construction are susceptible to acid rain corrosion, although the rate and intensity vary widely. Metal structures are the most vulnerable, and carbon steel which is used extensively in concrete reinforcement and structures leads them all. However, many other materials are also susceptible - limestone; concrete, marble, zinc, nickel, paint and some plastics. Sulphuric acid formed from sulphur dioxide is the most destructive acid rain but the compound has to be oxidized to sulphur trioxide (SO_3) to form acid by reacting with water, and the presence of nitrogen dioxide enhances the oxidation process. However, nitrogen dioxide also reacts with water to form nitric acid which is often a component of acid rain. Both exposed structures and those immersed in acidic waters or buried in acidified soils are vulnerable to acid rain.

6.8.1.2. *Impact of pollutions on agriculture and forestry*

Acid rain can alter soil chemistry significantly, increasing the acidity and destroying many microorganisms which break down organic matter and transform into nutrients for plants. Also, acidified soil may release aluminium metal which is toxic to plant roots. The response of soils to acidification varies depending on the inherent soil chemistry. Soils rich in marble or limestone (carbonates) can effectively neutralize acid rain, while granite-rich soils which are devoid of carbonates have no neutralizing effect. Acidified soils can stunt the growth of plants and trees, and also lower their immunity and resistance to diseases, fungal attack, and invasion of insect pests, and many plants and trees die as a result of soil acidification. Acid rain settles on plants and crops, destroying the protective waxy coating of leaves and allowing acids to diffuse into them. The natural plant-atmospheric evaporation and gas exchange system is thereby interrupted, with a negative impact on crop yield and quality. Vegetables are particularly vulnerable to acid rain and many other crops (corn, potatoes, soy beans, etc.) are also susceptible to acid rain damage. Black carbon, in particular $PM_{2.5}$ and associated pollutants such as tropospheric ozone can damage crops and ecosystems, with negative impacts on agriculture, livestock, fish, etc. In addition to affecting precipitation, $PM_{2.5}$ deposited on leaves has been shown to affect crops and ecosystems by reducing the surface albedo, thus obstructing access to adequate sunlight as well as carbon dioxide, both of which are critical to the natural plant biological processes.

6.8.1.3.   *Impact of pollutions on aquatic life and wildlife*

Deposits of acid rain, black carbon and other associated aerosols can be toxic to aquatic life. Over time, acid rain increases the acidity of surface water at the catchment or during transport over acidified soil, decreasing the average pH to levels that impact negatively on aquatic life. Oceans are major long-term sinks for carbon dioxide, and may also become acidified. Soil and aquatic organisms have different tolerances to acidity and some die while others have reduced proliferation. Soft-bodied aquatic life - eels, angelfish, snails, shrimps, leeches, crayfish - are usually the early victims of fresh/seawater acidification, but long-term effects have been identified on most aquatic life and insects. All life stages are affected, from egg survival to the survival of young fry and adults. Many studies have demonstrated that surface water acidification can lead to a decline in, and loss of fish populations. Below pH 4.5 no fish are likely to survive. Other aquatic life such as amphibians are also affected: many invertebrate species that contain high concentrations of calcium, such as mollusks and crustaceans, are very sensitive to pH levels and are among the first to disappear during the acidification of wetlands. It should be noted however that atmospheric pollution is not the only source of freshwater/ocean acidification, in most situations, freshwater acidification is a combination of both natural and anthropogenic factors. Natural imbalance between precipitation and evapo-transpiration can cause the leaching of minerals from the soil, in particular, iron, aluminium and mercury (podzolization). Action of atmospheric carbonic acids, use of nitrogen fertilizers, in particular, those that are not sulphur-free, invasion of catchment by livestock can all cause changes in freshwater acidification.

There is also evidence that oxygen concentration is declining in increasing areas of global ocean and coastal waters due to ocean warming which reduces solubility of oxygen in water and the rate of oxygen resupply from the atmosphere to the ocean interior. Increasing discharge of nutrients (nitrogen and phosphorus) and organic matter, primarily from agriculture, sewage and the combustion of fossil fuels into coastal waters is causing increased acidification and accelerated consumption of oxygen by microbial respiration. Deoxygenation of oceans and coastal waters can have a wide range of biological and ecological consequences, including creation of low-oxygen 'dead zones' which cannot support marine life (Breitburg *et al.*, 2018). Wildlife and domesticated animals and birds all feed on plants, vegetation, and fruits, all of which are exposed to the effects of atmospheric pollution, in particular, acid rain and ground-level ozone, hence they are all potentially susceptible to the negative effects of acidification and toxic aerosols. They are affected indirectly because their food chain is disrupted or corrupted. For example, animals and birds that feed on freshwater fish find fewer fish in acidified lakes.

6.9.   OUTLOOK ON ENVIRONMENTAL IMPACT OF ENERGY-RELATED POLLUTIONS

There are few signs that energy-related environmental pollution will decrease significantly for decades. Even if stratospheric pollution could be eliminated completely today, the negative effects of past pollution, particularly on climate could persist for decades or even centuries. It appears that the most urgent option is to promote adaptation strategies for the current global population who are having to deal with some of the consequences of climate change in real time. Also, the world needs to unite and introduce enduring policies that strengthen current climate mitigation strategies, thereby lowering the risks for future generations. These include strategies for coping with the current negative impacts of climate change and ambient pollution; strong policy support for decarbonization of energy; more aggressive efficiency improvement across the energy value chain, from production to end-use; and promotion of public awareness on how they can reduce their carbon footprint through lifestyle changes. Tropospheric pollution is much easier to deal with because most of the causes are local and most of the needed policy instruments can be developed by local governments, and some are already in place, for example the increasingly stringent emission standards for power plants and vehicles, refuse collection and disposal, faster transition to modern energy worldwide, etc. Reducing local pollutions will help clean up urban atmospheres and reduce pollution-related deaths and other negative impacts on ecosystems. However, prospects, particularly in the developing world are not very bright, in spite of numerous and diverse interventions by many organizations to move the regions towards the United Nations' goal of modern energy for all by 2030. The problem with stratospheric pollution is much more complex because the main effects are global warming and climate change and the negative impacts are largely independent of the sources of emissions. The developed world is moving away from energy-intensive production but they are becoming increasingly dependent on the emerging world for vital primary products (primary metals and non-metals, chemicals and petrochemicals, polymers, fertilizers, intermediate industrial components, solar PV cells, electric vehicle and storage batteries, etc.). Many of the industries are assembly plants that depend critically on imported components and a very large proportion of consumer goods come from the emerging regions. While this is consistent with globalization principles and would help clean up the local environment, most of the growth of future emissions will come from the emerging economies and the effects on global warming and climate change will be universal. Furthermore, any disruption of the supply chain could cripple their economies.

Chapter 7

Pathways to a sustainable environment

7.1.   INTRODUCTION

There is ample scientific evidence in support of the fact that human activities are influencing the climate system, concentrations of greenhouse gases are the highest they have ever been, and their effects, together with those of other anthropogenic drivers, are evident throughout the whole spectrum of the climate system. Many repercussions of climate change have had widespread negative impacts on human and natural systems. Extensive scientific data indicate (with a high degree of confidence) that these emissions have been the dominant cause of the observed global warming since the mid-20th century. The Earth's surface temperature is projected to rise by up to 2-3°C over the 21st century under all possible emission scenarios, and it is very likely that heat waves will occur more often, they will be more intense, and will last longer; flooding will be more frequent and severe; oceans will continue to warm and acidify; storms and tornadoes will be more devastating because they derive their energy from warming oceans; and extreme precipitation events will become more intense and frequent in many regions of the world. The thrust of global mitigation efforts has been to reduce energy related pollution emissions, but the International Panel on Climate Change (IPCC, 2014) has concluded (with a high degree of confidence) that current and planned efforts are inadequate, as is evident from its statement quoted below:

> *"Without additional mitigation efforts beyond those in place today, and even with adaptation, warming by the end of the 21st century will lead to high to very high risk of severe, wide-spread and irreversible impacts globally (high confidence). Mitigation involves some level of co-benefits and of risks due to adverse side effects, but these risks do not involve the same possibility of severe, widespread and irreversible impacts as risks from climate change, increasing the benefits from near-term climate change mitigation efforts...........Global emissions show no signs of peaking.........Global emissions in 2030 need to be approximately 25 percent and 55 percent lower than in 2017 to put the world on a least-cost pathway to limiting global warming to 2°C and 1.5°C respectively"*

7.2.   PATHWAYS TO A SUSTAINABLE ENVIRONMENT

Many independent and partisan organizations publish outlooks over two to three decades on global energy demand and use, based on historical antecedents, and various assumptions on the future pathways of the main determinants of energy demand: global economy, population growth and demographics, global energy market dynamics, emerging technological innovations, urbanization, emerging policies, etc., all of which are subject to significant uncertainties. The energy industry is in transition, shaped by growing global economy and population, rising prosperity and urbanization, new disruptive technologies, government policies, social preferences, challenges of producing more energy with less carbon emissions, unforeseen shocks that can severely disrupt either the global economy or the energy supply chain, etc. The net effect of any of these variables and many more on future energy demand and consumption is impossible to predict today. This is evident from significant changes which are often made in projections and forecasts from year to year by the same organizations, and the evolving strategies of exploring different potential scenarios in energy forecasts. Nevertheless, even though there may be differences in specifics, there is general consensus on the likely growth pathways of global energy demand and energy-related emissions.

Considering the inertia of response of the Earth and its ecosystem to climate change, current mitigation efforts may have some near-term benefits particularly in urban areas but the major positive impact on climate change will become evident in the later decades of the 21st century and beyond. The world's primary energy demand is growing, in response to strong economic and population growth, and all recent projections show that the trend will continue. Fossil fuels have been the major sources of primary energy for centuries and the dominance is unlikely to diminish in the foreseeable future. Around two-thirds of the total global atmospheric pollution comes from the use of energy, and fossil fuels account for over 90%. In effect, fossil energy-related pollution will continue to rise as long as energy demand continues to increase, even if all current and planned global mitigation actions are effective. Perhaps the most realistic plan is to continue to reduce the growth rate so that it becomes negative over the next few decades, and there are many options (Figure 7.1).

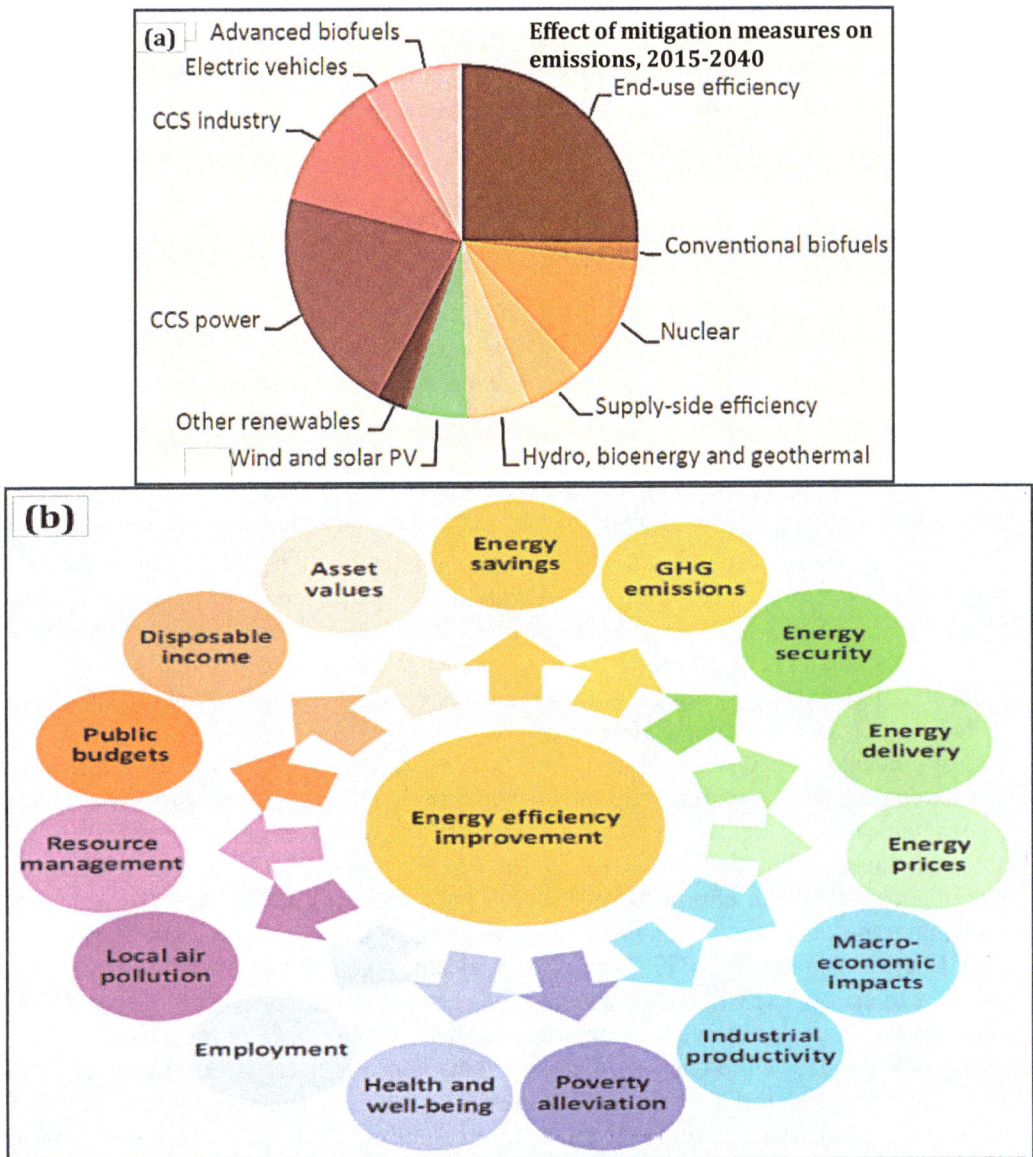

Figure 7.1 (a) Global cumulative emissions reductions by mitigation measure,   (113 Gt CO₂eq abatement) (b) Multiple benefits of energy efficiency improvements (*IEA, 2015; 2017b*).

Many international organizations have developed potential pathways to a sustainable environment. Five key options that could totally decouple fuel demand and the economy within a decade and help put the world on the path to a sustainable environment are:

- Reduction of energy use

  - Efficiency improvement in production, distribution and use
  - New technology innovations

- Decarbonization of energy

  - Efficiency improvement
  - Change in energy mix
  - Decarbonizing power generation
  - Decarbonizing industry
  - Decarbonizing transportation
  - Decarbonizing buildings
  - Carbon capture and storage

- Energy and waste recycling

- Institution of strong national and global policy instruments

- People's lifestyle changes that lead to a reduction of personal carbon footprints

7.3. REDUCTION OF ENERGY USE

Reduction of energy use in spite to the exponential growth of the global economy and population is one of the most challenging issues in the world today, especially when around eighty percent of the world's population currently have no adequate access to modern energy. The International Energy Agency has projected that current and planned policies for the reduction of global energy use are grossly inadequate, more aggressive strategies are required to reduce demand by around 30% from the 2017 level by 2040 (Figures 7.2 and 7.3). Energy-related emissions mirror the trends in energy consumption which in turn reflects global economic, population and prosperity trends. Increasing energy efficiency and decreasing energy intensity will be the key drivers of future anthropogenic emissions abatement across all regions of the world. In spite of virile economic growth and associated rise in energy demand, improved energy efficiency will continue to slow down the rate of growth of emissions, and all available projections show that emissions could fully decouple from energy demand by 2040. Improving efficiency and the resultant decrease in CO_2 intensity of energy use will help stem emissions growth in spite of growth in population and GDP, as amply demonstrated by China: although the country's GDP rose by about 1,000% from 1990 to 2015 while energy efficiency gains kept the rise in CO_2 emissions to about 300%. For the same reason, emissions in OECD countries have remained relatively flat for the last two decades and are projected to decline by about 20% by 2040. However, emissions from the non-OECD countries will rise by around 50% over the same period despite a 40% gain in efficiency across the emerging economies even though the growth rate of less than 1% will be much lower than the 3%/year from 1990 to 2015.

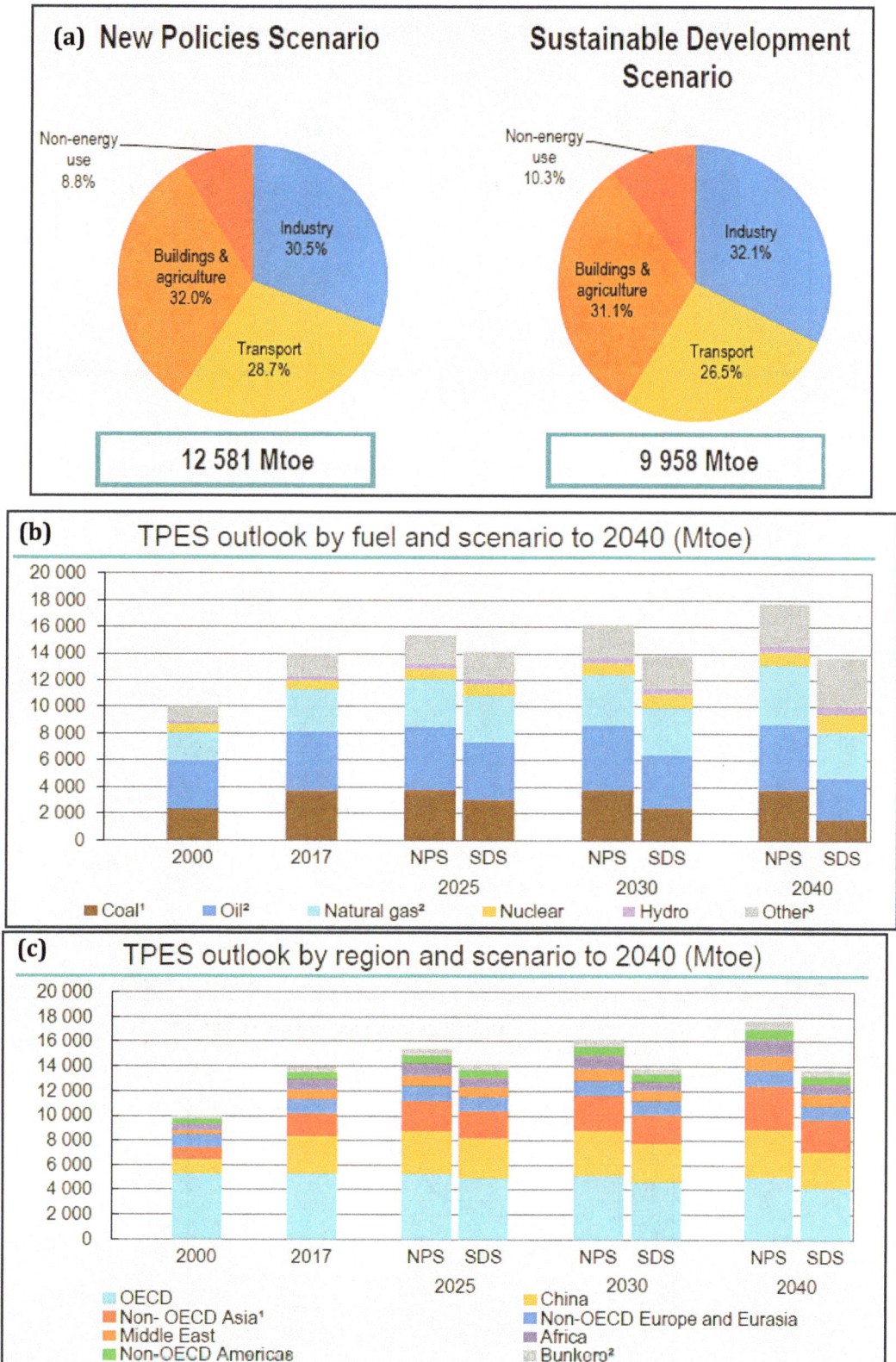

Figure 7.2 New Policy and Sustainable Development Scenarios on global primary energy supply by (a) economic sector, (b) fuel and (c) region *(IEA, 2019g).*

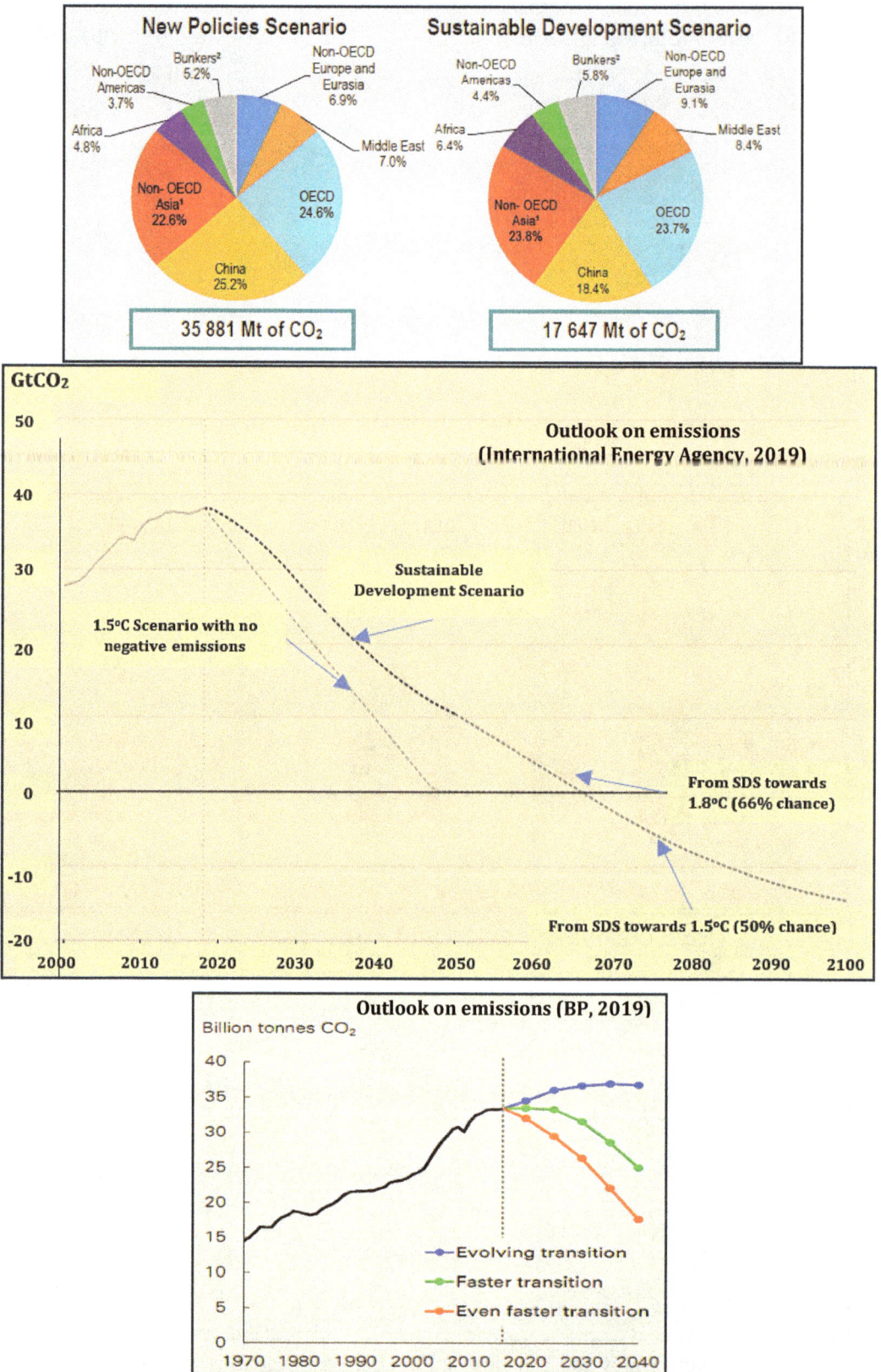

Figure 7.3   Outlook on emissions in different scenarios *(IEA, 2019h; BP, 2019a)*.

The prospects of lowering energy-related emissions depend largely on ability to reduce demand through improved efficiency and fuel switching and opportunities vary across sectors of the global economy. Events in two regions which contributed more than half of the total energy-related emissions in 2017 - Asia (41%) and North America (15%) - will largely determine the global outlook. China alone accounted for 28% while the United States contributed about 14.5%. In the same year, non-OECD economies accounted for 65% of total energy-related emissions. China's outlook to 2050 shows that a combination of efficiency improvements, technology implementation, fuel substitution, and emission control policies will hold emissions level flat in the next thirty years, overall, and in all sectors of the economy, with the exception of industry where an annual growth rate of around 0.6% is projected (EIA, 2018). However, Non-OECD Asia will remain the world's largest emitter of energy-related pollution in 2050, partly because the region will host most of the energy-intensive primary industries, but also because of the expected growth in their economies, population and prosperity. China's economy is slowing down and also transiting from high-intensity manufacturing to less energy-intensive production. Furthermore, growth of the renewable and nuclear energy sectors has been rising, hence a gradual decrease in energy demand rate is expected, and energy intensity will also start to decrease. Coal will account for a lower share of the primary energy demand and the share of natural gas, nuclear and renewables will rise significantly. Overall, there should be a significant fall on CO_2eq emissions of the country by 2040. Significant shifts from coal to natural gas, renewable energy and nuclear power are projected for the country, and the mitigation effect on CO_2eq emissions is expected to be substantial because of the size of the country's economy. On the contrary, energy demand by India continues to grow at more than 2.5 times that of China, representing more than a third of the global increase, and most of the growth will be filled by coal and nuclear power. However, the contribution of the country to global energy-related emissions will be no more than half the contribution of China for the next two to thee decades.

7.3.1.   Efficiency improvement in energy production and use

Energy efficiency is simply the ratio of output to input, typically expressed as a percentage. A power plant that converts only 40 kilograms of 100-kilogram feed coal into electric power has only 40% thermal efficiency. Energy intensity (a prime indicator of production efficiency) is the amount of primary energy required for a given output and is often expressed as energy/GDP, energy/capita, kilometers covered/liter of gasoline, etc. When it declines, it means that the world is able to produce more GDP for each unit of energy consumed. Efficiency rises either through re-organization of existing methods or as a result of infusion of new technologies, and higher values mean lower energy consumption and associated emissions per unit output. Carbon intensity (CO_2 emission/unit energy output or use) is primarily a function of carbon content of the fuel that produces the associated energy use (although it is also related to energy intensity), hence responds to fuel switching. Increasing efficiency over the next two decades or so and switching to lower carbon-intensive fuels should drive down energy intensity and carbon intensity, in spite of strong growth in the main energy demand drivers: population and economic growth. Energy efficiency has been evolving as a key resource, with multiple benefits for economic and social development across all economies, and has emerged as a major driver for uncoupling energy consumption from economic growth. The International Energy Agency projects that efficiency improvements can reduce emissions drastically in spite of projected increase in energy consumption (Figure 7.4). Apart from the obvious benefits of enhanced energy efficiency: reduced energy demand and lower greenhouse gas emissions - investment in energy efficiency can provide many different benefits to many different stakeholders. Whether by directly reducing energy demand and associated cost (which releases funds for other investments), or facilitating the achievement of other objectives (such as more efficient internal combustion engines and household energy appliances that reduce atmospheric aerosols, the enormous multi-faceted potential of energy efficiency improvement has become a prominent topic of global research interest in recent times.

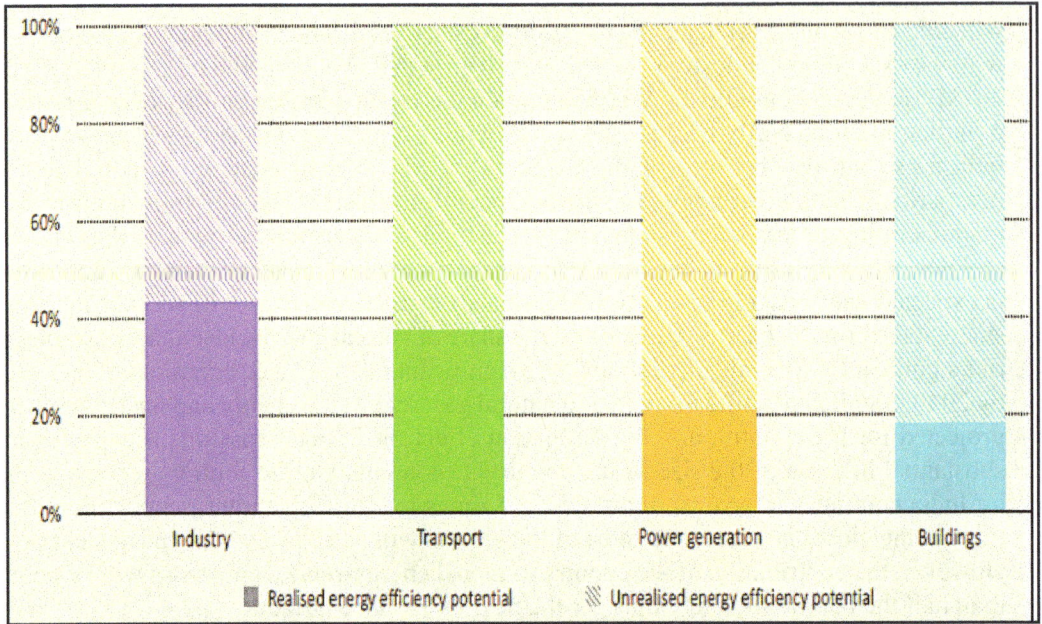

Energy demand, energy intensity & GDP in the EWS, 2000-40

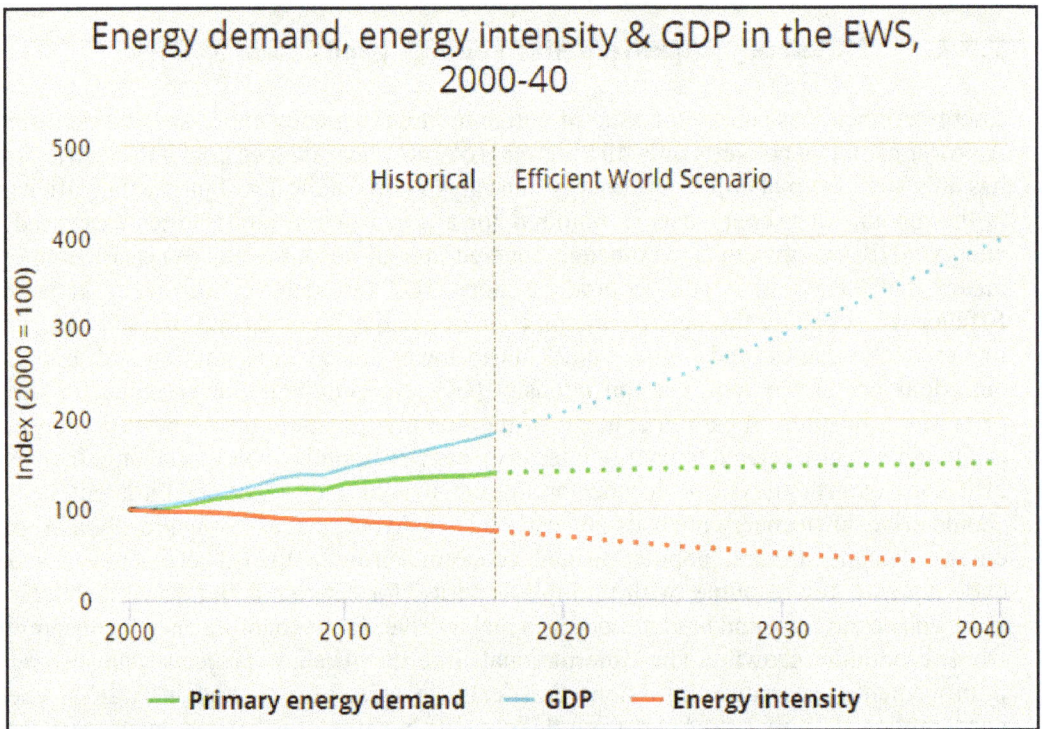

Figure 7.4a Long-term energy efficiency economic potential by sector and scenario *(IEA, 2017c).*

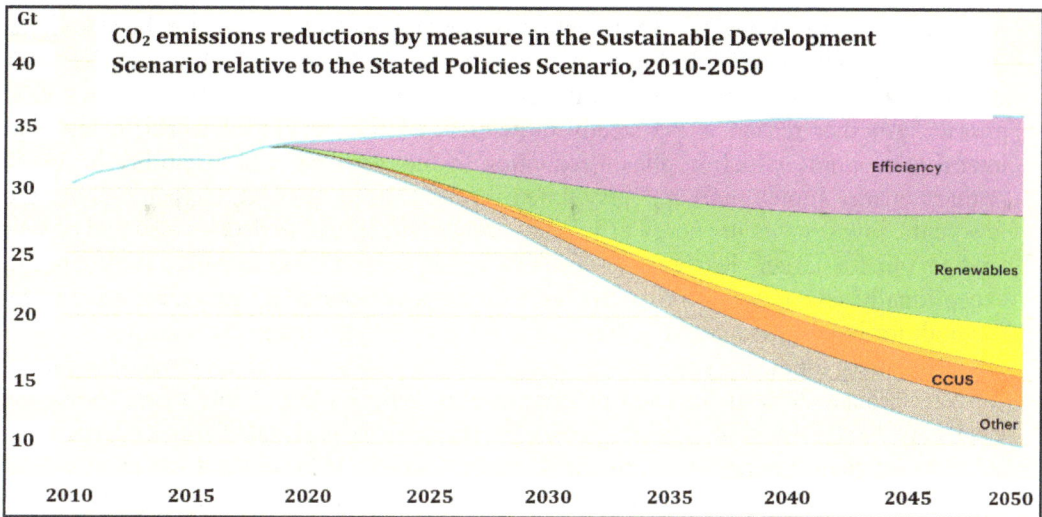

Figure 7.4b   Projected effects of climate mitigation policy scenarios *(IEA, 2020).*

Global energy intensity has declined at an average rate of 2.1% per year for the last ten years or so and the benefits, also known as *energy productivity bonus*, are both enormous and multi-dimensional. It is estimated that, without energy efficiency and intensity improvements, the world would have used 12-15% more energy than it did in 2018, equivalent to the total energy consumption of the European Union and worth over USD 2 trillion. Energy efficiency is regarded as the most effective environmental mitigation effort and the biggest contributor to reduced energy use and emissions. It is estimated that a combination of the decline in energy intensity and the change in energy mix towards lower carbon energy have been responsible for stalling growth and steadying emissions at around 32-33 billion metric tons over the last five years, with energy efficiency and associated fall in energy intensity accounting for the offset of around 77% of the impact on global emissions from GDP growth. Rising energy efficiency and associated decline in energy intensity make big contributions to strengthening energy security since many nations that depend on external sources for primary energy would need to import significantly less, with substantial economic and social benefits. Energy efficiency is considered a major resource that remains largely untapped and believed to hold the key to future sustainable environment. The International Energy Agency describes energy efficiency as "the first fuel of a sustainable goal energy system" and has developed a model that shows the enormous potential of energy efficiency gains in mitigating emissions over the next twenty years or so (IEA, 2019o; 2019i). The model known as *Efficiency World Scenario* (EWS) shows that a strong growth rate in energy efficiency across the global economy could reduce global energy-related emissions by more than 40% of the emissions cuts needed to reach the sustainable climate goals without new technologies. The model also delivers the energy efficiency Target One of the United Nations Sustainable Development Goals: *energy equity for all.* Under the EWS, the amount of global GDP produced for each unit of energy could double over the next two decades, for only a marginal increase in energy demand. However, this would require strong policy actions across all regions, starting from now and focusing on many of the two-thirds of final energy use not currently covered by mandatory energy efficiency policies. Also, under the Organization's *Efficient World Strategy*, decline in energy efficiency needs to be consistent at around 3% per year: it has been decreasing steadily over the last two decades by around 2% per year. Furthermore, the rate has declined consistently over the last five years to 1.2% in 2018. The decline in improvement rate is believed to be due to a combination of demand and supply factors. For example, the use of energy-intensive fossil fuels, such as coal for electricity generation has risen steadily over the same period, particularly in the emerging world where efficiency is not a pressing priority. Another reason is the steady movement by the developed world to less energy-intensive economy which implies a reduction in fuel demand, and instability in the global primary energy supply market. Every time oil price falls as has been happening over the last few years, investments in energy efficiency improvement tend to decline also and consumers tend to opt for larger, less energy-efficient goods such as automobiles. Furthermore, the exceptionally severe weather over the last few years has caused a significant spike in energy use both for heating and cooling, much of it filled by energy-intensive fossil fuels.

The energy industry produces all primary energy - fossil, nuclear, renewable fuels - and converts around 78% of the total primary energy supply (TPES) into other forms namely electricity, heat, refined oil products, coke, but the industry is also the largest energy user, and the least efficient. Energy losses by the sector are around 30% (the end-use energy output is only about 48.7%). Significant losses occur at the production stages, during transportation, conversion, power generation, transmission, and distribution. Many power plants and distribution infrastructure across the world are decades old and, while some are frequently upgraded by retrofitting, many others remain grossly inefficient. For example, most coal-fired power plants worldwide operate at below 40% efficiency even though technologies that deliver 45-80% have been available for decades, and power transmission/distribution losses range from 10 - 40%, again in spite of the availability of more efficient distribution technologies. Higher efficiencies in the energy production and processing sector could resonate in the end-use sectors, and global energy savings multiplication factors could be significant - 4.7 for electric power generation, 2.7 for heat generation, and 1.07 for coal and petroleum products (Bashmakov, 2009). Industry is the largest consumer of end-use energy, accounting for 84% of the final energy use of coal, 26% of petroleum products (for energy and as raw materials), 46% of natural gas, 40% of electricity, and 43% of heat. Transportation consumes 62% of liquid fuels final use, while the building sector accounts for 46% of final natural gas consumption, 76% of combustible renewables and waste, 52% of electricity use, and 51% of heat (IPCC 2014b).

Efficiency is low across the total energy value chain, from production to end-use, and there is a growing awareness that improving efficiency in energy production and end-use is the most promising mitigation option that helps decouple growing energy use and energy-related emissions in the shortest possible time. Emissions growth rate can be reduced significantly through improved efficiencies across all sectors, from production to transportation, conversion and use. Improved energy efficiency over the whole value chain is now regarded as a central pillar of pollution mitigation, an energy resource, and the easiest and least expensive way of conserving primary energy resources. For example, a power plant operating at 40% efficiency is wasting 60% of the primary fuel input, and an LED lamp would deliver the same light (lumens) as an incandescent lamp for 80% less energy consumption is saving much more than 80% energy if losses in the production chain are taken into account. A simple illustration is the production of one unit of electric light. About 320 units of thermal energy is required to produce one unit of useful electric light. In effect, energy equivalent to 319 units is lost in generation, transmission, distribution, wiring and fittings. Investment in more efficient power generation, transmission and distribution technologies could reduce fuel input by around 30%, while conversion to more efficient compact fluorescent and LED lighting could save much more, up to 80%. Projections indicate that up to 40% of future global primary energy requirements could come from improved efficiencies in production and utilization, which implies a reduction of fossil fuel use and potential anthropogenic greenhouse gas emissions by an equivalent amount, enough to set the world on the path to a sustainable environment (McKinsey, 2010; WEC, 2016c; IEA, 2017d, 2017e and 2017f).   E n e r g y efficiency has immense potential for reducing process energy and carbon intensities and is emerging as a key area of innovation. Improved energy intensity has been the biggest factor responsible for the downward trend of the global anthropogenic emissions over the last two decades or so. Furthermore technology innovation has been dynamic and is playing a key role in creating and actualizing new opportunities. Energy efficiency has been driving down energy intensity for many years although there are significant variations between countries and regions, with OECD countries and China accounting for most of the achievement. The most important implication of this development is the fact that the world is generating more value from its energy use thereby making huge savings in primary energy use, and holding back the growth of anthropogenic emissions. Efficiency improvement is impacting positively on the economies across all regions of the world. In emerging economies, energy efficiency gains are limiting the increase in energy use associated with economic and population growth, thereby making more energy available and improving access by their populations. The worldwide improvement in energy efficiency is helping to decouple both energy use and emissions from global economic and population growth. Global total energy use peaked in 2007 and has been declining since then in response to rising efficiency. Improved efficiency has also impacted significantly on the third dimension of the global energy trilemma - energy security (the others are energy access and environmental sustainability, see Chapter 2). As a result of declining demand for primary energy, propelled by increased efficiency, many countries were able to cut coal, oil and gas imports by as high as 30% in 2016, thereby improving internal energy security and freeing funds for other urgent needs. All sectors of the primary energy value chain have efficiency improvement potential: coal, oil and gas exploration, industrial production and transportation, power generation, energy transmission and distribution, domestic and commercial utilization. For example, electric motors, lighting and home appliances account for approximately 70% of the total electricity consumption in the industrialized nations but efficiency is between 10 and 50%. Estimates vary widely but it is believed that between 30% and 50% of primary energy produced worldwide is wasted through inefficient production processes, transmission and utilization.

7.3.1.1. Efficiency in fossil energy production

The recovery efficiencies of coal mines, oil and gas deposits are currently low, sometimes limited by available technologies, but more often by the typical inertia that plagues the adoption of new technologies; at the production stage, oil is spilled and gases are flared; and significant amounts of products are lost during transportation. These and many other energy sectors present very high opportunities for efficiency improvement. Oil wells are being pressurized with carbon dioxide primarily to improve production efficiency, but this also helps sequestrate anthropogenic emissions. Fracking technology has been available for decades, but it is only in the last decade or so that its use became widespread and this has greatly boosted oil and gas production efficiencies in some countries. Also, the large quantities of gas being flared worldwide could be processed for power generation or liquefied petroleum gas to boost overall production efficiencies. Energy conserved through the improvement of efficiencies of production and consumption translates to lower energy demand, lower fuel consumption and lower energy-related harmful emissions. It also means that more fossil fuels are conserved. Unfortunately, in spite of the large number of efficiency-improvement technologies already available, deployment has been slow and the enormous potential of improved energy efficiency in reducing primary energy consumption and pollution emissions is still largely untapped.

7.3.1.2. *Efficiency improvement in power generation*

Around 65% of the total global electric power generation is powered by fossil fuels and over 40% of global primary energy use is for power generation. However power plants operating around the world are only 33% to 40% efficient, in particular, coal-fired power plants, the most polluting and most of which utilize sub-critical technology, whereas much more efficient ultra- and super-critical technologies have been available for decades. In effect, 60-70% of the carbon energy locked in the coal is wasted, in addition to transmission and distribution losses which could be 10-40% of the generated power. Efficiency improvement translates to lower coal consumption per kilowatt-hour of generated power and lower emission: one percent increase in efficiency of conventional pulverized coal-fired plants results in 2 to 3% reduction in CO_2eq emissions and highly efficient modern coal plants emit up to 40% less CO_2eq than the older plants. Many of the coal power plants operating in the developed world are old and nearing end of life of around 40 years and few new ones are being built. However, most plants in the emerging world are only 10-14 years old and are generally more efficient because they adopted more modern technologies. Newer power plants and many retrofitted installations now operate at around 40% efficiency or above; adoption of pre-gasification of coal could raise efficiency to around 60%; while combined heat and power complexes have even higher efficiencies. Even at the higher levels of efficiency there is still a tremendous amount of energy lost in the generation process which could be reduced through the adoption of more advanced combustion optimization technologies and control systems. The economic and ecological benefits of recovering and utilizing more of the energy in the input coal are also enormous, and many technologies are already available, for example, cogeneration plants, energyplexes and other advanced power generation technologies could return efficiencies as high as 80% (see Afonja 2017).

7.3.1.3. *End-use efficiency improvement: industry*

Every sector of the economy has significant potentials for exploiting the benefits of enhanced energy efficiency. Industry accounts for over half of the total end-use energy and there is extensive scope for efficiency improvement. About 40% of the global electric power generation and 65% of industrial power are used to run electric motors in a very wide variety of applications ranging from driving industrial equipment to domestic fans microwaves and washing machines, with efficiencies as low as 30%. Most electric motors deployed in industry and incorporated in domestic appliances are fixed speed motors which consume substantially more energy than required at light loads, and variable speeds are obtained through gear reduction units. Variable speed motors are much more efficient than fixed speed motors and could reduce the world's demand for electricity substantially. However, the relatively high cost has restricted use even in industry. Virtually every industrial process that utilizes electrical motors has a potential for improved efficiency. These include food and consumer product manufacture, machinery and automotive industries, pulp and paper, metallurgical, chemical and petrochemical industries and service industries. There is considerable technology inertia in the electric motor industry, with little change over the last few decades, for example, wider use of variable-speed motors will stimulate research and development, and drive down costs. Efficiency regulation of electric motors is currently low, policies which support higher efficiencies could greatly improve efficiency: an improvement of just 10% could make a significant difference to energy consumption because of the very large population of electric motors worldwide. Industrial energy intensity (amount of energy required per unit product) has been declining in recent years (about 20% since 2000) due to wider adoption of newer, more energy efficient technologies across all sub-sectors and regions. There has also been a notable shift in the regional pattern of industrial energy demand in favor of emerging regions which now use 66% of global industrial energy. Five energy intensive sectors - chemical and petrochemical, iron and steel, aluminium, cement and pulp and paper - account for 67% of the world's total industrial energy consumption, and many are located in emerging countries. Very few innovative energy-saving technologies have emerged in these sectors in recent years but the fact that many of the facilities are new or have been recently expanded/upgraded, provided some opportunity to adopt more efficient technologies. However, opportunities for efficiency gains in the industrial sector are still very high and application of current best available technologies (BATs) could reduce energy use in the energy-intensive industries by 10% to 25%. Options include process and plant modification, waste and heat energy recycling, more efficient process control, and greater adoption of proven but under-utilized technologies.

7.3.1.4. *End-use efficiency improvement: transportation*

Improvement in fuel economy of transportation vehicles is considered one of the most cost-effective ways to reduce GHG emissions. The average kilometers per liter of fuel is expected to double by 2040. This should reduce demand for oil and transport-related pollution in 2040 substantially. However, the population of vehicles is growing exponentially and is expected to double by 2040, and the bulk of the growth will be in the emerging world where vehicles are much less fuel-efficient and pollution control is lax. Furthermore, the increasing demand for personal transportation is fueling the growth in highly-polluting single-stroke, 2-3 wheeled motorcycles which are used very widely for personal and commercial transportation. Furthermore, growth in diesel-fueled commercial trucking, aviation, and shipping will be dynamic in the regions. The impact of energy efficiency in transportation is well below the potential, due largely to the instability in the global oil market. Significant improvements in fuel efficiency of conventional vehicles combined with low global oil prices have led to faster increase in sales, in particular, less efficient large passenger vehicles such as sport utility vehicles and family vans in the last decade or so, thereby dampening the global rate of improvement in passenger vehicle fuel efficiency, which has been evident over the last two years or so. Another problem is the relative lack of improvement in the truck sub-sector which accounts for around 43% of total oil consumption for road transport, and uses the bulk of diesel, the most polluting of fossil end-use fuels. While many countries have mandatory efficiency standards for passenger vehicles, only a few have fuel economy standards for commercial trucks in place.

It is widely accepted that electric vehicles (EVs) hold great potential in decarbonizing the transport sector (assuming access to lower carbon electricity source), and global sales have been strong, rising from around 2 million total population in 2016 to about 5 million in 2018. However, in spite of the high relative advantages of electric vehicles compared with combustion-powered vehicles (better fuel economy, zero anthropogenic tailpipe emissions and lower carbon footprint), the penetration of electric vehicles into the global auto market has been very slow: the global population of EVs in 2018 was only around 0.5% of the population of nearly one billion vehicles. Although growth in the EV market is expected to be strong over the next two or three decades, EVs will still account for no more than 15% to 35% of the total vehicular population in 2040 (based on several recent projections and different levels of optimism), and even this modest growth will depend on the extent to which battery size and cost can be reduced and charging/swapping bays can be proliferated, and, crucially, consumer interest which may be dampened significantly by increasing fuel economy in conventional vehicles and cheaper fuels. Furthermore, the sterling environmental credentials of electric vehicles depend critically on the carbon intensity of manufacturing and operating energy. China which currently accounts for nearly half of the total global population of EVs, projected leader of future growth, and the main source of electric batteries also has one of the highest energy carbon intensities in the world. Most of the growth in EVs will be accounted for by China, Europe, North America, and Japan, while nearly all the growth in vehicle population will be in emerging countries. Furthermore, the electric vehicle market is very sensitive to global oil prices which have been low for some years, thereby fueling the growth of internal combustion engine vehicles at the expense of EVs. However, some countries, particularly in Europe have introduced punitive taxes on ICEs and are offering strong incentives to promote the EV market. It should be noted however that every EV vehicle on the road contributes to emissions mitigation especially when low-carbon electricity is available for manufacture and operation. The most obvious, immediate and important impact is on the highly problematic urban pollution and the negative effect on human and ecosystem health.

7.3.1.5.   *End use efficiency improvement: buildings*

Buildings (residential and commercial) accounted for around 21% of final energy consumption globally in 2017 and, although demand is projected to increase by 32% by 2040, the share of end-use energy will still be around 21%. The projected increase will be driven by rising population, urbanization and increasing wealth in emerging countries. Substantial potential exists (up to 25%) for improvement of the energy efficiency of buildings, particularly in heating and cooling which is the largest end-use energy sub-sector, accounting for one-third of all energy consumed in buildings (heating accounts for around 60% of household energy use in cold climates) (IEA, 2017d). A major area of efficiency improvement is the upgrade of the building thermal envelope through design and use of best available technologies (BATs). Potential areas of innovation include insulated walls, windows, cool roofs, improved natural lighting, efficient lighting and appliances, and smart control of household energy use. The impact of improved efficiency in the building sector is cutting across all regions, and it is estimated that household energy efficiency gains saved consumers 10-30% of their annual energy spending in 2016.

The energy utilization efficiency of buildings also known as Energy star rating (ESR) is becoming a critical variable for both home builders and buyers in many countries. New buildings feature effective wall and roof insulation; double glazed windows; compact fluorescent/LED lighting which can provide the same light power for 10-20% of the power consumption of a conventional resistance bulb; smart control systems which switch off lights in unoccupied rooms; energy-efficient heating and air conditioning systems; and energy star-rated appliances. Also, consumers are becoming increasingly energy-conscious and look critically at energy efficiency ratings of appliances before they buy. Furthermore, the "smart home" concept also known as "Internet of Things (IoT)" is spreading rapidly and involves connecting homes and appliances to the Internet to facilitate the ability to control home appliances remotely on cell phones, which means that home heating and other appliances can be switched off when every one is out and reactivated shortly before return. Also, temperatures can be monitored and controlled remotely. Consumer education will also play an important role in mitigating energy use in the building sector, and many home and appliance buyers now list energy star rating as one of the leading factors that determine choice from options. Consumers need to be aware of the potential cost savings that could come from energy conservation and should be sensitized on how to access the potential gains. Already, packaging of energy-efficient light bulbs and appliances carry details of potential energy and cost savings. Also, energy marketing companies in some developed nations offer detailed analysis of day-to-day energy use of each customer including use for heating, cooling, lighting, etc., which serves as a useful domestic energy management tool. Many also offer lower unit pricing and bonuses for low energy consumption, and strong incentives for using more energy in low-peak periods such as nights and weekends.

7.3.1.6.   *End-use efficiency improvement: Heat generation*

Heat is the largest energy end-use. Providing heating for homes, industry and other applications accounts for around half of total global final energy consumption (larger than 27% for transportation). Also, around 65% of this demand comes from sources fueled by coal, fuel oil, gas, solar, geothermal or traditional biomass. The share of renewable energy (mainly biomass) has been low, less than 10% in 2018 and growth over the next few years will be insignificant. Fossil-fueled equipment and less-efficient conventional electric heating technologies still dominate the sub-sector, accounting for around 80% of new sales in 2019. Decarbonizing heat production is an important mitigation option, particularly in the building sector, but remains a challenge. However, sales of heat pumps and renewable heating equipment such as solar hot water systems have increased, representing more than 10% of overall sales in 2019. To be in line with the Sustainable Development Scenario, the share of clean heating technologies - heat pumps, district heating, renewable and hydrogen-based heating - needs to more than double to 50% of sales by 2030 (IEA-2020). Much of the expected progress in decarbonizing heat energy over the next decade or so will be in China and the European Union where modern biomass-fueled heat generation plant deployment is growing. Other potential sources are renewable electricity, solar, and geothermal energy. Some combined systems which produce both heat and electricity, thereby improving operating efficiencies drastically, are also being deployed. It is estimated that energy efficiency and switching to clean final energy sources (including decarbonized electricity and district energy that produces both heat and electricity) could cut fossil fuel consumption for heating and cooling in half by 2060 compared with today.

7.3.1.7.   *End-use efficiency improvement: the digital revolution*

The information and communications sector (ICT) has literally exploded in the past decade or so: Internet connectivity is growing at an exponential rate. There were around 4 billion Internet users in 2017, over half of the global population, and Internet traffic could double in the next few years (Statistica, 2018; Internet World Stats, 2018). About 5 billion connected devices were in use in households worldwide and may double by 2020. Deployment of network-enabled devices in offices and homes has increased exponentially - broadband connectivity, wireless mobility, cloud computing, tele and video conferencing/medicine, e-commerce, social media, video-enabled devices, smart devices, smart systems and sensors, cell phones, personal computers, servers, laptops, game consoles, printers, etc. Many home appliances - TVs, microwave machines, washing machines, home security surveillance systems - now have Internet connectivity. There were, around 17 billion networked devices in 2016, expected to rise to about 27 billion by 2021 (3.5 devices per capita) (CISCO, 2018). Projections indicate that network-enabled devices will be around 500 billion by 2030 (WEC, 2017). Edge devices - routers, routing switches, integrated access devices, multiplexers, metropolitan area networks (MAN), wide area networks (WAN) - all of which provide authenticated access to enterprise or provider core networks are now indispensable in industry, commerce and homes, and the world is experiencing an exponential growth in the deployment of digital technology in all other sectors of the economy. Adoptions of smart control systems and meters are leading to significant performance and efficiency gains in the power industry. Around half a billion smart meters which can be monitored and controlled remotely are helping to track and control electricity use in real time. Virtually every industrial process has advanced smart control systems which help optimize production and energy use.

There has been a surge in energy demand to power ICT systems, accounting for more than 10% of total final global energy consumption in 2016. The energy required to meet the projected growth in network-enabled devices is expected to double in the next decade. Most of the energy (up to 80%) is consumed when devices are in standby modes (ready and waiting but not performing the main function). The aggregate electronic devices in offices and homes: modems, routers, TVs, PCs, servers, set-top boxes, games consoles, printers, security equipment, etc., constitute more than 40% of ICT electricity demand, and standby energy accounts for 1 to 2% of global electricity consumption and approximately 10% of residential electricity use. There is a significant scope for reducing this wastage and associated emissions, and equipment manufacturers are doing a lot to achieve this goal. For example, a recent European legislation limits energy consumed by an appliance to 1 watt when in standby mode compared with older equipment which consume 12-15 watts: this needs to spread to other countries. Furthermore, energy savings will become substantial as old appliances are replaced because new models are generally more efficient and many feature innovations that switch off idle equipment.

7.3.2.   New technology innovations

Technology innovations have played a key role in improving efficiencies across the whole spectrum of energy production and use, notably in the development of new energy-efficient equipment, appliances and processes. For example, the average car today would cover around twice the distance per liter of fuel compared with the average car twenty years ago despite the fact that they are much heavier. Many new technologies targeting efficiency improvements in other areas of transportation, in power generation, manufacturing and building are emerging all the time and many more are incubating, including such technologies as the capture and use of anthropogenic carbon dioxide. The main goal of innovation is to develop energy-efficient and low carbon-intensive technologies that are self-sustaining and can be proliferated irrespective of negative factors such as oil price volatility and shifting policy instruments. The impacts of technology innovations vary significantly across sectors of the global economy. While sectors such as the light vehicle industry have benefited tremendously, sectors with the least innovation progress toward energy decarbonization are those where policy incentives are weak and long-term perspectives are lacking, notably the heavy industry, freight transportation and aviation. One major issue that may blur the potentially significant impact of technology innovation on environmental mitigation is the fact that, historically, low carbon technology has been driven mostly by the industrial economies whereas most of the growth in energy use and emissions will come from the emerging economies which are increasingly hosting the energy-intensive production processes (primary metals, cement, chemicals, petrochemicals, fertilizers, etc.) and which have different energy equity/security priorities. Furthermore, up to three-quarters of vehicles operating in many emerging countries are imports of cars and trucks that fail to meet emission standards in developed countries.

7.4.   DECARBONIZING ENERGY

One of the main thrusts of global effort to mitigate energy-related environmental pollution is to decouple energy demand and greenhouse gas emissions, and the power sector offers the best opportunities. Global warming potentials of emissions are determined by the carbon intensity of a process which in turn is determined by process efficiency, carbon intensity of the energy source and carbon footprint of raw materials. There are three key ways of decarbonizing energy and they are discussed below.

7.4.1.   Energy production efficiency improvement

As discussed earlier, the processes of producing primary energy and conversion to end-use energy are the least efficient in the supply and use chain, accounting for around 40% of total primary energy consumption. Potential ways of improving efficiencies in mining, transportation, conversion and distribution have been discussed in the last section. Improved process efficiencies in all the sub-sectors translate to lower carbon intensities of the various end-use energy sources.

7.4.2.   Change of energy mix

The carbon intensities of primary energy sources vary very widely. Lignite, the youngest coal emits about 1050 metric tons of CO_2eq per gigawatt-hour of electricity produced. Comparative figures for other primary energy sources are hard coal, 880; oil, 740; natural gas, 500; solar PV, 85; biomass, 45; nuclear, 30; hydroelectric, 26; wind, 26. Clearly, increasing the share of renewables in the world's primary energy mix has a huge potential in decarbonizing energy. However, various issues that have so far limited more aggressive deployment have been discussed in Chapter 4. A combination of efficiency improvement and switch to lower carbon energy can also rapidly decarbonize power generation and the energy end-use sectors: industry, transportation, building. It is widely believed that efficiency improvement and radical increase in renewable energy use particularly for power and heat production can flatten and reverse energy-related emissions within a decade, leading the world to a carbon-neutral environment within two decades. Much of the efficiency improvement can be achieved by adopting simple technologies that are already available (best available technologies, BAT), such as retrofitting a coal-fired power plant with more efficient burners, or enriching combustion air with oxygen. Improved efficiencies in the end-use sector and switching to lower carbon sources of primary energy (natural gas, renewables, nuclear) drive down energy intensity and the carbon footprint. However, one of the main challenges is the promotion of widespread deployment of a wide range of technologies which are already available and many new technologies that are at different stages of development and will emerge in the near future. Transformational change in the energy sector can be a very long process due largely to '*status quo*' mentality. This explains why many power plants all over the world are still operating at 30-35% efficiency when simple retrofitting technologies that could lift efficiencies to 45% and above, save substantial operating costs and reduce emissions have been available for decades. Power plants have a life span of 30-40 years (many power plants worldwide are close) and, while many undergo retrofits over their life span to improve operating efficiencies, most of the investments in recent times have been in new power plants, featuring new advanced technologies that not only improve efficiencies, but also facilitate compliance with increasingly stringent worldwide environmental control regulations. Strong environmental mitigation policy instruments have been the main drivers of change in the power industry. The International Energy Agency estimates that stronger deployment of technologies that are familiar and available at commercial scale today can reduce significantly future emissions from the energy sector (IEA, 2017d). Coal is the backbone of power generation in many countries and has been responsible for more than 40% of global energy-related emissions growth since 2000: around 6 Giga metric tons (half of the total emissions from the power sector today) come from inefficient, low-technology coal power plants. Upgrading from sub-critical operation to super-critical and ultra-super-critical boiler technologies which have been available commercially for more than a decade could decarbonize fossil electricity substantially. Also, investments in renewable energy power generation technologies (in particular, wind and solar which have only 2-3% of the carbon footprint of a typical coal power plant) which are readily available commercially and are becoming increasingly competitive could further mitigate pollution from the power sector. Industry, transport and buildings end-use sectors also offer substantial energy efficiency improvement opportunities which translate to lower carbon emissions, through the infusion of new technologies and fuel switching

7.4.3. Carbon capture, use and sequestration (CCS/CCUS)

Atmospheric pollution has become a major issue because human activities are altering the natural carbon cycle by increasing the concentration of carbon dioxide in the atmosphere. One key way to return the world to carbon neutrality is to suck as much as possible of the excess carbon dioxide from the atmosphere and either use it or store it in a benign way. This can be achieved by capturing as much of the gas as possible, for example during coal, oil and gas production, during the conversion processes, and in end-use processes that generate significant anthropogenic gases. Other options include benign storage, carbon trading, and enhancement of the natural carbon cycle.

7.4.3.1. *Carbon capture/use/storage*

Carbon capture involves trapping emissions at source for use or storage in a form that poses minimal threat to the environment. Technologies are already available and many are under development for the capture and safe storage of carbon dioxide in major industrial operations, in particular, power generation. Use for captured carbon (CCU) is currently limited, mainly to oil and gas production enhancement and many industrial processes - fire extinguishers, carbonated drinks, welding, pneumatic pressured gas systems, refrigeration systems, chemicals production, water treatment processes, metals casting, etc. Carbon capture and sequestration (CCS) involves capture and storage in suitable locations or conversion to benign form. It is currently one of the available technologies that can significantly reduce GHG emissions, and a key technology option to decarbonize the power sector, especially in countries with a high share of fossil fuels in electricity production. Carbon capture, use and storage (CCUS) is an emerging technology for capturing and storing as much of energy-related emissions as possible at source by capturing flue emissions from fossil fuel powered generation and other industrial processes. Technologies are already available and many more are under development for the capture and safe storage of carbon dioxide and particulates produced in major industrial operations, in particular, petroleum production, power generation, iron and steel, cement and chemicals/petrochemicals production. Carbon capture technologies either separate CO_2 from waste gases before or after combustion (pre- and post-combustion capture). In pre-combustion capture, coal is gasified to produce synthetic gas (syngas), a mixture of hydrogen and carbon dioxide, further reformed with steam to produce almost pure hydrogen which is then used for driving a turbine to produce electricity. The process not only reduces CO_2 production but also offers an opportunity to capture the little amount in the reformed syngas at high efficiency because the gas is at high pressure, thereby virtually eliminating pollution from coal-fired power plants. Gasified coal has comparable or even better carbon footprint credentials than natural gas because it contains neither carbon dioxide nor methane which are the predominant compounds in natural gas. Large-scale coal gasification installations are operating or under construction in many parts of the emerging world primarily for electricity generation, or for production of chemical feedstocks. The hydrogen obtained from coal gasification can be used for various purposes such as making ammonia, powering a hydrogen economy, or upgrading fossil fuels. Coal-derived syngas is being used to power fuel cells and can also be converted into transportation fuels such as gasoline, diesel or methanol fuel/additive through additional treatment. Also, syngas can be liquefied for use as a fuel in the transport sector. Post-capture technology processes flue gases from coal or natural gas combustion to recover CO_2, and NO_x gases, but the process is less efficient compared with pre-capture technology. A third option (oxy-fuel combustion technology) is designed to minimize production of carbon dioxide in fossil fuel combustion by using oxygen-enriched air or recycled flue gases instead of the conventional air for combustion.

Over twenty large-scale CCS projects are in operation or under construction in different parts of the world, with a combined capacity to capture up to 40 million metric tons of CO_2eq per year. These projects cover a range of industries, including gas processing, power, fertilizer, steel-making, hydrogen production, and chemicals. Most carbon capture demonstration projects are in North America and the world's first large-scale adoption of carbon capture technology in the power sector was commissioned in Canada in 2014. Several demonstration plants are operating in other countries and some commercial projects are under construction, particularly in the United States. Other countries - Japan, Australia, South Korea, others in the Middle East - are also actively considering CCUS projects. Many CCUS projects are integrated with oil and gas production because the captured gas is either used to pressurize oil wells to enhance production, or stored permanently in exhausted wells. About 30 million metric tons of GHG gases are captured annually, with ExxonMobil accounting for around 40%. However, the International Energy Agency (IEA, 2019g) estimates that this needs to rise to around 24 billion metric tons per year (800% scale-up), accounting for 7% of the cumulative emissions reductions needed globally by 2040 for positioning the world on the path to a sustainable environment. Another promising option involves sucking out carbon dioxide from the atmosphere and storing underground or converting to useful products.

The viability of CCUS technology is not in doubt, especially as a very promising method of significantly reducing energy-related GHG emissions, and a key technology option to decarbonize the power sector and industry. The potential is particularly high in countries with a high share of fossil fuels in electricity production. However the total global capture capacity in 2019 was 40 million metric tons, equivalent to only 0.1% of the global emissions in 2019. There are formidable obstacles to widespread deployment: retrofitting existing plants or including carbon capture in new plant design could drive up capital and operating costs by up to 30%, and many emerging nations that will host most of the expected new fossil-fueled power plants cannot absorb the added costs. Another major constraint is what to do with the captured gas, since current options are very limited. Apart from limited use, finding suitable storage sites, preferably near the source is a major problem, while transportation over long distances for safe disposal (in disused mines and other suitable underground sites) is too expensive and often impractical for wide adoption. Furthermore, technologies for capturing carbon dioxide from combustion effluents and from the air and conversion into useful products or benign state are at early stages of development. Progress in resolving these issues will largely determine the reductions in anthropogenic emissions that can be achieved over the next 2-3 decades through carbon capture. Currently over 40 carbon capture projects are operating around the world, under construction or consideration but they are mostly tied to petroleum production which uses the gas to pressurize wells and enhance oil production.

Many research and development projects are also ongoing, targeting the enhancement of the natural carbon cycle by promoting the use of the excess anthropogenic carbon dioxide in the atmosphere by plants to produce carbohydrates and proteins, using the Sun's energy, thereby moving the world towards carbon neutrality. The Intergovernmental Panel on Climate Change (IPCC) has determined that achieving the ambitions of the Paris Agreement to limit future temperature increases to 1.5 degrees will require more than just an acceleration of efforts to reduce emissions; it may also require the deployment of technologies to actually remove carbon from the atmosphere. There are many options under development but the most mature carbon dioxide removal technology is bio-energy with CCS, or BECCS. The technology involves enhancement of the natural carbon cycle by growing biomass which extracts carbon dioxide from the atmosphere as it grows, harvesting and processing (with carbon capture) to chemical products or other forms of energy. Possible applications of BECCS include: dedicated or co-firing of biomass in power plant; combined heat and power; cement plants, pulp and paper mills; lime kilns; ethanol plants; biogas refineries; and biomass gasification plants. Certain biomass conversion processes, including fermentation and gasification technologies generate high-purity carbon dioxide streams as an intrinsic part of the process, thus providing lower-cost capture opportunities. The IPCC found that in pathways with limited or no temperature overshoot, up to 400 Gigatonnes of BECCS could be required this century. Currently, there is one large-scale BECCS facility in operation, at the Illinois Industrial Carbon Capture and Storage facility in the United States, capturing and storing one million metric tons of carbon dioxide per year, a very long way from the IPCC target.

7.4.3.2. *Carbon trading*

Carbon trading (emissions trading, cap-and-trade) is a market-based tool designed to limit GHG emissions mainly from the industrial and commercial sectors of the economy. The system is designed to penalize high emitters and compensate low polluters. A limit (cap) is set for allowable emissions and any excess incurs a tax while emissions that are lower than the limit attract credit. Those that are over the limit can buy credit from low-polluting enterprises to offset the excess emissions. Trading gives companies a strong incentive to save money by cutting emissions in the most cost-effective ways. Caps are divided into units, typically metric tons and distributed to companies freely or by auction. Companies that cut their pollution faster can sell allowances to companies that pollute more, or save them for future use. Caps are determined by the scheme's governing body which may be mandatory or voluntary and revised upwards or downwards regularly to moderate supply and demand forces which primarily control pricing. Member companies that do not have enough allowances to cover their emissions must either make reductions or buy another company's spare credits. The system has been very effective in controlling GHG emissions in the European Union and is spreading to other regions. In the EU system, a cap is set on the total amount of certain greenhouse gases that can be emitted by installations covered by the system, and is reduced over time so that total emissions fall. Within the cap, companies receive or buy emission allowances which they can trade with one another as needed. They can also buy limited amounts of international credits from emission-saving projects around the world. After each year a company must surrender enough allowances to cover all its emissions, otherwise heavy fines are imposed. If a company reduces its emissions, it can keep the spare allowances to cover its future needs or else sell them to another company that is short of allowances.

The International Energy Agency (IEA) has set a cap on emissions and a timetable for achievement if the 2°C or below emissions target is to be achieved. Working with the World Bank and the International Monetary Fund, targets have been set for specific industries. The iron and steel industry which accounts for 7-9% of all direct fossil fuel emissions and also the world's largest industrial source of energy-related emissions has been set a target of 65% reduction standard by 2050. However a recent evaluation report shows that most companies in the industry are far behind target. Many cleaner iron and steel technologies are already available (direct reduction, carbon capture, natural gas injection, heat and materials recycling) or under development but most new iron production technologies and innovations are too limited in output compared with the traditional blast furnace process which produces around three quarters of the annual global output, or too expensive to deploy because they would raise production costs by 20% to 30%. While a few companies are embracing innovations, have set targets that achieve carbon neutrality by 2045 and appear to be on course, most others either have no targets of are well behind in developing and adopting low-carbon technologies. However increasing carbon prices particularly in Europe are forcing companies to comply or slash production and suffer major financial losses. Unfortunately, relatively few countries mostly in Europe and East Asia are enforcing carbon pricing, while the major pollution emitters - China, United States, Russia - lag behind because of lax policy instruments.

7.4.3.3.   *Reforestation*

Aforestation and reforestation also have significant potential for removing carbon dioxide from the atmosphere. Land is both a source and a sink of natural and anthropogenic greenhouse gases (GHGs) and plays a key role in the exchange of energy, water and aerosols between the land surface and the atmosphere. Changes in land cover (deforestation, aforestation, reforestation) directly affect the Earth's surface temperature by altering the natural balances of moisture and heat exchange with the atmosphere. Land exposure through human activity will alter the natural albedo and cause the global land mass to absorb more of the Sun's energy than it radiates, resulting in global warming, draught and desertification. These will increase the vulnerability of land ecosystems, biodiversity and society to climate change and weather and climate extremes. Although it is difficult to separate natural and anthropogenic fluxes, it is estimated that net carbon dioxide emissions of 5-7 Gigatonnes per year come from land use and land use change (IPCC, 2019). In effect, degraded and desertified land areas warm the world and affect the climate, while climate change exacerbates land degradation, particularly in low-lying coastal areas, river deltas, drylands and in permafrost areas, with severe consequences for life in the areas. Sustainable land management can contribute significantly to reducing the negative effects of these multiple stressors It is estimated that human activities have destroyed about half of the world's population of trees that existed before the rise of human civilization. Planting billions of trees across the world in land areas that were previously degraded (reforestation), sparsely vegetated and not used for agriculture (aforestation) is perhaps the most effective and cheapest way of reducing the concentration of carbon dioxide in the atmosphere. Trees are very effective at taking carbon dioxide out of the atmosphere through photosynthesis, in some places offsetting human emissions of carbon dioxide by 30 percent. Carbon dioxide is a primary input for plant photosynthesis and a precursor to plant carbohydrates and proteins which feed both humans and animals.

About 15 billion trees are cut down yearly across the world to make room for agriculture, construction of infrastructure (buildings, roads, dams, etc), and for use by industry. Reforestation not only provides a natural sink for carbon dioxide, it results in cooling due to enhanced evapotranspiration which in turn can result in cooler days particularly in the tropical regions, and a reduction in frequency, intensity and duration of heat-related events such as heat waves. It is estimated that a worldwide planting program could remove two-thirds of all the emissions from human activities without encroaching on crop land or urban areas (land and oceans already absorb about 55% of CO_2 produced by human activity). A recent study (Bastin *et al.*, 2019) estimates that about 11% of the global land area could host around 1.2 trillion native trees, which means increasing the estimated global population of trees by about 40%, and the tropical areas are particularly suitable. This proposal may sound ambitious but presents an essential and achievable vision and a potentially important blueprint for country and global public policies as well as private sector initiatives. Although forest restorations could take up to a hundred years to bring about the full effect of removing around 200 billion metric tons of carbon dioxide from the atmosphere (bioenergy crops and shrubs take only months), future generations can look forward to a much less hostile environment and the devastating effects of pollution-induced climate change. The right trees planted in the right locations could capture around 200 Gigatonnes of carbon dioxide, equivalent to about two-thirds, of all anthropogenic carbon dioxide burden since the Industrial Revolution. Tree planting initiatives already exist, notably the Bonn Challenge, a German initiative endorsed and extended by the New York Declaration on Forests at the 2014 UN Climate Summit. It has become a global effort to bring 150 million hectares of the world's deforested and degraded land into restoration by 2020, and 350 million hectares by 2030. It should be noted however that as trees mature, the capacity to sequester and use carbon dioxide from the atmosphere declines, hence sustainable forest management strategies which include harvesting and replanting economic trees will be vital.

Natural peat bogs (ponds and marshlands filled with rotted plant matter) are regarded as natural high-carbon ecosystems estimated to hold over 40% of all stored carbon, more than the carbon stored in all other vegetation types, including the world's forests. While mature vegetation will eventually reach saturation and cease to absorb carbon dioxide, peat bogs can continue to sequester for centuries. However, peat bogs are currently being drained and cleared for agriculture and peat has many uses as fuel, in oil spill control, etc. Environmentalists have advocated that global carbon-curbing plans should include conservation of natural pit bogs. It should be noted however that peat bogs are potential precursors to coal formation although the process takes around 300 million years. There are many other carbon sequestration technologies under development, notably the use of carbon dioxide to fertilize ocean algae which are valuable bioenergy feedstocks, machines that pull out carbon dioxide from the atmosphere for sequestration or use, conversion of carbon dioxide to benign natural carbonate ores, photo-electrochemical production of synthetic natural gas (syngas) from water and carbon dioxide using solar energy, or reacting carbon dioxide with hydrogen under very high pressures to produce methyl alcohol and many other fuels.

7.4.4. Waste and energy recycling

The world generates about 3 million metric tons of waste daily, and the rate is projected to double by 2025. Nearly half is organic waste and about a third is made up of metals, plastics and paper (Figure 7.5). Around 80% is used for landfills and the balance is recycled, incinerated to generate electricity and heat, or used for composting. The United States, Japan, China, and many countries in Europe have extensive waste-to-energy (WtE) programs and most claim substantial reductions in anthropogenic emissions. In 2014, the United States recycled and composted about a third of the 258 million metric tons of municipal solid waste (MSW) generated, resulting in a reduction of over 181 million metric tons of carbon dioxide equivalent emissions, comparable to the annual emissions of nearly 40 million passenger cars. Over 33 million metric tons were combusted with energy recovery (which reduced fossil fuel use), while 133 million metric tons were landfilled. The United Kingdom Waste and Resources Action Program (WRAP) claims a more modest but still substantial reduction of 10-15 million metric tons a year of emissions, amounting to nearly 50 million metric tons between 2010 and 2015 (WRAP, 2010; 2018).

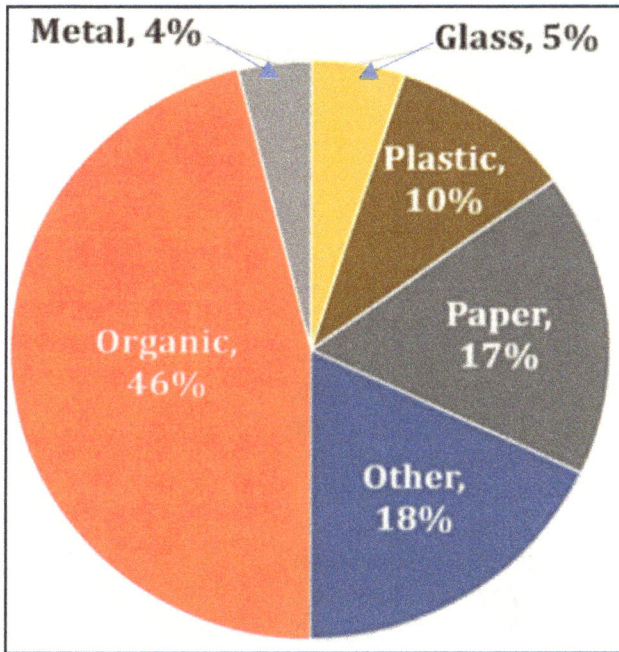

Figure 7.5 Global waste generation by type *(WEC, 2016b)*

Recycling is the process of converting discarded waste to useful products, sometimes different from the original products, with significantly less energy than would have been required for making the same products from virgin raw materials. There is also the additional advantage of conserving natural raw materials and the associated production and processing emissions. Furthermore, recycling slows down the need for exploiting low-grade ores, a more energy-intensive process that releases higher emissions. Recycling requires significantly less energy per kilogram of material produced than primary production, and also decreases the negative impact of mining, processing and transportation of ores. The process offers higher environmental benefits and lower environmental impacts compared with other methods of waste disposal. Different products offer different opportunities and gains: metals are inherently recyclable and the same product can be manufactured from recycled material, with significant gains. For example, producing aluminium soft drink cans from recycled material saves around 95% of the energy required to make the same product from virgin bauxite ore and saves four metric tons of bauxite ore and the associated mining and processing emissions per metric ton of recycled aluminium. Every metric ton of recycled aluminium reduces CO_2eq emissions by about 14 metric tons compared with production from virgin ore (EPA, 2011). Recycling one metric ton of paper saves about 17 trees, and also saves one metric ton of CO_2eq of emissions. Many other metals and materials - iron, steel, copper, lead, silver, paper, glass - are recyclable, with less dramatic but very significant energy savings and GHG emissions reduction when recycled feedstock is used (Table 7.1).

Table 7.1 Environmental effects of recycling *(BIR, 2008; UCO, 2014; EPA, 2015).*

Material	Energy savings	Air pollution savings
Tin	98.9%	99%
Lead	98.7%	99%
Aluminium	95%	92%
Copper	63%	65%
Steel	56%	58%
Glass	5-30%	20%
Paper	50-70%	73%
Plastics (reuse)	70%	50%
Plastics (recycle)	70%	30%

Polymers are products of fossil fuel-sourced petrochemicals and, since invention in the 1950s, have taken over virtually every aspect of human life, from packaging to clothing. One of the most attractive properties is that they are unreactive, which makes them suitable for food and chemicals storage. Unfortunately, this property also makes it difficult to dispose of polymer waste. Furthermore, there are many different types of plastics made by different chemical processes, consequently plastics offer less but still significant opportunities for recycling compared with metals and paper, primarily because of the diversity of types with very divergent thermo-chemical characteristics. Most plastics cannot be recycled to produce the original product, but can be reprocessed into other useful products, with significant savings in energy and reduction in anthropogenic emissions.

Waste-to-energy (WtE) or energy-from-waste (EfW) is the process of generating energy in the form of electricity or heat from the primary treatment of waste either by direct combustion, or by conversion to biofuels such as methane, methanol, hydrogen or other synthetic fuels. Modern incinerators can convert heat to electricity at 14-28% efficiencies, or up to 80% if configured as part of co-generation which combines electricity generation with space heating. Also, strict emission regulations have forced manufacturers of incinerators to produce plants with low emissions of nitrogen and sulphur oxides, as well as particulates. Apart from the fact that conversion of waste to useful energy makes economic sense, the more important advantage is the avoidance of uncontrolled incineration or dumping of waste, particularly plastic waste which releases hazardous gaseous/particulate products into the atmosphere when incinerated, or fails to degrade for many years when used for landfills. When exposed to ultraviolet rays of the Sun, plastics can degrade and break up into fine particles, causing new environmental issues. Landfill is the most popular disposal system worldwide, but decomposition of the organic matter releases significant amounts of carbon dioxide and methane into the environment and some plastics may not degrade for decades.

Waste-to-energy technologies can contribute to global climate mitigation by significantly reducing emissions from other methods of waste disposal. Many developed countries are adopting incineration to produce power and heat, a system which cuts CO_2eq emissions by more than 50% compared with landfill systems. New, more efficient waste gasification technologies are emerging, and adoption could cut emissions by two-thirds. Also, using the energy generated by WtE plants will reduce the demand for energy from fossil fuels and eliminate the associated emissions. A modern WtE plant can produce carbon emission savings in the range 100 to 350 kg CO_2eq per metric ton of waste processed, depending on waste composition and amount of heat and electricity produced. Even greater savings up to 800 kg CO_2eq per metric ton of waste can be achieved if WtE completely replaces landfilling (WEC, 2016b). Emissions (gas-phase and particulate) from WtE plants which feature incineration are comparable to coal-fired power plants, but those which adopt gasification produce significantly lower emissions. In any case, strict environmental regulations in many developed countries have made it mandatory for both coal and incineration plants to adopt a series of process units for cleaning flue gas and disposal of post-combustion residue in an efficient and environmentally friendly manner. Waste recycling requires significantly less energy per kilogram of material produced than primary production, and also decreases the negative impact of mining, processing and transportation of ores. Recycling slows down the need for exploiting low-grade ores, a more energy-intensive process that releases higher emissions. Fossil fuels, mostly natural gas and coal are the precursors to plastics production as well as the main sources of energy for production. Considering the exponential increase of plastic use in the last five decades and projections of even faster rate in the coming decades, their share of global fossil fuel and primary energy use will continue to increase in the foreseeable future and disposal of plastic waste will remain a major global problem. Enormous energy is wasted in many energy conversion and industrial production processes, mainly through the discharge of hot liquid and gaseous waste. It is estimated that around three-quarters of all the energy produced is wasted as discarded heat (waste process heat; heat losses in all combustion systems, power transmission lines, industrial equipment, domestic appliances, etc.). Improvement of process efficiencies reduce waste heat because waste heat recycling is one of the primary options. Much of the heat can be recovered and recycled for preheating feed materials thereby reducing significantly the additional heat input. The heat can also be used to generate electricity or heat buildings. Many modern industrial processes now feature waste heat capture and reuse systems: power generation, primary metals production, etc. Many large industries, in particular primary metals and materials industries recover and recycle substantial heat which is used to fuel captive power plants and provide heat and power to neighbouring communities, or feed power to grids. Recycled waste heat is emerging as a valuable renewable energy resource and an effective way of reducing carbon intensity of processes

One major constraint to recycling is the development of policy-driven appropriate strategies for collection, sorting, pretreatment, etc. Motivating communities and individuals to support recycling is also an issue. Many developed countries have systems for collection, mostly managed at the municipal level. However, most of the collected waste is either used as landfill or exported to emerging Asian countries where they create new environmental issues because they are usually sorted manually to retrieve useful components or incinerated, often in the backyards of small enterprises in Asia, thus releasing potentially dangerous aerosols into the atmosphere and contributing to ambient pollution. This is happening because policy instruments in the developed countries which promote local processing are weak, enabling local contractors to export their collections without sorting, thereby transferring an environmental problem from one region to another.

All products have environmental impacts, from the extraction, upgrading and transportation of raw materials to manufacture, distribution, use and disposal, and there are enormous potentials for mitigating environmental pollution through recycling. Raw materials and lifecycle energy associated with the manufacture of new products are conserved, with significant reductions in emissions. However, while municipal governments need to play a leading role in protecting the environment from indiscriminate disposal of waste, there is also considerable scope for the people who are the main consumers of industrial products and also the main sources of recyclable waste. A people-targeted waste management model has been evolving and is already well embedded in some countries, mostly in the EU-28: *refuse, reduce, reuse, recycle (R⁴)* (Figure 7.6). For example, they can refuse to use disposable plastic shopping bag by buying reusable shopping bag; they can limit the number used on each shopping trip; they can reuse the bags in many ways; they can deposit the bags in the municipal recycling system. Adoption of this model by many more municipalities around the world could conserve materials and energy, reduce emissions, and mitigate environmental damage caused by waste dumping.

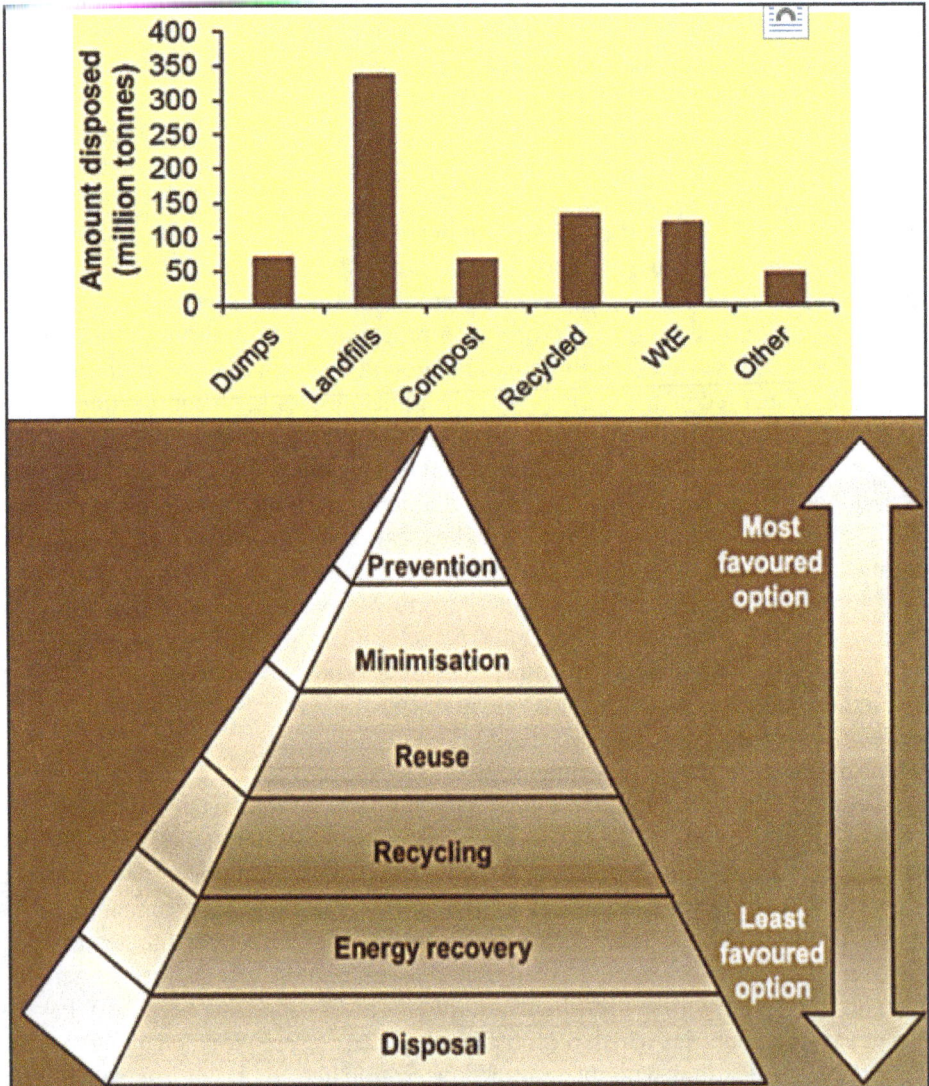

Figure 7.6 Global waste generation by disposal technique *(WEC, 2016b)*

Chapter 8

Mitigating climate change: power of We the People

8.1.   INTRODUCTION

There is little doubt that the Earth's average temperature has been increasing steadily in the last hundred years or so, with potentially severe negative consequences. What is not so clear is the extent to which all the observed changes can be attributed to human activities. However, there is strong scientific evidence that nature's control systems in terms of regulating the energy balance of the Earth and its environment, are being compromised significantly by anthropogenic emissions from various human activities. It is inevitable that climate change will make tornadoes and heat waves much more likely and much more severe, energized by rising global temperature and warming oceans. It is noteworthy that Europe-28 which is the most successful region in energy decarbonization and environmental pollution reduction has also been one of the most hard-hit regions by severe weather in recent years, reinforcing the fact that climate change knows no national or regional boundaries and the negative effects will resonate across all regions of the world irrespective of the source. The debate is ongoing on who is liable for degrading the environment and what should be the consequences, a development which is discussed briefly in the next section.

8.2.   LIABILITY FOR CLIMATE CHANGE

Global debate is strengthening on who should be held responsible for degrading the environment and what should be the price to be paid, and the focus is on the global energy companies, with lawsuits beginning to emerge. The first issue is that no science has proved conclusively that human activities are entirely responsible for climate change. As discussed earlier, there are many natural phenomena and processes that have been causing climate change and extreme weather on and off well before human civilization such as slight changes in the Earth's orbit of rotation and variabilities, are likely to continue even without the intervention of human-generated emissions. Therefore it would be difficult to apportion blames between mankind and nature. There is however no doubt that human activities especially in the last century or so have been making the consequences of climate change more intensive and more widespread. The energy sector is perhaps the world's largest employer of labour; crude and refined oil exports dominate global trade and are the lifelines of the economies of many countries across all regions of the world. It is difficult to imagine that any country could contemplate a reduction in primary energy output, in particular, coal, natural gas and oil because of potential damage to the environment. Furthermore, the ever increasing demand for modern infrastructure and consumer products across all regions of the world will continue to drive the demand for even more energy. However, most countries can enact potentially powerful and resilient policies that regulate the excesses of both producers and users of energy.

While it is true that energy production and use account for two-thirds of anthropogenic pollutions significant emissions and pollutions also come from other human activities, especially agriculture and land use. There is little doubt that energy companies have prioritized profit at the expense of environmental sustainability over time and need to be held accountable through more compensative and punitive tax policies, but the entire world is in fact liable: the agricultural sector which accounts for around 15% of pollutions is virtually off the radar; consumers who are the ultimate beneficiaries of all the products of energy and agricultural technologies are either unaware that their activities, choices, preferences, and excesses propel energy demand growth and account for most of the pollutions from energy use, or are unprepared to review and modify their choices in an environmentally sustainable manner. Not many are willing to modify their lifestyles to optimize energy use (moderation and optimization of home and transportation energy use, discretional use of consumer goods, support for recycling, etc.). Livestock production and rice cultivation are the main sources of pollution from agriculture, mainly methane which is one of the most potent of the greenhouse gases but not many people are prepared to give up meat or reduce consumption for the sake of the environment. Soil fertilization is the other main source of pollution (nitrogen oxides) from agriculture but more than half of the world population would starve without it. Perhaps the best way to make the world pay for environmental degradation and moderate their lifestyle choices is to upgrade the current carbon taxes into *attribution taxes* which should cover energy and agricultural production and spread to every consumer good and agricultural product.

8.3 ADAPTATION AND MITIGATION STRATEGIES

It is clear that human activities that damage the environment will not reduce due to benevolence or voluntary action: it would require powerful political and social pressure in order to make any significant impact on current global efforts to reposition the world on the pathway to a sustainable environment. The issues of anthropogenic environmental pollution can be viewed from two different perspectives which require different mitigation efforts: *local*, and *global* pollution. Most pollutions (gaseous, aerosol, solid waste) are local, emanating from power generation, transportation, industrial, commercial and domestic energy use, land use, refuse management, and the negative effects are also local initially. Gaseous and aerosol pollution reside in the first layer of the atmosphere (troposphere) around 25 kilometers above the Earth's surface just for a few weeks before they either drop down or rise higher to the second atmospheric layer that controls climate (the stratosphere). Even in this short period of residence in the troposphere they can cause substantial changes in the local weather (cloud systems, precipitation, etc.). They cause smudge which obscures view; they cause or exacerbate human respiratory ailments and other diseases that are responsible for an estimated 750 million deaths worldwide annually (WHO, 2018). Although national policies can help control major sources like transportation and power generation, local policies are the most effective.

Depending on wind dynamics, some of the gaseous and fine particulate contents of pollution may rise to the next layer of the atmosphere (stratosphere) where they mix with the natural greenhouse gases that control temperature on earth thereby increasing their concentration, heat trapping capabilities and the amount of heat reflected back to earth, causing global warming. This is the layer that also hosts natural ozone which protects the world from the dangerous cancer-causing rays of the Sun. Some of the anthropogenic gases especially the fluorocarbons which emanate from refrigeration and air conditioning react with the ozone thereby reducing concentration and causing 'holes' in the protective layer through which some of the damaging rays can reach the Earth's surface. The greenhouse gases (both natural and human-sourced) are well-mixed and fairly evenly dispersed across all regions of the world, hence the negative impact of global warming is gradual and global, irrespective of the source of pollution. Also, it may take decades, centuries or even millennia for the world to experience the consequences.

It is clear from the discussion above that, even if anthropogenic emissions were to stop today, the negative effects of past emissions will continue to impact the world for a long time. Perhaps the most current and urgent issue is how to manage the immediate consequences in real time and find ways of reducing human-related environmental pollution going forward. All available projections show that fossil fuels will remain dominant in the global primary energy mix for decades to come, in response to the vibrant global economy, growing population and prosperity, increasing demands of human development needs, and the prime role of fossil fuels in the global economy. The most urgent reality is that mankind must seek to manage energy and land use in an environmentally sustainable manner by developing coping mechanisms for dealing with the undesirable impacts (that are playing out in real time already) on survival and health of people and the ecosystems, and fast-tracking movement towards a carbon-neutral world. Two options that have gained prominence in recent times are *adaptation* and *mitigation*. Adaptation and mitigation are complementary strategies for managing environmental pollution: while adaptation focuses on minimizing the negative consequences of climate change on people, mitigation seeks to develop and promote ways of moderating human actions that cause climate change.

8.4.   ADAPTATION TO CLIMATE CHANGE

Adaptation options are largely localized and within the jurisdiction of local and state governments, hence the impact is more immediate compared with most mitigation options that depend on both intra- and international actions, and take much longer time before benefits become evident. It is difficult to quantify the impact of climate change (natural or anthropogenic) on human and ecosystems for many reasons: the nature of hazards, exposure and vulnerability can vary widely across regions and socio-economic stratifications; and there are different non-climatic stressors that can influence vulnerability, exposure and ability to adapt to changing situations. Adaptation measures are designed to minimize potential consequences of phenomena such as extreme weather and ambient pollution on life. Potential adaptation strategies include improving the design of structures to withstand extreme weather, multi-level response strategies to disasters, provision of early warning systems, emergency evacuation plans, provision of storm and heat shelters, mobilization of first responders, use of respiratory masks, etc. Others include effective plans for emergency medical and food supplies and building of awareness on what people can do to reduce personal carbon footprints, such as embracing and supporting recycling, energy conservation in transportation and across the home front, and a drastic change in individual lifestyle. However, the world is largely unprepared to cope with the potential consequences of climate change. For example, the intense heat waves of recent years particularly in Europe and North America, believed to have shattered records and caused many deaths have stimulated action on coping strategies. The level of preparedness in Europe was low considering that less 5% of homes in Europe which has historically had a temperate climate are air-conditioned, compared with over 90% in the United States, and relatively few public transportation systems have air conditioning. Many potentially vulnerable utilities were severely disrupted: expanded rail lines; disruption of control systems; system collapse due to unusual spikes in electricity demand and lack of backup strategies; traffic gridlocks due to signal failures; power outage; inadequate first response systems, etc. Current measures include critical reviews of emergency response measures for public transportation and utilities, creation of public cool rooms, extension of public swimming pool hours, strengthening of local and national emergency response strategies, closure of schools, activation of help lines, and public awareness campaign.

Adaptation policies need to be targeted to be effective: for example, urban areas hold more than half of the world's population and most of its built-up assets and economic activities, hence most adaptation efforts usually target urban communities. A high proportion of greenhouse gas and aerosol emissions is also generated by urban-based activities and residents and mitigation efforts often focus on these areas. However, design of urban coping strategies needs to take account of the social stratification: while the wealthy often can take care of themselves, the majority of urban dwellers are usually low to middle income people many of who depend a lot on public welfare and utility systems. Rural areas have different vulnerability to disasters, for example, obstructions that could slow down tornados are relatively few and the impact of touchdowns can be very severe. Also, while many of the more prosperous urban dwellers can afford personal transportation which facilitates quick evacuation from impending weather disaster areas, and air conditioners in the event of heat waves, there are few escape or coping options for poor rural people. Furthermore, most disaster response systems are based in urban areas and help can be slow in arriving in case of disasters. Rapid urbanization and migration have peaked in many regions, and poverty/extreme poverty rates in rural areas are falling, with the exception of sub-Saharan Africa where rates are rising. Many in the region live on subsistence agriculture, and access to fertile land and adequate rainfall are crucial. Most of these people are already subjected to many other non-climatic stressors and any disruption of their livelihood, for example, as a result of draught can be devastating. However, many rural communities devise their local adaptation strategies since there is often very limited institutional support.

8.5.   MITIGATION OF CLIMATE CHANGE

The international Energy Organization (IEA) and many other independent organizations including the Intergovernmental Panel on Climate Change (IPCC) have projected that the rise in carbon emissions needs to stop by 2030, followed by sustained decline in order to place the world on a path towards the goal of limiting global temperature rise to 2°C or below by the end of the century. This would require the decoupling of energy use from carbon emissions and a major effort to slow down energy demand and emissions in spite of the rising trends of the global economy and population, the main drivers of energy demand. This goal was adopted at the Paris-2015 conference on the environment coordinated by the United Nations in which countries submitted plans for action. However, all indications suggest that current and proposed mitigation efforts are grossly inadequate and much greater action than currently in place or planned would be required to avoid the potentially disastrous consequences of climate change. Furthermore, the commitment of many nations to even the inadequate planned country actions is very much in doubt. The International Energy Agency (IEA, 2019f; 2019i) has proposed three potential scenarios for energy use and related emissions over the next two decades: Emerging Transition Scenario (ETS), Faster Transition Scenario (FTS), and Even Faster Transition Scenario (EFTS). Several other organizations have also developed different scenarios which generally fall under these three categories. The ETS projection (also known as New Policies Scenario, NPS or 'business as usual') already takes into account all existing, expected and anticipated technologies and policy instruments. It assumes that government policies, technology and social preferences continue to evolve in a manner and speed consistent with the recent past. Projections by IEA and several other international organizations show that, under this scenario, emissions will not peak and start to reduce by 2040. A faster transition scenario (FTS) which demands stronger policy instruments and greater adoption of innovative technologies that promote efficiencies could peak emissions but there will be no significant reduction from current level by 2040. The EFTS postulates a scenario which could peak emissions within the next decade and reduce the total by about 45% by 2040 relative to current levels. This would place the world on a viable path towards the Paris-2015 target. Achievement would require a wide range of improvements and changes: more rapid efficiency improvement across the entire spectrum of energy production and use; decarbonization of energy through much higher renewable energy share of power generation and carbon capture, use and storage (CCUS); stronger market penetration of electric vehicles; and much more energized societal mitigation through awareness campaign and lifestyle choices that reduce personal carbon footprints.

The Even Faster Transition/Sustainable Development Scenario (EFTS/ SDS) which is needed to meet energy and climate goals, outlines an integrated approach to achieving internationally agreed objectives on climate change, air quality and universal access to modern energy. The projected emissions in the New Policies Scenario will be about 36 Gt in 2040 and need to drop to around 18 Gt in the Sustainable Development Scenario. The EFTS/SDS model could limit the global temperature rise to below 2°C, possibly as low as 1.5°C. The Agency estimates that the EFTS/(SDS) model would lead to efficiency gains that reduce energy-related emissions by about 42% by 2040 while energy decarbonization (fuel switching to renewables, natural gas, nuclear; heat and materials recycling; and carbon capture) will account for about 51% (IEA, 2019j). Potential pathways to achieving this goal are shown in Figures 8.1 to 8.3.

In a recent study on energy outlook by BP (2018), three different scenarios were considered: Evolving Transition (ET), based on current and expected policies and trends in previous years; Faster Transition (FT); and Even Faster Transition, energized by much stronger technology and policy support (EFT). The last two scenarios reflect possible implications of different judgements and assumptions. The key conclusion of the study in respect of CO_2eq emissions is that, in the ET scenario emissions will continue to grow through to 2040, increasing by about 10% above current levels. This is inconsistent with the sharp decline that is necessary to fully decouple energy use and emissions from the global economy by 2040 in order to meet the Paris-2015 climate goals which in any case have been found to be inadequate. The EFT scenario projects a further reduction in emissions by around 50%, consistent with the International Energy Agency's 'Sustainable Development Scenario', also known as IEA-450 Scenario. None of the scenarios published by any organization so far takes full account of all possible uncertainties, and all are cautious about placing high probability of happening as projected in view of the many uncertainties surrounding energy demand forecasts (see Figures 8.4 to 8.6). In summary current and anticipated policies will not achieve the Paris-2015 goal which requires halving current emission levels by 2040 (the IEA-450 Scenario). Much stronger and effective policies that would drive down emissions, significantly improve efficiencies across all sub-sectors of energy production and use, and promote much faster decarbonization of power generation appear to be the surest pathways to a healthy, sustainable global environment.

The International Renewable Energy Agency (IRENA, 2018) has shown that a profound and radical transformation of the power sector beyond current plans and policies would be crucial, one that enhances efficiency and fast-tracks decarbonization of energy. Cumulative emissions must be reduced by around 500 Gigatons (equivalent to around 15 years of annual global emissions at the current rate) by 2050 to meet the sustainable environment target. In order to achieve this goal, electricity which is expected to increase its share in total energy end-use from 20% currently to 50% by 2050 must be heavily decarbonized preferably by fast-tracking the share of renewable energy from around 25% currently to around 86% of power generation (the most optimistic of recent projections predict 30-35%) and its share of primary energy consumption from the current 15% to around 65%. This could reduce emissions in the energy sector by as much as 60%. However, this is a highly ambitious proposition largely because most of the growth in energy demand and associated emissions over the next three decades or so will come from emerging regions that depend primarily on fossil fuels especially coal for power generation. Potential pathways to these ambitious targets have been discussed in Chapter 7 and are summarized in the following section.

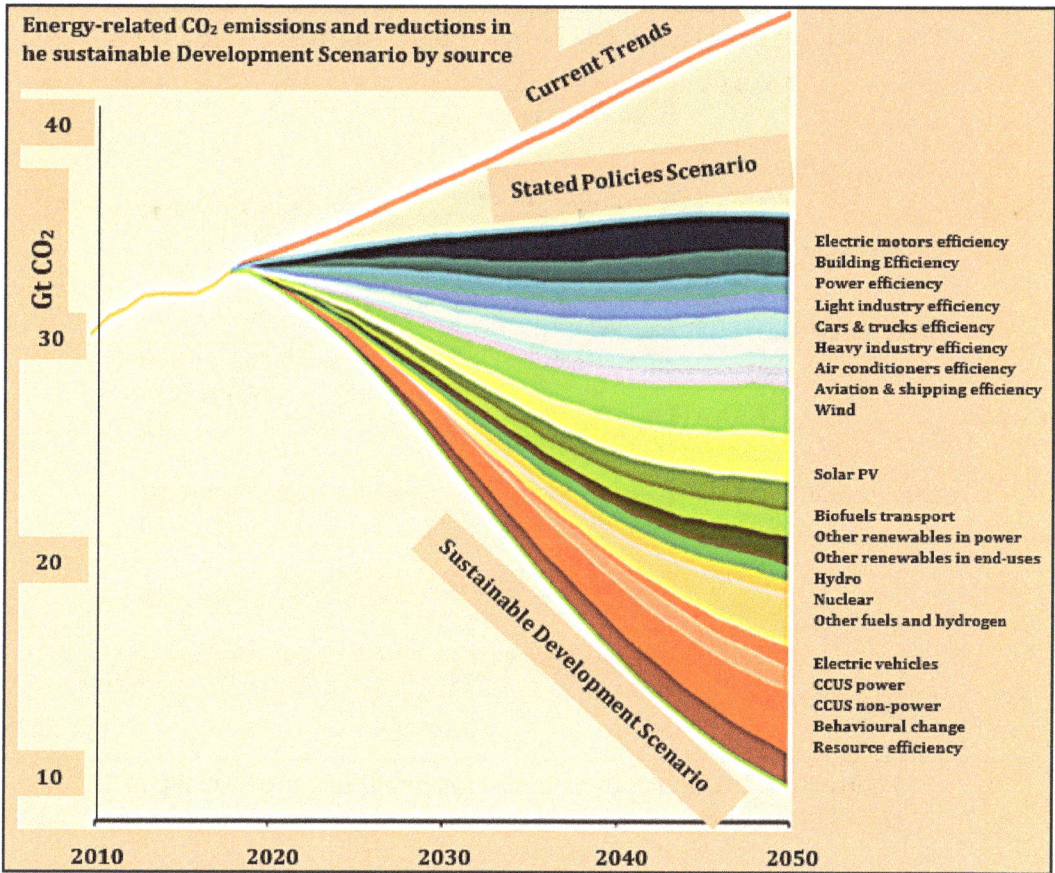

Figure 8.1 International Energy Agency Sustainable Development Scenario
(IEA, Energy Outlook, 2019g).

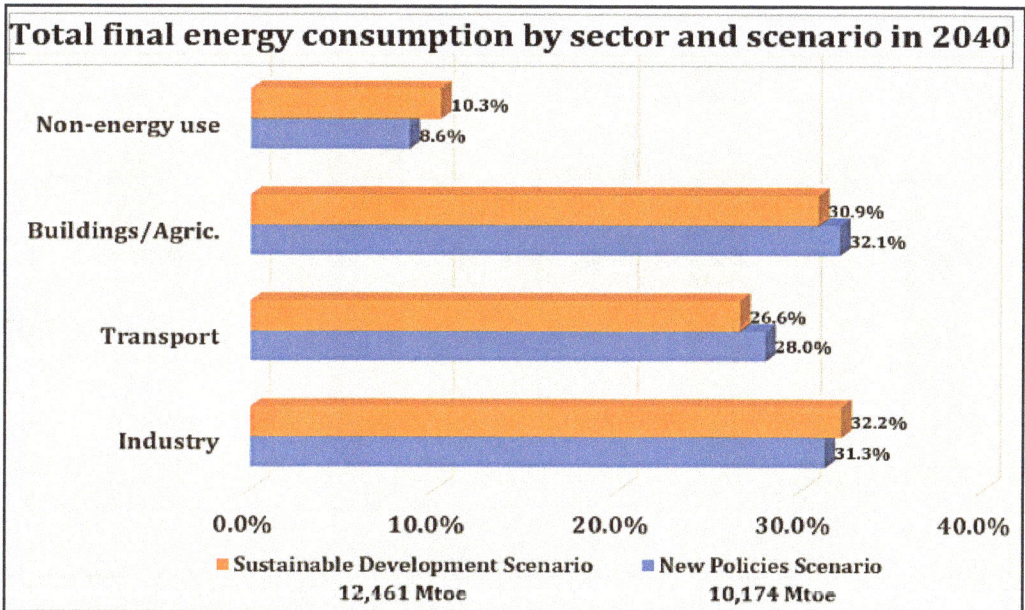

Figure 8.2 Outlook on total final energy consumption by scenario
and sector *(Data from IEA, 2017g).*

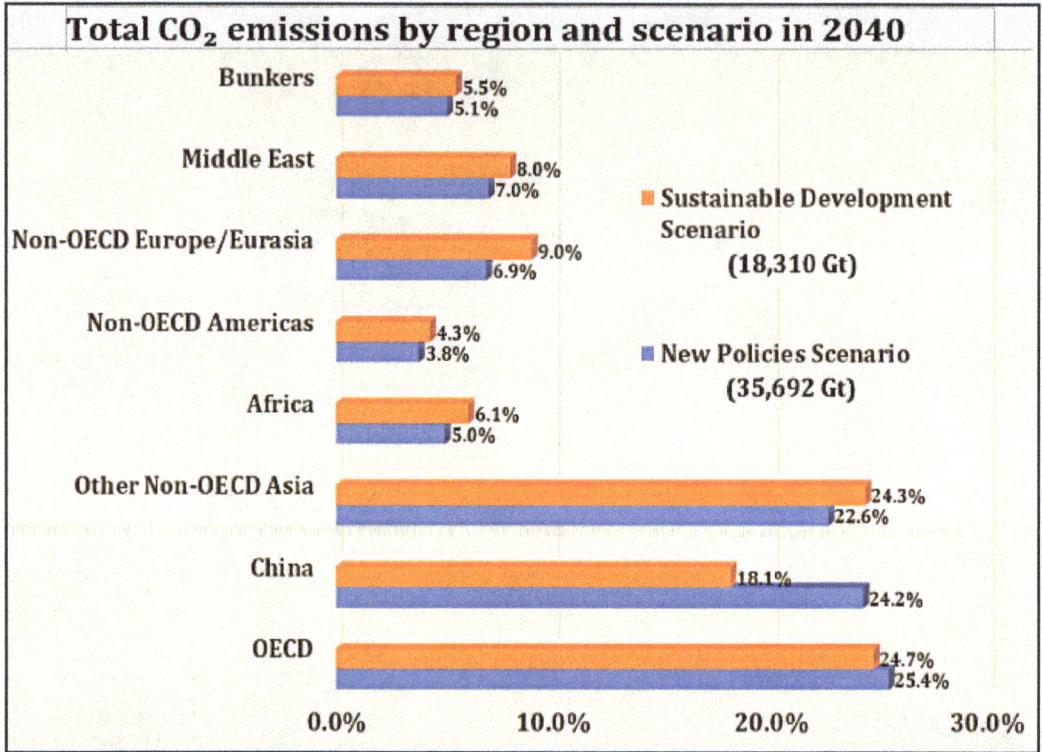

Figure 8.3 Two scenarios on global total final energy consumption and CO_2 emissions outlook to 2040   *(Data from IEA, 2017a).*

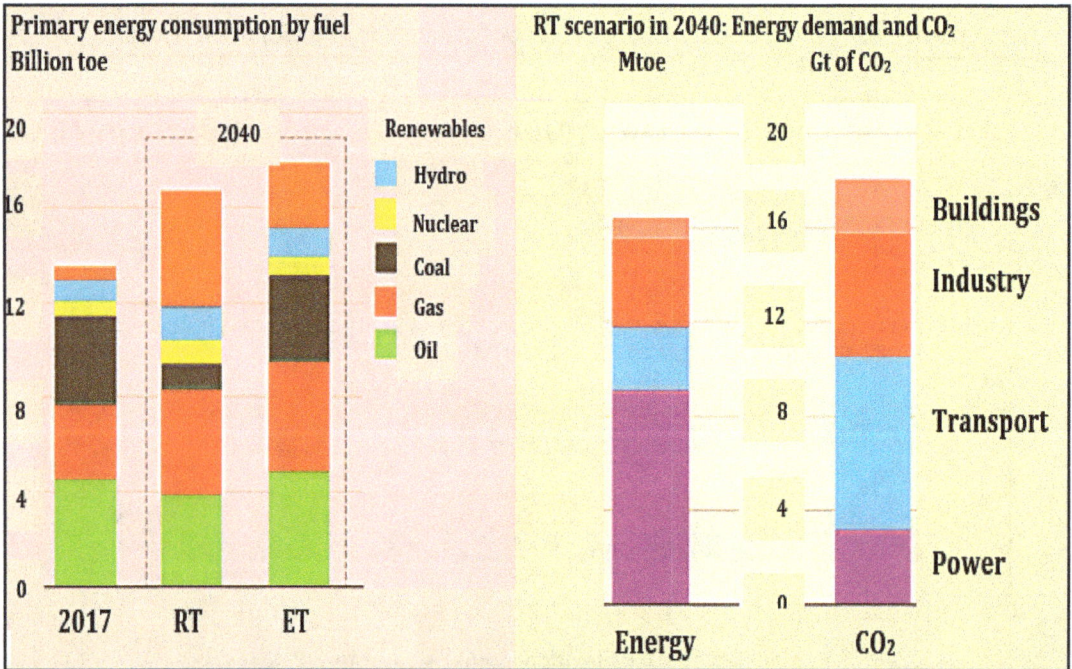

Figure 8.4 Outlook on primary energy consumption by fuel and economic sector under different scenarios. (ET and RT are Emerging and Rapid Transition resp.) *(BP, 2019a).*

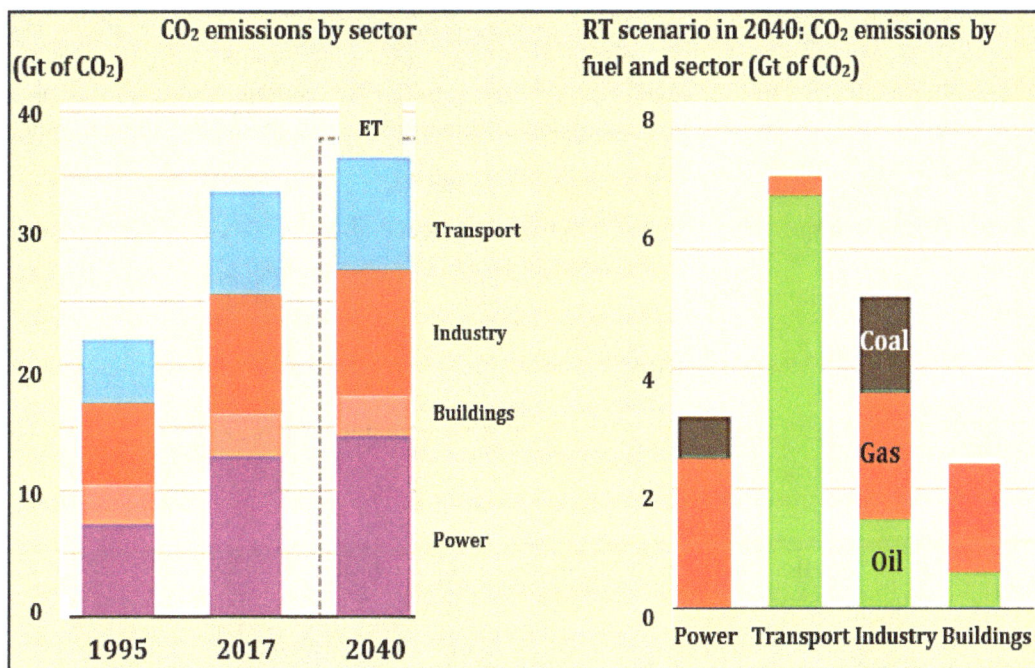

Figure 8.5 Outlook on CO_2 emissions by fuel and economic sector under Emerging Transition (ET) and Rapid Transition (RT) scenarios. *(BP, 2019a).*

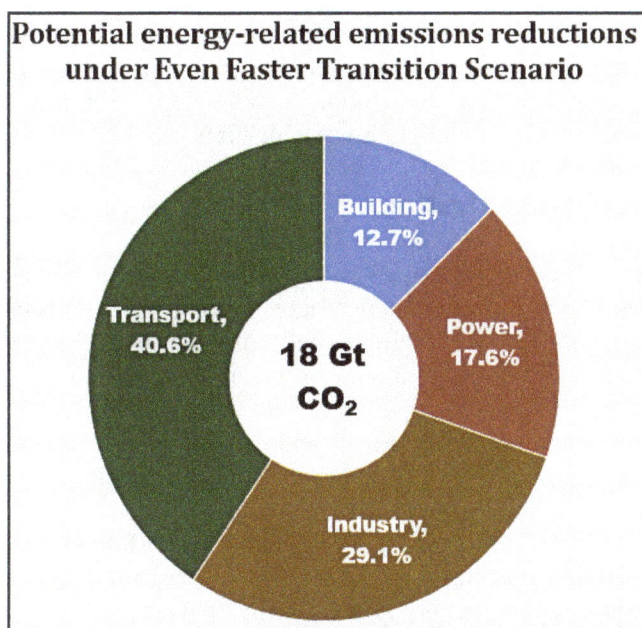

Figure 8.6 Outlook on CO_2 emissions by economic sector under Even Faster Transition Scenario that is compatible with sustainable development. *(BP, 2019b).*

Mitigation involves development of aggressive and effective policy and technology strategies for reducing the growth rate of environmental pollution and, while there are many options, a detailed analysis of the potential benefits of many mitigation options shows that four options will be the key drivers of future energy-related anthropogenic emissions' reduction across all regions of the world:

- Efficiency improvement in energy production and end-use.

- Decarbonization of energy.

- Promotion of waste-to-energy recycling.

- Strong and resilient policy instruments that survive political cycles.

- Mobilization of societal action.

The first three options have been discussed in some depth in the last chapter and the two others are discussed below.

8.5.1.   Strong and sustained policy instruments that survive political cycles

Society tends to ignore or resist changes that target the status quo and some compelling force is always necessary to stimulate compliance. This is the primary driver of the advertisement industry which largely employs motivational strategies. However, statutory regulations and strong policy instruments that employ both motivational and enforcement strategies are required at three different levels of political governance to achieve any significant results in the global quest to mitigate environment pollution: municipal, national and international. The effectiveness of environmental mitigation actions depends critically on resilient policy instruments, (both supportive and punitive) which transcend political cycles.

8.5.2.   Municipal mitigation

Municipal governments can effectively institute and enforce regulations that help pollution abatement, in particular, on activities that have direct impact on their areas of governance. Some of the regulations may be national but many are designed to address specific local issues, notably transportation pollution control and refuse management. Imposition of punitive taxes on or outright prohibition of polluting vehicles is already having a significant impact on efforts to clean up urban environments and regulations on waste disposal and recycling of reusable items such as plastics, metal products, paper and electronic waste, plastic waste, compostable household waste, are becoming increasingly common although actions need to be expedited and more widespread. It is estimated that a trillion single-use plastic bags are disposed of worldwide annually (around 2 million every minute). More than 480 billion of plastic bottles (water, soft drinks, etc.) were used in 2016 across the world, projected to double by 2040 (Euromonitor International, 2017). Fewer than half of the bottles were collected for recycling and only 7% of those collected were turned into new bottles, the balance being discarded or used in landfill. Most of the discarded bottles fragment and end up in rivers and oceans, causing serious environmental and ecological problems. It is estimated that between

5 million and 13 million metric tons of discarded plastic have ended up in the world's oceans, ingested by fish and other aquatic life, and some of it is already showing up in the human food chain. Dead whales and other sea animals with up to ten kilograms of ingested plastic have been found on shores in different parts of the world. Apart from being potentially toxic, ingested plastics cannot be usefully assimilated into the food chain. Another problem with plastics is the fact that they do not degrade easily and, although exposure to strong ultraviolet rays from the Sun can embrittle some types, causing them to break up and possibly end up in the atmosphere as aerosol, or in the oceans, other types can remain intact for several hundred years. Many municipalities especially in Europe and the USA have achieved phenomenal success in restricting the use of one-cycle shopping bags simply by introducing a small tax. A recent policy by the United Kingdom to impose just 5 pence on plastic shopping bags provided to customers by a few major retailers has had a dramatic effect, cutting demand by 85% in just six months, and the few large supermarkets that were involved in the pilot project reported very substantial cost savings. Only 0.5 billion (less than 100 million a month) of plastic bags were purchased by customers in the first six months since the introduction of tax, compared with more than 7 billion (more than 600 million a month) in the preceding year when it was free (Defra, 2018). Most shoppers now carry reuseable shopping bags. The policy is being extended to more polybag outsources; refundable deposits are being introduced for bottled drinks and cans; and drinking straws are increasingly prohibited. Many other countries in most regions of the world have reported similar dramatic gains and others particularly in Europe and North America also have in place or are introducing policies that help stimulate the R^4 culture: *refuse, reduce, reuse, recycle* (see Chapter 7). However, this culture needs to spread across the world much more rapidly, particularly in China and North America, the two largest sources of plastic pollution.

8.5.3. National mitigation

Mitigation of energy-related emissions requires strong national policies in addition to local efforts, particularly in relation to enforcing emissions standards in power generation, industry and transport. Introduction of stringent power generation and vehicle emission standards and support policies for electric vehicles are all having significant effects in cutting down emissions, forcing polluting vehicles off the roads and inefficient power plants to upgrade or close down. Other policy interventions promote renewable power generation, in particular, solar and wind, and establish carbon taxes whereby the government sets a price that emitters must pay for each metric ton of greenhouse gas emissions they emit beyond a set limit. Carbon taxes include an emissions tax which is based on the quantity an entity produces and a tax on goods or services that are generally greenhouse gas-intensive such as a carbon tax on fossil fuel-sourced energy. This system which has been discussed in the last chapter drives businesses and consumers to take steps, such as switching fuels, adopting new technologies to reduce their emissions and avoid paying the tax, or trading carbon.

Carbon trading is an effective policy instrument and is becoming the instrument of choice for controlling greenhouse gas emissions. The scheme is one of the prime and very effective GHG emissions control strategies used by The European Union which leads the world in the development of strong energy-friendly policy instruments. All 28 member states are obliged to adopt emission laws and regulations irrespective of the internal political dynamics and harsh and enforceable penalties are in place for contravention. The United States also has strong national policies on the environment but enforcement powers are limited due to the largely independent status of states, most of who have their own policies based on local exigencies. For example, it is difficult to enforce the stringent national emission codes in states that depend heavily on coal for power generation and employment, although economic forces are at play, forcing coal power plants to convert to cheaper natural gas or close down and driving coal mines to bankruptcy. China, one of the largest sources of energy-related pollution also has strong mitigation policies but the pressures of power demand from the very large primary industry and growing middle class population have resulted in many policy reversals. Many coal plants have been closed, but many more higher-efficiency plants are under construction. India, another significant polluter has strong renewable energy policy but coal remains the primary energy source for its growing primary industry and teaming population who want access to modern energy.

As discussed earlier, forests are important parts of the natural carbon cycle and act as effective carbon sinks (biosequestration). It is estimated that 4 billion hectares of forest ecosystems (about 30% of the global land area) store large reservoirs of carbon, together holding more than double the amount of carbon in the atmosphere (FAO, 2005). Forests also minimize disruption of the natural albedo (heat exchange between land and the atmosphere), and act as barriers that slow down potentially damaging winds. Although the climate protection role of forests is in no doubt, it is complex to determine how much of the forest carbon sink and reservoir can be managed to mitigate atmospheric CO_2 buildup, and in what way. It is estimated however that each intact hectare of the world's tropical forests across Africa, Amazonia and Asia traps around 0.6 metric tons of carbon a year, which adds up to about 5 billion metric tons, equivalent to about a sixth of the total global energy-related emissions in 2018. Many countries, municipalities and non-governmental organizations across all regions of the world have strong reforestation programs, not only for environmental reasons but also to restore economically valuable trees and regain valuable land devastated by excessive deforestation. Reduction of current global deforestation rates by 50% by 2050 would be a major contribution to a sustainable environment.

8.5.4.   International mitigation

Many international organizations are playing key roles in promoting global discourse on climate change and energy-related emissions. The United Nations (UN), International Energy Agency (IEA), World Energy Council (WEC), Intergovernmental Panel on Climate Change (IPCC), Energy Information Administration (EIA), ExxonMobil, BP, World Health Organization (WHO), The World Bank, United Nations Environmental Protection Agency (UNEP), European Union (EU), and many more have been providing extensive information and data which have become vital resources for country and international actions. The World Health Organization has brought into full global focus the consequences of ambient outdoor and household pollution, which have a more immediate and severe impact on the health of life on earth than climate change. The 1992 United Nations' conference in Rio de Janeiro was perhaps the first major global initiative to address the problem of environmental pollution. The meeting developed an International Environmental Treaty (United Framework Convention on Climate Change (UNFCCC) which has become a major stimulant and guideline for country action. The thrust of the Treaty was the stabilization of greenhouse gas concentrations in the atmosphere at a level that would prevent dangerous anthropogenic interference with the natural climate system. The parties to UNFCCC treaty have met annually since 1995 as Conference of the Parties (COP) to assess progress in dealing with climate change. The meeting in Kyoto in 1997 developed a Kyoto Protocol which established legally binding obligations for developed countries to reduce their greenhouse gas emissions. Six greenhouse gases were identified which, if reduced could drive down global warming significantly and reverse ozone depletion. From 2005 the conferences have also served as the Meetings of Parties of the Kyoto Protocol (CMP). The latest meeting was held in Paris in November 2015, attended by 196 parties and the Paris Agreement, a global treaty on the reduction anthropogenic emissions was adopted. Signatories were obliged to establish National Greenhouse Gas Inventories and develop strategies for control and removal by 2020. The Paris-2015 Protocol set a goal of limiting the increase of the world's average temperature to no more than 1.5-2 degrees centigrade above pre-industrial levels by the end of the twenty first century.

8.5.5.   Societal mitigation (Lifetime choices)

Society is the ultimate consumer of energy and its products. It is difficult to imagine any aspect of human life and development that is not intimately dependent on energy mostly supplied by fossil fuels: prosperity, health, education, communication, transportation, etc. Also, there is hardly any product that does not have a fossil fuel footprint in terms energy use and precursor materials, notably primary metals and materials that are indispensable for civil, mechanical electrical, and structural applications. Other areas of the global economy which depend critically on fossil fuels especially as precursors are information technology, communication, transportation equipment (airplanes, railways, automobiles), appliances, clothing, household goods, sporting goods, drugs, fertilizers. Unfortunately, prosperity has been promoting excess consumption on all fronts, spiking energy demand and enhancing emissions. Gradual moderation of the 'consumption syndrome' could make a significant impact on global anthropogenic pollution.

Carbon footprint calculation has become a powerful tool for assessing the impact of human (including personal) behavior on global warming. Every consumer good or fuel use has a carbon footprint: the foods and goods that humans buy everyday, bottled drinks, polymer shopping bags, household energy and water use, transport energy, recreational facilities, appliances, consumer goods, drugs all have carbon footprints which must be accounted for in calculating personal contribution. For example, ten liters of petrol or diesel burnt in a car, or of oil used in home heating contribute 23-27 kg of carbon dioxide to the atmosphere; one transatlantic flight a year contributes nearly two metric tons of emissions per person; the manufacture of three empty one-liter plastic bottles of water/soft drink or twenty plastic shopping bags releases about 1 kg of CO_2.eq into the atmosphere. Also, significant emissions are associated with the end-use disposal of most consumer goods. In effect, reducing personal carbon footprint can reduce people's contribution to global anthropogenic emissions significantly. The world average of personal carbon footprint was about 5 metric tons/person/year of CO_2.eq in 2018, but there were wide variations between countries. For example, the value for Qatar was 38.2, Canada 16.1, Australia 16.8 USA 16.1, China 8.0, UK 5.6, Uganda 0.1 (Crippa *et al*., 2019; Wikipedia, 2019). With over 7 billion people in the world today, (projected to rise to nearly ten billion over the next twenty to thirty years), a reduction of annual personal contribution to emissions by just 20% could have reduced the world total by around 7 Gigatonnes of CO_2.eq, equivalent to over one-fifth of global energy-related emissions in 2018.

Potential areas of personal carbon footprint reduction are many, and both the environment and personal finances benefit: rationalizing energy use across the entire spectrum of personal activities such as opting for energy-efficient homes and appliances when buying or leasing homes; upgrading the energy efficiency of existing homes through improvement of insulation; replacement of inefficient appliances and lighting fixtures with energy-smart systems; switching off idle appliances, opting for energy-efficient internal combustion or electric vehicles; planning journeys such as shopping to eliminate unnecessary trips; pooling vehicles; using public transport when feasible; moderating consumption of goods and energy by responding to needs rather than wants and impulse, etc. On line shopping which is becoming increasingly popular has the potential to cut personal carbon footprint significantly since just one delivery van can eliminate the need for several hundred individual trips to shops, saving significant transportation energy and associated emissions, and use of thousands of plastic shopping bags. Furthermore, the traditional market place/mall models will be replaced gradually by much less energy-intensive warehouse-online model. Voluntary adoption of reusable bag shopping and support for recycling will eliminate the energy required to manufacture several hundred disposable bags and plastic bottles (from fossil fuels) per person per year and the associated carbon emissions, thereby further reducing personal carbon footprint. Also the natural gas and coal that would have been precursors in the manufacture of the products would be left in the ground.

8.6.   OUTLOOK ON PARIS-2015 TARGET

The focus of the Paris 2015 Agreement was the decision of 196 countries to work together to limit global temperature rise to well below 2 degrees Celsius above pre-industrial levels, and to strive for 1.5 degrees Celsius, in order to mitigate the risks and impacts of climate change. Many institutions and organizations have tried to define feasible pathways to achieving this goal, now known as 2DS (2-Degree Scenario) or 450S (450 Scenario which targets 1.5 degrees Celsius, limiting the concentration of CO_2eq to 450 parts per million (ppm) in 2040, compared with 403 ppm in 2016 which was 40% higher than in the mid-1800s. A recent analysis by the International Energy Agency (IEA) concluded that current trajectory of policies and actions fall far short of what is needed to achieve this goal in a long time, and much stronger and aggressive global policies are needed to reduce currently projected emissions of around 37 GT in 2040 by around 50% (see Figures 8.1. to 8.3). The Intergovernmental Panel on Climate Change (IPCC), the leading international body for the assessment of climate change recently reviewed progress on country commitments (NDCs) at the Paris-2015 meeting and concluded that virtually all countries fall far short of goals (IPCC 2018), (Figure 8.7).

Global greenhouse gas emissions show no signs of peaking; CO_2 emissions from energy and industry increased in 2017 following a three-year period of stabilization; and total global GHG emissions, including land-use change reached a record high of 53.5 $GtCO_2eq$. Emissions need to decline by approximately 25% and 55% over the next ten to twelve years in order to put the world on a least-cost pathway to limiting global warming to 2°C and 1.5°C respectively. Without this drastic reduction in emissions, global warming which has increased by about 1°C (average) above pre-industrial levels will likely reach 1.5°C between 2030 and 2052, with high risks for natural and human systems, the magnitude of which will depend on geographical location, levels of development and vulnerability, and on the choices and implementation of adaptation and mitigation options. The International Energy Agency (IEA) has proposed much wider deployment of innovation technologies, many of which are already available or in the pipeline, and stronger policy support to meet global climate ambitions. Aggressive infusion of new technologies is needed at both the supply and demand sides of the energy system: for example, the power sector would need to be heavily decarbonized by reducing its share of energy-related emissions (which stood at about 41% in 2018), to no more than around 7%, and with carbon capture by 2040. Industry which contributed 25% emissions in 2018 will need to reduce levels by at least 20%; the electric vehicle market penetration will need to grow to around 15 million a year compared with the current level of around 2 million, leading to 90% of all cars on the road being electric by 2060; and most of the electricity that will be required to power the electric cars should come from lower-carbon, non-fossil renewable sources. Furthermore, the building sector which contributed 8% (24% when emissions from electricity and heat are reallocated on the basis of sectoral use) will account for around 40% of the total global electric power demand in 2040, hence its contribution to energy decarbonization will largely depend on the source of the electricity and the extent to which the use of traditional biomass is reduced. Achievement of these targets will depend critically on strong and resilient local and international policy instruments.

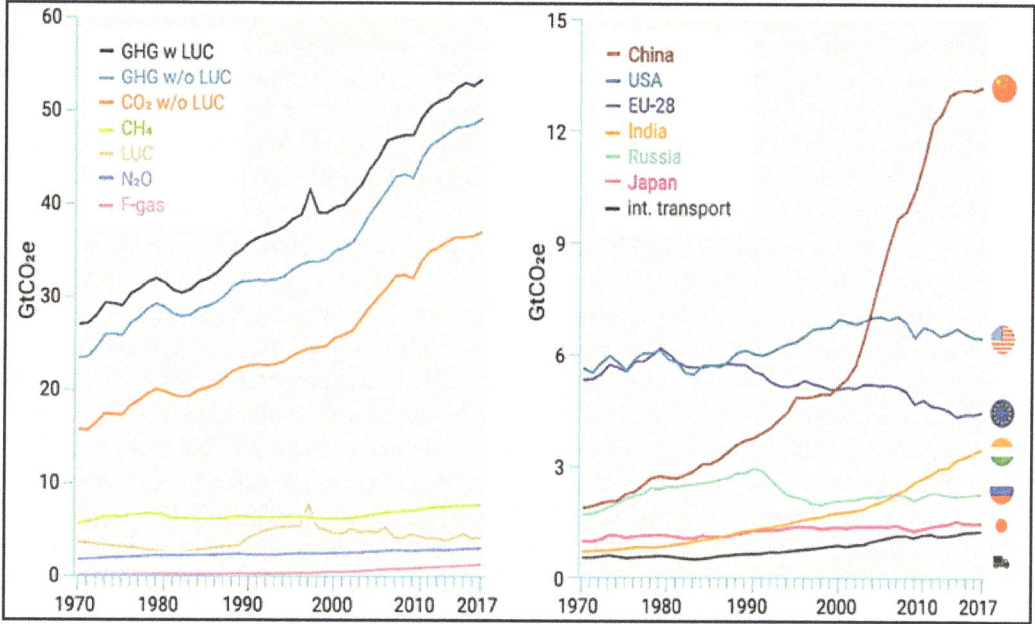

Figure 8.7 Global greenhouse gas emission trend and levels for major emitters per type of gas. (LUC = Land-use change) *(IPCC, 2018).*

One of the major problems with energy use and emissions outlook projections is the amount of uncertainty about major variables. Variations in global oil market are particularly disruptive: when prices are low (as they are currently), they tend to promote energy use and depress incentives for investments in efficiency upgrade or lower carbon technologies. This is evident from the latest report on global energy and CO_2 status report (IEA, 2019d): contrary to projections, global energy consumption in 2018 increased at nearly twice the average rate of growth since 2010 driven by a strong global economy and extreme weather in some parts of the world which increased the demand for heating and cooling energy. As a result, improvements in global energy efficiency and energy intensity have slowed down dramatically. Demand for all fuels increased, led by natural gas, and higher electricity demand accounted for over half of the growth in energy needs. This unanticipated rise in global primary use increased emissions by around 2% over the 2017 level. Although deployment of renewables grew at double digit pace, fossil fuels still met 70% of the growth and accounted for about 81% of the global primary energy use in 2018. Asia accounted for most of the growth in global energy demand, with India and China accounting for more than 40% of the increase. Emissions which had remained flat for the previous three years, grew significantly by 560 million metric tons, equivalent to emissions to 200 million additional cars on the roads. The trend of growing emissions is apparently as a result of the combined effect of growth in energy demand and weaker energy efficiency improvement efforts, and cuts across all regions, including most major economies, with a few exceptions such as the United States, United Kingdom, Japan, and Mexico.

Several international bodies including the Intergovernmental Panel on Climate Change, and the International Energy Agency have concluded that the emerging scenario NPS) which takes into account commitments in the Paris-2015 Agreement, current and anticipated policies, probable technology innovations, will be inconsistent with a sustainable environment. The IEA has proposed a faster transition/sustainable development scenario (SDS) that could put he world on the pathway to the 2 Degree goal or even lower (see Section 8.5). One major achievement of international action on pollution mitigation coordinated by the United Nations is the mobilization of country and regional actions. Many countries, particularly in the industrialized world have put in place voluntary measures to reduce environmental pollution. These measures are often referred to as Intended Nationally Determined Contributors (INDCs). Europe-28 leads the world in terms of regional action to balance the energy trilemma: energy equity, energy security and energy sustainability. In 2008 the European Union (EU) committed to climate and energy goals to be reached by 2020. These targets known as '20-20-20' targets aim to achieve a 20% cut in emissions of GHG compared with 1990 levels; a 20% share of the final energy consumption coming from renewables; and a 20% increase in energy efficiency. Already, the Union has achieved an average of 25% renewable power generation and some EU countries have set even higher goals. For example, Denmark leads the world in renewable power generation: in 2016, wind and solar accounted for 44% (42% wind and 2% solar), and the country expects to achieve around 70% by 2022. Germany set a national emissions' reduction target of 40% and a 55% cut in emissions by 2030. Also, the share of renewable energy (mainly solar and wind) in the country's electricity consumption should rise from around 30% in 2017 to 65% by 2030, and coal-fired power generation will be phased out completely. On the other hand, Poland is one of the top ten producers of coal in the world, and the second largest user in Europe, after Germany. Coal-fired power plants currently generate more than 80% of electricity in Poland and the coal industry is a major employer. Furthermore, coal is used extensively in home heating, not only in Poland, but +also in many other European countries. According to a recent report by the WHO, 33 cities in the country featured in the list of the 50 most polluted cities on the continent. For Poland and many other countries which still rely heavily on coal, (China, India, Australia, Germany, South Africa, and some states in the US), energy security remains the prime goal and balancing the energy trilemma is a major struggle. The Polish government acknowledges that coal will continue to play a prime role in the country's primary energy mix in the foreseeable future, but the country has developed extensive plans to cut environmental pollution by increasing energy efficiency and decarbonizing the transport system. Many of the country's power plants are old and are being replaced by much more efficient coal-fired power plants and nuclear power could also play a significant role in the country's future energy supply mix.

France derives over 70% of its electricity from nuclear power from 58 nuclear reactors, the second largest nuclear power fleet in the world after the United States. The country is also the world's largest exporter of electricity to neighboring countries due to the very low cost of generation. Much of the country's reactor fleet is reaching the end of its lifetime (average, 30 years) and there has been no clear policy on de-commissioning or replacement, nor a feasible trajectory to renewable energy in such a short time, considering that renewables account for only 15% of the country's power generation fuel mix currently. In spite of the country's significantly low-carbon electricity mix arising from nuclear power generation, France released the Energy Transition for Green Growth Act in 2015, which set goals to cut the share of nuclear power from 78% presently to 50% by 2025, while also reducing greenhouse emissions by 40% in 2030. Environmental issues have become very prominent in French politics and this explains the inherent policy instability. For some years now, the country's energy policy has not survived political cycles, for example, the current government has reversed the previous government's decision to reduce nuclear power generation capacity and determined that nuclear is "the most carbon-free way to produce electricity, along with renewables." Also, a progressively increasing tax regime on CO_2eq emissions was introduced to fast-track decommissioning of coal-fired power plants. This decision appears the most pragmatic considering that the French nuclear industry employs over 200,000 people and the pace of renewable energy deployment is slow. However, considering the antecedents, it is unclear whether this policy will survive for long.

The United States and China lead the world as potential INDCs - they are also the two leading sources of CO_2eq pollution: together, they accounted for about 45% of global emissions in 2018. However, both countries have committed to INDCs and are introducing strong policies to meet set targets. In 2015, the United States Environmental Protection Agency (EPA) introduced a Clean Power Plan (CPP) to reduce emissions by 26-28% below 2005 levels by 2025 (EIA, 2016). Around 80% of emissions in the US is energy-related, the balance coming from other sources such as agriculture, land use and forestry. Two of the largest sources of energy-related emissions are the transportation and electric power sectors. The CPP targets power plants which are the largest sources of carbon pollution, accounting for around one-third of all greenhouse gas emissions. Emission performance rates (Best System of Emission Reduction, BSER) were established for existing fossil fuel-fired electric generating units (EGUs) - electric utility steam generating units and stationary combustion turbines. The CPP reflects the different needs of different states and each state is given the flexibility to choose how to meet the set goals. The CPP, if implemented, was projected to reduce U.S. emissions by 0.5 billion metric tons by 2040 (EIA, 2018). One major flaw in the United States' CPP action plan which was signed into law in 2015 was its obvious focus in eliminating coal-fired power plants. It would be near impossible for existing plants to meet the stringent emission control standards, and, for the many states which depend heavily of coal for power generation and coal industry employment, the CPP was unacceptable. In any case, most states already have local emission mitigation policies, designed for their specific local conditions. Many states have instituted legal actions against the federal government, and succeeded in stalling implementation, even before the recent action by the current administration to repeal the law and reverse or temper several other environmental mitigation policies, notably automobile efficiency regulations; policies targeting inefficient lighting; regulations on gasoline-ethanol blending; emissions from power generation plants; and fracking in the petroleum industry. However, even without the CPP, various state policies, rising use of renewables, increasingly competitive natural gas pricing, and negative economic forces in the coal industry have been driving down the use of coal in power generation, with a decline of about 16% between 2010 and 2017. A further decrease of about 35% is expected by 2030, after which it levels off. Nearly 300 coal power plants (40% of the country's coal power generation capacity) have closed down in the last ten years and, in spite of the current administration's efforts to save the coal industry, fifty plants closed in the last three years due to economic forces and many of the leading coal companies have collapsed. In summary, the CPP has the potential to reduce the US power generation emissions by about 30% through 2050, but its future is in doubt, considering the fierce resistance by many states and efforts by the current administration to cancel the initiative and reverse many other national policies targeting environmental pollution. It should be noted also that the United States which accounts for the second highest level of global energy-related pollution in 2018 and championed the Paris-2015 Accord has withdrawn from the Agreement, reversed policies designed to discourage the use of coal for power generation, and is on course to reverse many efficiency improvement standards and environmental pollution mitigation policies that are in place. This demonstrates clearly the political sensitivity and vulnerability of environmental issues, and the urgent need for resilient policies that survive political cycles. Fortunately, most states in the country have continued to pursue vigorously energy decarbonization and pollution reduction goals in line with the Paris-2015 Protocol. Nevertheless, the withdrawal of the country that has contributed more to global energy-related pollution than any other in the last century, and played a pivotal role in evolving the Paris-2015 Agreement will likely have a cooling effect on the commitment of many other countries who are already facing the challenge or reconciling energy equity and security with environmental sustainability.

China which surpassed the United States as the world's largest CO_2eq emitter in 2008 has also set a goal of 20% emission reduction and 20% of non-fossil energy use by 2030, and the main drivers will be solar and nuclear energy. Although the country's economy grew by around 7% in 2017, emissions increased by just 1.7%, clearly mitigated by increasing renewables deployment, faster coal-to-gas/nuclear power switching and replacement of many existing coal power plants with more efficient plants. Many other European OECD countries have also set ambitious INDC targets. In fact, 146 national climate change panels presented draft INDCs at the United Nations Climate Change Conference in Paris in 2015. However, a recent progress assessment by the Intergovernmental Panel on Climate Change (IPCC, 2018) indicates that most INDCs are unlikely to meet or exceed the planned targets, considering the intra-country economic and political dynamics. Asia accounted for two-thirds of the growth in global emissions in 2018, due largely to the growth in the economies of the region, and all recent projections show an even faster growth over the next two decades or so. In spite of declared efforts to decarbonize energy in the region, it is difficult to see a clear path towards achieving this goal, considering that coal is a major primary energy and economic resource that fuels power and carbon-intensive primary materials production in most countries in the region. All recent projections indicate that the region will account for most of the growth of energy consumption, fossil energy use, and energy-related emissions over the next two decades or so.

Events in the emerging economies, in particular Asia Pacific will largely determine the extent of success in global energy-related emissions' abatement over the next two decades or so. The region has accounted for most of the emissions' growth in the last decade and this situation is expected to be sustained over the next two decades. Most countries in the region including Japan and Australia depend on coal for 70 - 85% on of power generation and accounted for over half of the global energy-related emissions in 2018 (Australia is also a major exporter of coal). The region also accounted for three-quarters of total global emissions from industry from metals, minerals and chemicals/petrochemicals production. Furthermore, Asia has caught up with the Americas which had been historically the largest producer of transport emissions, with each contributing about 2.5 Gt CO_2eq to the global total of 8 Gt CO_2eq in 2016. Since the region will account for most of the future growth in vehicle population, electricity demand and energy-intensive industrial production over the next two decades, projections show that nearly all the growth in energy-related emissions will also come from there. While several countries in the region have policies targeting decarbonization of energy, the strong drive for energy equity and security will make any significant achievement unlikely. Furthermore, the region has become the world's center for energy-intensive primary industries and depository of toxic solid waste from the developed world. As discussed earlier, the environment has no hard country or regional boundaries, most energy-related GHG emissions end up in the stratosphere where they enhance the heat trapping capabilities of greenhouse gases and warm up the Earth. The gases in the stratosphere are fairly well-mixed and the net effect across the globe is fairly uniform, irrespective of the origin. While local efforts to decarbonize energy will help to clean up the local environment, no country or region will be immune to the negative impact of emissions from any other region on the global climate. It is vital therefore that international efforts focus more on assistance to Asia-Pacific region and all other emerging economies.

In spite of the commendable global effort to mitigate energy-related pollution, it does not appear the world is on course for achieving the Paris-2015 goal. All recent projections show that emissions will grow by around 30-35% over the next two decades reaching around 36-37 Gt CO_2eq in 2040, in spite of the numerous country and regional mitigation policies in place and anticipated. While the growth rate will be significantly lower compared with the last two decades, the impact will not be sufficient to reach the Paris-2015 goal. Projections indicate that, taking into account current and planned mitigation policies, global average temperature will likely rise by up to 3°C in 2100, with severe consequences for future generations. Much stronger policies that can reduce emissions by a further 50% over the next two decades will be required to fully decouple energy use and emissions, and set the world on track for limiting global warming to no more than 2°C (preferably lower) above the pre-industrial level by the end of the century.

A recent publication by the International Energy Agency (IEA, 2019g) examines in-depth the complexity of reconciling the divergent dimensions of the tumultuous global oil markets, geopolitical tensions, carbon emissions and climate targets, and the goal of providing electricity for the 850 million people around the world who currently lack access. The publication presents a set of scenarios that explore different possible futures, the actions or inactions that bring them about and the interconnections between different parts of the system. The report projects what would happen if the world continues along its present path (Current Policies Scenario), determines that even the New Policies Scenario that takes account of proposed and other likely new policies will be inadequate in stopping and reversing energy-related environmental pollution: the scope and momentum of proposed new policies are not enough to offset the effects of an expanding global economy and growing population. The rise in emissions will slow down but, with no peak before 2040, the world falls far short of shared sustainability goals. The report also presents an updated feasible pathway to a sustainable environment (Sustainable Development Scenario) which requires much more rapid and widespread changes across all parts of the energy system. Considering the complexity of the world's current and future energy needs, there are no simple or single solutions: multi-dimensional actions will be required, notably rapid adoption and scale-up of multiple low-carbon energy technologies and improved efficiencies across the whole spectrum of energy production and use, which lead to much more efficient and cost-effective energy technologies that must spread across all regions of the world to make any significant, positive impact on climate change.

8.7. OUTLOOK ON THE FUTURE OF LOW-CARBON ENERGY

Renewables (solar, wind, hydro, biofuels, geothermal, tidal) are at the center of the transition to a less carbon-intensive and more sustainable energy system and the rate of deployment has grown rapidly in recent years, driven by policy support and sharp cost reductions for solar photovoltaics and wind energy in particular. Use is primarily for power generation currently, accounting for about 25% in 2019. The power sector is the single largest source of carbon emissions within the energy system and provides the brightest opportunity for reducing carbon emissions over the next twenty years or so. However, electricity accounts for only a fifth of global energy consumption, and the role of renewables in other sectors of the global economy, in particular, the transportation and heating sectors is critical to the energy transition. Renewable power capacity is projected to grow by about 50% over the next five years, led by solar PV which accounts for around 60% of the expected growth, with onshore wind representing about a quarter (IEA, 2020). Projections indicate that solar and wind could provide more than half of the additional electricity in the Emerging Transition Scenario and almost all the growth in the Sustainable Development Scenario. However, in spite of the truly exceptional rates of growth of renewable energy deployment, the very low starting base and the ultra-fast pace of growth of global power demand, particularly in the emerging regions, limit the pace at which the power sector can decarbonize. The fuel mix in the global power system has remained flat, with shares of both non-fossil fuels and coal (36% and 38% respectively) unchanged from their levels twenty years ago, in spite of a significant shift towards greater electrification. This highlights the urgent need to decarbonize electricity through much faster deployment of renewable power generation. However, the most optimistic projections predict renewable contribution of around 30-35% to global power generation capacity. One major issue is the fact that the developing world has been driving the growth in electricity demand, accounting for 81% of the growth in 2018-2019, led by China and India. Although renewable energy use increased in both countries by about 25% over the period and the region accounted for over half of the global growth in generation from renewables, coal remained the primary power generation fuel, driven by the strong growth in power demand.

Coal remains the fuel of choice in the emerging regions because it is the cheapest way of meeting the exponentially increasing demand for modern energy and, considering the large number of existing coal-using power plants and factories for primary metals and materials, those under construction, and increasing deployment of natural gas, emissions from the region will likely continue to neutralize the worldwide gains from renewable energy use in the foreseeable future. Furthermore, most of the key inputs into renewable energy come from the region, powered by fossil fuels: PV solar cells, electric vehicle and power storage batteries, copper, and most of the minerals required for manufacturing these products. As discussed earlier, the negative impact on climate will be global. The intermittency and weather sensitivity of renewable energy also constitute significant restraints to the rapid growth of deployment since most still need to be configured with fossil power plants for continuity. Furthermore, there is increasing resistance to deployment of renewable energy, in particular, hydro and wind from conservationists and communities due to potential enviro-ecological degradation and noise, and many projects have been stopped as a result of legal action in the developed countries. Concern is also rising about the rate at which valuable resources needed to feed a hungry world (land, irrigation infrastructure) are being diverted to the production of biofuels, solar and wind farms.

The tenuous future of nuclear energy has been discussed in Chapter 4. While the excellent credentials as a low-carbon energy source are not in doubt, any significant change in its current low contribution to global primary energy is unlikely. The developed countries which have accounted for over ninety percent of the global capacity are retiring nuclear power plants and, with the exception of the Russian Federation, most are not commissioning new projects. Also, safe and permanent disposal of nuclear waste which has been piling up in temporary storage sites all over the developed world for around six decades is becoming increasingly problematic. Most recent projections indicate a steady decline of nuclear power in the primary energy mix of most of the countries and many current plants will be decommissioned over the next two decades or so. On the contrary, interest in nuclear power generation is increasing rapidly in the emerging world: China, India, Egypt, and over fifty plants are under construction or planned. Most recent projections on primary energy show that the increasing deployment of nuclear energy in the emerging regions will be largely neutralized by the declining use in the industrial world, and nuclear energy contribution to future global primary energy demand will remain largely flat.

8.8. OUTLOOK ON INTERNATIONAL ENVIRONMENTAL MITIGATION ACTION

The 1997 Kyoto Treaty which established legally binding obligations for developed countries to reduce their greenhouse gas emissions was not ratified by many countries. The USA, the world's second largest emitter did not ratify the Treaty because it would cause serious harm to its economy, and also because it did not cover developing countries (80% of the world) including China, the leading emitter in the world. Some countries which ratified the Treaty initially, for example, Canada, also withdrew in 2012 because the country was unable to achieve the 6% reduction in emission from the 1990 level. Rather, emission was 17% higher in 2012. Nevertheless, global emissions in 2012 were nearly 23% lower than 1990 levels compared with the set target of 5%, probably not because of the Treaty but because many developed countries including the United States already had policies in place to decarbonize energy, largely by substituting gas for coal. Also, many developed countries adopted the 1987 Montreal Protocol to eliminate ozone-depleting gases but implementation has not been universal. For example, use of fluorocarbons in refrigeration and air conditioning has been widely discontinued in the developed world but is still prevalent in the emerging economies.

The United Nations Conference on Climate Change comprising Conference of Parties (COP) and the Meeting of Parties to the Kyoto Protocol (CMP) held in Paris in November 2015 was attended by 197 parties which adopted the Paris Agreement, a global commitment to the reduction anthropogenic emissions. The Agreement outlined strategies for strengthening the global response to the threat of climate change, and set a target of keeping a global temperature rise this century well below 2°C above the pre-industrial levels, while pursuing efforts to limit the increase to 1.5°C. Several other clauses in the Agreement include the enhancement of adaptive capacity of member nations by strengthening their resilience and reducing vulnerability of their people to climate change. Policy instruments were developed to help strengthen the ability of countries, especially the developing countries, to deal with the impacts of climate change; these included the establishment of financial support and capacity building. The Paris-2015 conference also agreed on a goal of achieving zero net anthropogenic greenhouse gas emissions by the second half of the 21st century, and introduced a legal framework to compel signatories to the Agreement to adopt and domesticate the Treaty within their own legal systems. Signatories were obliged to establish National Greenhouse Gas Inventories and develop strategies for control and removal by 2020. However, many sections of the Treaty are promises, aims, goals and indefinite time frames which are not enforceable. Furthermore, as discussed earlier, the priority of most developing countries is to provide primary energy to their growing population and most lack the wherewithal to deal effectively with pollution control. In order to help mitigate this problem, the Paris protocol established a revolving fund to help developing countries adopt low-carbon energy technologies.

Zero net carbon means carbon neutrality and requires balancing the carbon released into the atmosphere with an equivalent amount sequestered, offset with the use of zero carbon energy technologies and extensive re-forestation that will suck more carbon dioxide from the atmosphere. Any shortfall could be compensated for by buying carbon credits. All parties to the Agreement are required to submit their nationally determined contributions (NDCs) and report regularly on their emissions, and on progress of their implementation efforts. The landmark accord entered into force in November, 2016, after having been signed or ratified by 197 countries. However, the agreement is non binding since there are no verification or enforcement mechanisms, and the recent withdrawal of the United States of America from the protocol and subsequent move to cancel the country's Clean Power Plan are major setbacks. However, all other major countries and many states within the U.S.A. have expressed the determination to move forward with the implementation of the Agreement.

There is little doubt that the world is making some progress towards mitigation of climate change: every country that signed the agreement now has at least one policy or law in place, but it is not clear to what extent the stated goals can be achieved. What is becoming increasingly certain is the fact that, even if all the nations succeed in achieving stated NDIC goals, the effect will not be sufficient to move the world towards the Paris-2015 goal. Global emissions started increasing again by 2.1% to 33.1 Gigatonnes (GT) in 2018, having stayed flat for the previous three years. Although renewable energy is projected to grow by around 2.6% per year over the period (the fastest-growing energy source), its share of global primary energy use in 2040 will still be no more than about 15% by all recent projections, but its use in power generation could be as high as 35-50%. Projections also show that fossil fuels will still meet nearly 80% of the global primary energy needs in 2040, and possibly remain dominant well into the 21st century. Emissions are projected to increase to about 36.5-37 GT in 2040, in spite of current and planned mitigation actions. However, a recent analysis by the International Energy Agency (IEA, 2017c)) and several other studies have shown that emissions need to peak by around 2030 and start to reduce to about 18 GT (about 55% of current level) by 2040 in order to put the world on the pathway to the achievement of the Paris-2015 goals.

8.9.   OUTLOOK ON VOLUNTARY COUNTRY MITIGATION (INDCs)

The main impact of stratospheric pollution is on climate change, and the effect on human life and the natural ecosystem is very gradual, often taking decades to become prominent. However, in spite of the relatively short life of tropospheric (ambient) pollution, the negative impact is localized, immediate, and severe. Smog and pollution-related human deaths are major issues in most urban areas in all regions of the world and the negative impacts on human health and natural ecosystems are well documented (see Chapter 6). The sources of lower atmosphere pollution vary across countries, hence abatement solutions will differ, and this has prompted many countries to develop local action strategies to mitigate pollution. The cumulative effects of local pollutions eventually resonate in the stratosphere and enhance global warming which is the primary cause of climate change, hence local actions are also impacting positively on stratospheric pollution. The extent and diversity of voluntary country mitigation actions were evident in the action plans (Intended Nationally Determined Contributors, INDCs) submitted by over 150 countries at the United Nations Paris-2015 conference. The problem of climate mitigation varies between countries and across all regions of the world because of different exigencies and priorities. However, the Intergovernmental Panel of Climate Change (IPCC, 2018c) has developed a series of multi-level and cross-sectoral climate adaptation and mitigation options which countries can choose from and still commit to the Paris-2015 goal. These include optimization and decarbonization of energy use across all sectors of the global economy through efficiency improvement, fuel substitution, carbon capture, recycling, and lifestyle optimization; improved management of ecosystems (natural/managed, avoided degradation and deforestation, restoration, sustainable agriculture, biodiversity management, etc. (Figure 8.8). If most countries choose any of the four portfolios of mitigation measures which strike different balances between lowering energy and resource intensity, rate of decarbonization, and emissions removal, global emissions will likely decline by about 45% from 2010 levels by 2030, reaching net zero around 2050. However, these options present different implementation challenges and need to be adapted to country realities and priorities. All pathways would require rapid and far-reaching transitions in energy, land and urban infrastructure (transport, buildings, etc.), and industrial systems, all of which would require a significant upscaling of financial and technology investments, actively promoted by strong and enduring policy instruments.

Breakdown of contributions to global net CO₂ emissions in four illustrative model pathways

Fossil fuel and industry   AFOLU   BECCS

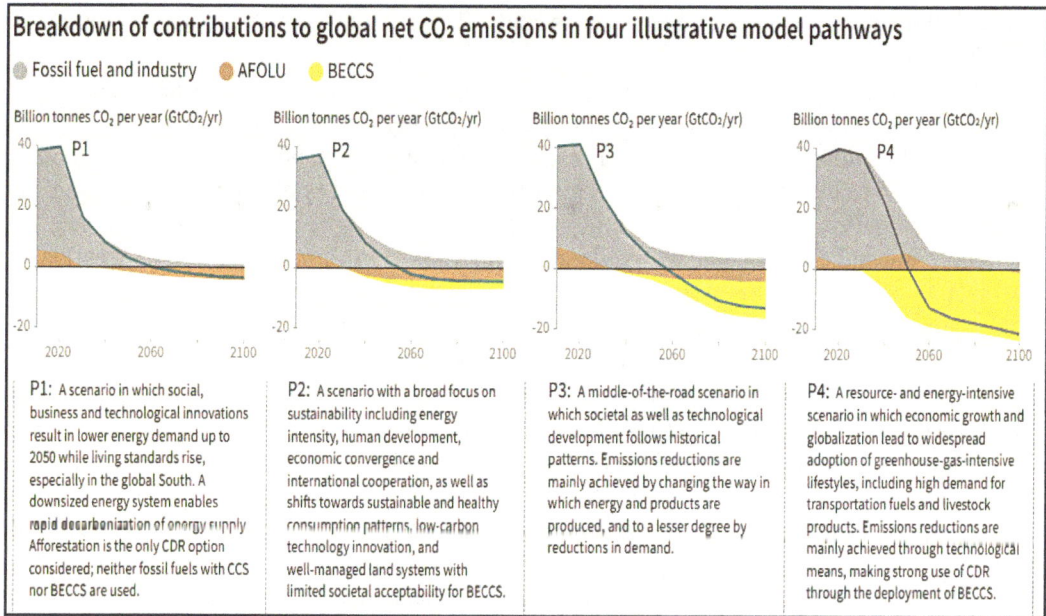

P1: A scenario in which social, business and technological innovations result in lower energy demand up to 2050 while living standards rise, especially in the global South. A downsized energy system enables rapid decarbonization of energy supply. Afforestation is the only CDR option considered; neither fossil fuels with CCS nor BECCS are used.

P2: A scenario with a broad focus on sustainability including energy intensity, human development, economic convergence and international cooperation, as well as shifts towards sustainable and healthy consumption patterns, low-carbon technology innovation, and well-managed land systems with limited societal acceptability for BECCS.

P3: A middle-of-the-road scenario in which societal as well as technological development follows historical patterns. Emissions reductions are mainly achieved by changing the way in which energy and products are produced, and to a lesser degree by reductions in demand.

P4: A resource- and energy-intensive scenario in which economic growth and globalization lead to widespread adoption of greenhouse-gas-intensive lifestyles, including high demand for transportation fuels and livestock products. Emissions reductions are mainly achieved through technological means, making strong use of CDR through the deployment of BECCS.

Figure 8.8   Potential pathways to a carbon neutral world *(IPCC, 2018c).*

8.10.   OUTLOOK ON SOCIETAL MITIGATION: POWER OF 'WE THE PEOPLE'

From early times, developments in energy technologies have been driven by human needs and the status of human development today could not have been achieved without the multi-dimensional innovations in energy technologies. Demand for energy is fueled by economic and population growth, both of which have been very dynamic for centuries. The global economy is projected to double over the next twenty years and the population will grow by around 26%. The world population of vehicles will reach around 2 billion by 2040, urbanization and prosperity will grow, particularly in the emerging regions, the rate of construction and infrastructure development will continue to rise, and many of the over 2 billion that do not currently have access to modern energy will gain access. It is inevitable therefore that energy demand will continue to grow to power these developments. Prior to 2000, global energy demand was growing roughly at the same rate as the economy and should double by 2040. However, immense progress in technology innovations' deployment have significantly decoupled energy from the economy and most projections predict a growth in energy demand of around a third. However, as has been the case for the last five decades, fossil fuels will still account for around 80% of the total global primary energy demand in 2040 and beyond. This is because, while renewable and nuclear energy can replace fossil fuels significantly in power generation, there is as yet no suitable substitute in many of the current demand centers that use fossil energy and precursors: production of primary metals, cement, polymers/plastics, carbon/nano materials, fertilizers, chemicals, petrochemicals, pharmaceuticals. Also, most of the growth in primary energy demand, human population/prosperity and vehicle population will occur in the emerging regions where coal is the main and probably the only locally available primary energy resource, yet these regions are emerging as the world's industrial zone for energy-intensive production. Furthermore, fossil fuel production and trading dominate the economies of many countries across all regions in terms of foreign exchange earnings and employment. Although coal use is declining in some countries, investments coal, oil and natural gas infrastructures remain strong in many others.

Primary energy supply and the global economy are closely interdependent: a major disruption of either could collapse the other with very severe consequences. The current near-total lockdown of the world as a result of the Covid-19 pandemic cut road and air traffic by around 85% in just a few weeks. The urban environment is cleaner and healthier, climate change mitigation benefits, but the negative consequences for the global economy will be monumental and could be enduring: already, hundreds of millions of jobs have been lost across all sectors of the global economy; oil market prices collapsed by around 80% in just weeks; the closely interconnected and interdependent global supply chain of industrial inputs and consumer goods is broken and the entire world economy is virtually on total lockdown. In a similar way, an abrupt major interruption of the world's primary energy supply (comprising over 80% fossil fuels) would collapse most sectors of the global economy - power generation, transportation, manufacturing, - and stall virtually every aspect of human development. The seismic impact of the pandemic on all aspects of the global economy will subside and the world will spring back. For example, human-related emissions declined by 15% in just weeks but have since risen by 10% since some economic activities resumed. However, this monumental shock is an invaluable opportunity for the world to rethink strategies in the process of re-opening the economy, in particular, the energy sector which needs to be done in a way that puts the world on a stronger footing towards an environmentally sustainable future. The International Energy Agency (IEA, 2020b) has recently released a Sustainable Recovery strategic plan which details options for the guidance of countries as they consider recovery plans which will shape investment, infrastructure and industry development strategies for decades to come. One key conclusion of the study is that fossil fuel use will rebound although the demand for oil, natural gas and coal could drop by 4-8% in 2020 but it is unclear whether the decline can be sustained. New investments in the sub-sector are currently on hold but they are vital to the sustenance of existing facilities. Furthermore, the fossil energy sector employed 50% of the around 40 million people around the world who worked directly in production, transport and distribution of fossil fuels, and crude and refined oil led global commodity trade in 2019. However, investments in renewable energy have remained strong and will likely grow at a much faster rate than previously projected because of the expected growth in demand for electricity to meet the new normals such as significantly increased home-based work, virtual meetings, e-commerce, and other digital services. The report presents comprehensive options that could rapidly transform the global energy system and provide a feasible pathway towards achieving long-term climate and sustainable development objectives.

The world needs to face the reality of continued dominance of fossil fuels in the global primary energy mix in the foreseeable future and the most pragmatic approach, is that society learns to manage energy production, conversion and use in an environmentally sustainable manner that seeks to make the world carbon-neutral in two to three decades, and there are many options which have been discussed in previous chapters. This section focuses on a key option which hardly features in global discourse on climate change mitigation: the need for people to reduce personal carbon footprints.

8.10.1.   Developing a carbon-neutral world

Carbon dioxide is the main component of anthropogenic greenhouse gases although it is the least potent in terms of environmental degradation per unit weight. Atmospheric carbon dioxide is also the primary source of carbon in proteins and carbohydrates produced by plants through the process of photosynthesis, both of which are indispensable to human and animal life. If the amount of carbon dioxide produced by human activities could be neutralized by the amount sucked out of the atmosphere by plants and other vegetation or the excess can be sucked out of the atmosphere, converted to useful products or stored in a benign state, the world would become carbon neutral. Carbon neutrality (net-zero carbon) is considered by experts as one of the best options that save the environment from continued degradation without a violent disruption of people's modern lifestyle and human development. Unfortunately land use and deforestation are destroying the natural vegetation sink, hence the current global drive for reforestation through tree planting needs to be energized. Some of the carbon dioxide produced by human activities is already being used in many ways: welding, fire extinguishers, fizzy drinks, et cetera but the proportion is negligible. However, a lot of technology innovations on decarbonizing energy, carbon capture, conversion, use/sequestration are emerging or incubating (see Chapter 7).

8.10.2. Developing strong, resilient policies

Governance is stratified in most countries of the world and policy intervention can by instituted at the various levels: vehicle pollution control, refuse management, and so on at the local council/county level, and policies on efficiency improvement across all sectors of the economy, power generation and industrial emissions control, building standards, and so on at the state/national level. Actions at the international level are purely advisory except in groups like EU-28 where they can be enforced, and this explains why the bloc is leading the world on environmental sustainability. On the contrary, there are as many policy centers as there are states in the United States and enforcement of policies at the federal level is tenuous. Strong, resilient and targeted policies especially at country levels are critical to the success of climate change mitigation. Policy resilience is very sensitive to political exigencies and many policies do not survive political cycles. This explains why environmental mitigation policies are unstable in many countries and some countries are announcing unrealistic, politically expedient measures such as banning sales of internal combustion cars in 2030 even though the market penetration of electric vehicles was less than 0.1% of the total global vehicle population in 2019, in spite of strong policy support. Furthermore, politicians largely project the views of their political base, even when they are contrary to personal views, which gives the people who elect them a potentially formidable tool to influence policies. Also, both climate change activists and sceptics frequently exploit the power of the news/social media to disseminate and spread often sensational, inaccurate and potentially inciting information to a gullible global population most of who cannot deal with the complexities of climate science.

8.10.3. The power of technology innovations

Just twenty years ago, no one could have imagined or predicted the phenomenal impact of technology innovations on human development: the Internet, cell phones, bio/nano materials, and the phenomenal improvements in all sectors of the global economy: fuel efficiencies of road and air transportation have doubled; carbon intensities of industrial production halved over the period; fracking technology has propelled the United States to global leadership in oil and gas production and export in just a decade; e-commerce, telemedicine, virtual meetings are now normal events. Technology innovations can be very disruptive and make a nonsense of predictions and projections, and many are incubating all over the world. This implies that predictions of what is likely to happen in any field of human development in ten years time should be regarded as speculative. For example, electric vehicles have great potential for reducing emissions in transportation and there is little doubt that most of the challenges restricting market penetration will be surmounted. However, they are competing with internal combustion (IC) vehicles which are backed by over a century of technology innovations which remain strong. There is little doubt that innovations that enhance performance, fuel efficiency, emissions reduction, comfort, and consumer appeal of IC vehicles will continue to emerge and pose a formidable challenge to market penetration of electric vehicles. Already, some of the latest models of IC vehicles return up to 20 kilometers per liter of fuel; some feature stop-start capability which is a major selling point of electric vehicles; and diesel vehicles are now fitted with catalytic converters that significantly reduce emissions; and consumer preferences in most regions of the world remain strongly in favour of IC vehicles. Also, electric vehicle manufacturers and policy supports are currently targeting the passenger/light vehicle sub-sector of transportation whereas the highest pollution emissions are from heavy commercial transportation. Furthermore, the downward trend in prices of oil in the global market is a disincentive to market penetration of electric vehicles. There are many technologies under development that target improved efficiencies in all sectors of the global economy, and for decarbonizing energy, and there is little doubt that many will mature continuously and could dramatically change the dynamics of global primary energy mix, consumption and anthropogenic emissions in ways that move the world faster towards carbon neutrality. This explains why there is always a significant amount of uncertainty attached to most of the scientific projections on climate change which often extend to the end of the century (80 years ahead).

8.10.4.   The power of We the People

We the people are the consumers or beneficiaries of energy and its products, and our choices and lifestyles drive the growth of the energy industry. We largely elect those that make policies and fund the energy industry through our investments. Therefore we hold a very formidable climate change mitigation option which hardly features in global discourse on environmental pollution and climate change. Three key ways in which this powerful mitigation effort can be mobilized are discussed below.

8.10.4.1.   *Reduction of personal carbon footprint*

The primary goal of the Paris-2015 Agreement is to limit the global average temperature increase to well below $2^{\circ}C$ limit by the end of the century. There are many potential pathways to achieving this goal which have been discussed in previous chapters. However, these options have decadal time frames: national policies and major energy transformations often take decades to change because of institutional inertia and the need for new infrastructure and technology innovations. However, many recent studies have shown that individual actions and lifestyle choices which are often missed in most climate mitigation options are perhaps the most effective and immediate (Gardner and Stern, 2008; Dietz *et al* 2009; Attari *et al* 2010; Girod *et al* 2014). The demand for energy keeps rising because the people are asking for more. This is evident from the way oil prices collapse every time there is a slow-down in the world's economy or any event that reduces road/air transportation intensity. It is estimated that around a quarter of the energy used by people is either wasted or deployed in unnecessary or wasteful activities. Also, studies have shown that people's acquisition instincts are driven more by want or impulse than need. It means that people's exercise of *prudence, moderation, discretion* and *lifestyle changes* could reduce carbon dioxide emissions beyond what is required to move the world in the direction of a sustainable environment, and there are many options some of which are listed in Figure 8.9. A recent study (Wynes and Nicholas, 2017) evaluated a wide range of lifestyle choices and calculated their potential to reduce greenhouse gas emissions. They identified four widely applicable high-impact actions with potential to contribute to systemic change and substantially reduce annual personal emissions, particularly in developed countries: having one fewer child would save nearly sixty tonnes of CO_2eq per year; avoiding one roundtrip transatlantic flight would save 1.6 tonne CO_2eq; living car-free (or one car less in the family) saves 2.4 tonnes CO_2eq per year; and eating plant-based diet saves nearly 1 tonne CO_2eq a year. However, the global distribution of anthropogenic emissions is grossly inequitable and there is wide variability between social stratifications: although the world average carbon footprint per capita per year is about 5 metric tons, there is a very wide disparity between the rich and the poor (carbon inequality): the poorest half of the global population (around 3.5 billion people) account for around 10% of total global emissions attributed to individual consumption; about 30% of emissions are attributable to the richest 10% of people around the world whose average footprint is about 11 times as high as the poorest half of the population, and 60 times as high as the poorest 10%; also, the average footprint of the richest 1% of people globally could be 175 times that of the poorest 10% (Gore, 2015).

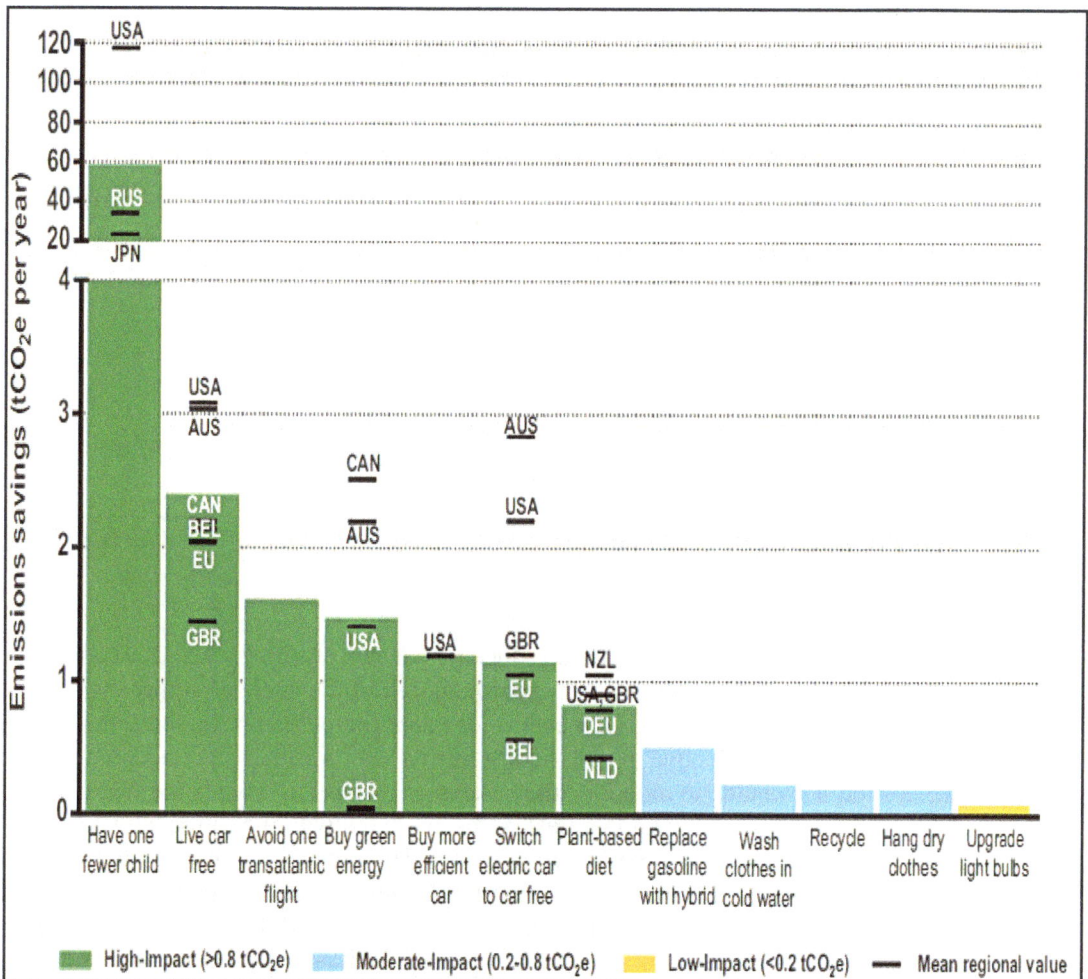

Figure 8.9a A comparison of the emissions reductions from various individual actions. *(Wynes and Nicholas, 2017).*

Percentage of CO₂ emissions by world population

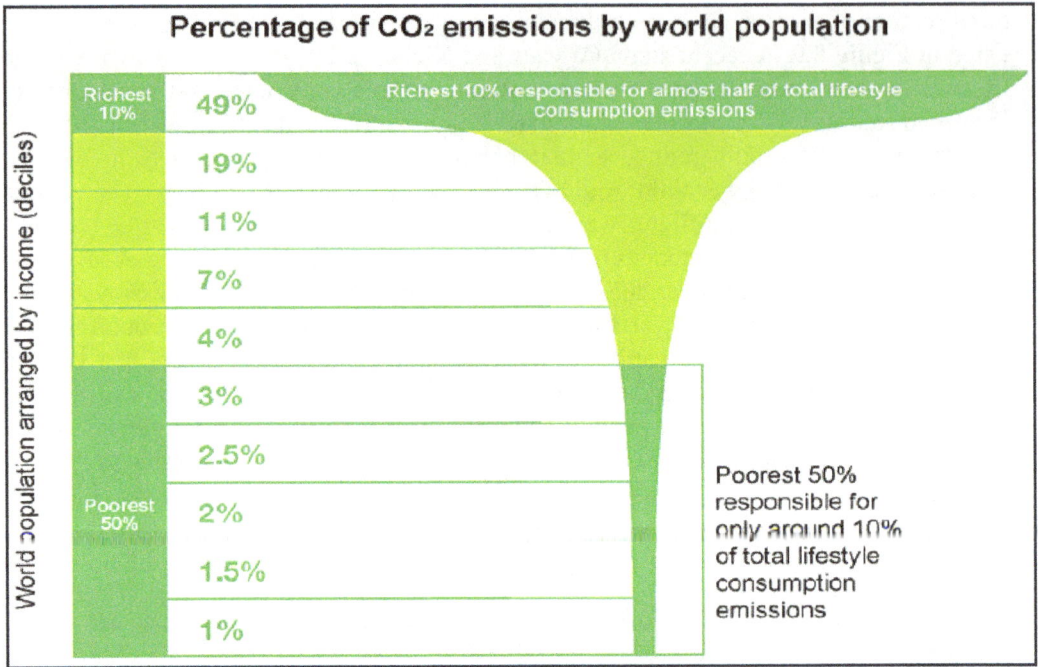

Figure 8.9b Global income deciles and associated lifestyle consumption *(Oxfam, 2020)*

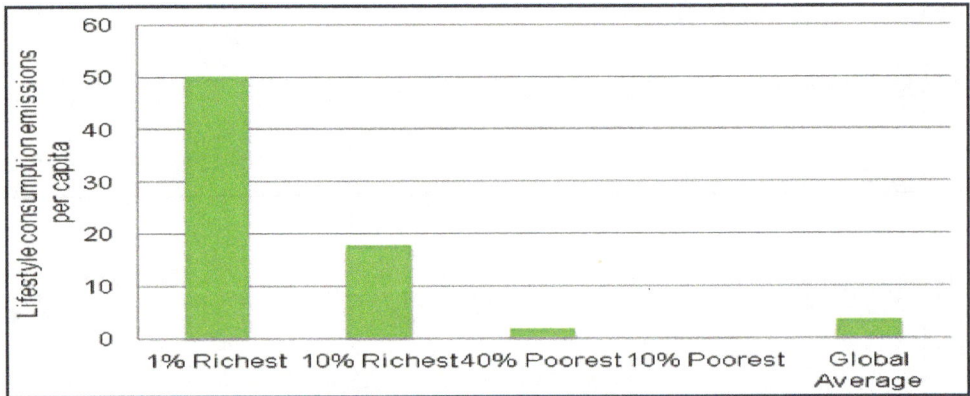

Figure 8.9c Lifestyle consumption emissions per capita from different global income levels *(Oxfam, 2020).*

If the current nearly 8 billion people could reduce personal footprint by just 20% on average, emissions would be reduced by 8 Gigatonnes, around half of the decline required to move the world to carbon neutrality by 2040. However, these data presented above show that lifestyle changes would be most effective if the richest across the world (but concentrated in the developed economies) decide to moderate their excesses. North America is the world's largest consumer of primary energy: 240 Gigajoules/capita compared with the European Union (140), OECD (73), Africa (15), and the world average (76). The power to reduce global emissions through lifestyle change and promote behavioural plasticity that spreads public support lies critically with two sectors of the global population: the wealthy and the youth. The wealthy are high-carbon individuals and can cut their carbon footprints drastically without any significant loss of comfort. Adolescents are poised to establish lifelong lifestyle patterns; they still have the freedom to make large behavioural choices that will structure the rest of their lives; their behavioural shifts have the potential to be more rapid and widespread; they are likely to suffer the greatest negative impacts of climate change; hence they are an important target group for promoting high-impact climate mitigation actions through lifestyle changes. People have many climate change mitigation options such as choosing wisely when shopping for homes, vehicles, appliances, electrical fixtures; switching off idle appliances; eliminating unnecessary auto journeys; pooling vehicles; and supporting municipal recycling efforts. Effective country and local policy instruments are necessary to stimulate demand reduction and promote recycling, especially of reusable items such as metal products, plastic products (shopping bags, drink bottles, plastic protective wear, packaging, etc.), paper and electronic items.

The contribution of recycling to individual emissions reduction in the analysis by Wynes and Nicholas is relatively low, but this is because emissions were not considered on a lifecycle basis which would have taken account of the energy carbon footprint. For example, plastic bottles and bags are produced from polyethylene terephthalate (PET) derived from fossil fuels, and processing requires enormous energy which produces large emissions: the processing of every metric ton into bottles produces around 3-4 metric tons of CO_2eq emissions. Recycling could save between 50-99% emissions compared with production from virgin materials (see Chapter 7). Also, the wealthy economies can do a lot to help promote individual and corporate lifestyle changes: the current global virus pandemic has changed the way the world operates: many more people now work from home; more are shopping on line; virtual meeting rooms, classrooms and even court sessions are proliferating; and the positive effects on energy consumption and environmental pollution have been dramatic. Many of these innovations will become new normal and the ultimate positive effect on climate change will be profound.

8.10.4.2.   *Activation of socio-political-economic power of 'We the People'*

Energy is produced, products and goods are manufactured because people buy them. Any prudent modification of people's lifestyle choices and excesses will force down the demand for both. While drastic steps may be unrealistic, such as abstinence from meat or dumping airplanes in favour of boats across the Atlantic, it is possible for people to make positive and significant impact on climate change without and violent disruption of progress in human development simply by adopting a keyword 'moderation' in lifestyle choices. The global energy industry is motivated by profit and it is highly unlikely that there will be a voluntary change in the current lackadaisical attitude towards climate change. The primary energy sector is the least efficient global economic sector and, while many innovations that can improve efficiency and reduce emissions are already available, only those that grow profit are favoured. For example gas flaring has been a perennial problem in the industry and technologies for gathering and utilization are already widely available. However, in spite of efforts by different world bodies, reduction has been slow and fluctuating in response to global oil prices: whenever oil prices rise, the incentive to invest in expensive gas gathering/use infrastructure declines. Countries have the power to enforce policies that could reduce gas flaring but this has been largely ineffective because local energy companies are the pillars of the economies of many countries and exert significant political influence. However, it should be noted that most fossil energy companies in the capitalist world are owned by the people through share/stock holding. While withholding investments from the sector will spell doom for the global economy, investors have a potentially formidable power to enforce climate-friendly actions in the industry, and this is already happening: major investors who believe in climate action are diverting investments and forcing coal mines to close, while major oil and gas companies are rethinking and reworking strategies towards more energy-efficient operations and a more sustainable environment. Investors are also motivating non-energy companies (Microsoft, Apple, Google, Amazon, and many others) to establish climate change funds and develop strategies towards carbon neutrality by reducing corporate carbon footprints, funding innovative startups that help companies reduce their carbon impact and operate more sustainably, and supporting projects in renewable energy, carbon capture and sequestration, tree planting/reforestation, etc. Many major companies are now committing to removal from the atmosphere as much carbon dioxide as they produce or buying equivalent carbon credits.

Politicians make policies that are largely motivated by the will of the people who therefore hold a very formidable instrument that could shape the deployment of effective, environment-friendly policies that transcend political cycles. The power of people's movement is aptly captured in the following quote from Time Magazine (December, 2019): *"Political leaders respond to pressure, pressure is created by movements, movements are built by people changing their minds"*. The youth have been particularly effective in this role: The French revolution of 1968 which eventually led to the collapse of the Charles de Gaulle government was led by the youth; the Tiananmen Square protest led by the youth in 1989 changed the course of events in China for ever; youth actions in the Middle East, Hong Kong, Latin America have altered the course of events; the current worldwide climate movement was initiated and led by a teenager (Greta Thunberg) who successfully organized a worldwide School Strike for Climate: "You will die of old age, we will die of climate change" recently and has moved to sensitize the world on climate issues including addressing the United Nations and meeting world leaders, featuring as Time Magazine Person of the Year in 2019. Young people from all over the world are raising their voices on issues which range from climate change to social issues such as inequality, corruption, freedom; they are mobilizing to tear down the toxic social structures built by past generations. Millennials (Generation-Y) are the largest living generation and the largest age group in the work force in many countries (and the Zoomers (Generation-Z) are poised to join), they are mobilizing across all regions and weaponizing their extraordinary appetite for the social media. The youth now form the progressive wings of political parties in many countries of the world and are becoming a force to reckon with; they have succeeded in installing many young country leaders and politicians; they are bringing social issues to the forefront and forcing society to confront the perils of inaction; they are making a significant difference on all these issues, evident in the words of the French President:

> *"When you are a leader and every week you have young people demonstrate with such a message, you cannot remain neutral. They helped me change."*
>
> ....*Emmanuel Macron, French President*

There is little doubt that the People's Movement, strongly energized by the 'youthquake' is one of the most powerful instruments that can change and reposition the world on a pathway to carbon neutrality and a sustainable environment.

Bibliography

Afonja, A. A., (2017). *Basic Coal Science and Technology*. SineliBooks.

Anderson, J. O Thundiyil, J. G, and A. Stolbach, (2012). "Clearing the Air: A Review of the Effects of Particulate Matter Air Pollution on Human Health." Journal of Medical Toxicology, Vol. 8(2), pp 166-175.

Attari, S. Z. *et al* (2010). "Public perceptions of energy consumption and savings". *Proc. Natl Acad. Sci.* **107** 16054-9.

Bashmakov, I. (2009). "Russian Energy Efficiency Potential. Scale, Costs and Benefits." Problems of Economic Transition, Vol. 52(1), pp. 54-75.

Bastin, J. F. *et al*. (2019). "The global tree restoration potential." *Science*, Vol. 365, issue 6448, July 2019, p. 76.

BCG (2018) "Batteries for electric cars: Challenges, Opportunities, and the Outlook to 2020." The Boston Consulting Group. bcg.com. Accessed 412/2018.

BIR (2008). *Report on the Environmental Benefits of Recycling*. Bureau of International Recycling.

Bond T. C. and C. S. Zender (2013). "Bounding the role of black carbon in the climate system: A scientific assessment ." *Journal of Geophysical Research*, Volume 118, Issue 13, pp. 5380-5552.

Boucher, O. (2015). Atmospheric Aerosols. DOI 10.1007/978-94-017-9649-1_2.

BP (2017). *Energy Outlook 2017*. bp.com

BP (2018). *Energy Outlook 2018*. bp.com

BP (2019a). *Energy Outlook 2019*. bp.com

BP (2019b). *BP Statistical Review of World Energy 2019*. bp.com

Breitburg, D. *et al*. (2018). "Declining oxygen in the global ocean and coastal waters." Science 05 Jan. 2018: Vol. 359, Issue 6371. DOI:116/science.aam7240.

CISCO (2018). *Global – Device Growth Traffic Profiles*. cisco.com. Accessed 3/17/2018.

DEFRA (2018). "Retailers report 83% reduction in plastic bags." Update from Department for Environment, Food and Rural Affairs, Defra, UK.

Dietz T. *et al* (2009). "Household actions can provide a behavioral wedge to rapidly reduce US carbon emissions". *Proc. Natl. Acad. Sci.* **106** 18452-6.

ExxonMobil (2018). *Energy Outlook 2018*. exxonmobil.com.

ExxonMobil (2019). *Energy Outlook 2019*. exxonmobil.com.

EIA (2015). *International Energy Outlook 2016*. Energy Information Administration. eia.gov.

EIA (2016). *International Energy Outlook 2016*. Energy Information Administration. eia.gov.

EIA (2017c). *International Energy Outlook 2017*. Energy Information Administration. eia.gov.

EIA (2018) International *Energy Outlook 2018*. Energy Information Administration. eia.gov.

EIA (2019). *International Energy Outlook 2018*. Energy Information Administration. eia.gov.

EPA (2011). "Reducing Greenhouse Gas Emissions through Recycling an Composting." epa.gov.

ERL (2017) "Reduction of one's carbon footprint ". *Environmental Research Letters. IOP Publishing.*

Euro-Int. (2017). " Euromonitor International, 2017 global packaging trends report."

FAO (2006). "Global Forest Resource Assessment 2005. Food and Agriculture Organization of the United Nations.

Flanner M. G. *et. al.* (2007). "Present-day climate forcing and response from black carbon in snow." *Journal of Geophysical Research*, Vol. 112, Issue D11.

Gardner, G. T. And Stern, P. C. (2008). "The sort list: The most effective actions US households can take to curb climate change". *Environ.: Sci. Pol. Sust. Dev.* **50** 12-15.

Geyer R. *et al.* (2017). "Production, use, and fate of all plastics ever made.) Science Advances 3(7): e1700782. July 2017.

Girod, B. *et al* (2014). "Global climate targets and future consumption level: an evaluation of the required GHG intensity". *Environ. Res. Lett.* **8** 014016.

Gore, T. (2015). "Extreme Carbon Inequality: Why the Paris climate deal must put the poorest, lowest emitting and most vulnerable people first" Oxfam International Media Briefing, December 2, 2015. Oxfam.org.

Grant K. *et al.* (2013). "Health consequences of exposure to e-waste: a systematic review." http://dx.doi.org.

Grossman E. (2011). "Radioactivity in the ocean: diluted, but far from harmless." *Yale School of Forestry & Environmental Studies.* e360.yale.edu/.

IAEA (2016). *Annual Report 2016.* International Atomic Energy Agency. iaea.org.

IAEA (2017). *International Status and Prospects for Nuclear Power 2017.* International Atomic Energy Agency. iaea.org.

IAEA (2019). *International Status and Prospects for Nuclear Power 2019.* International Atomic Energy Agency. iaea.org.

IAEA (2019b). *Energy, Electricity and nuclear power.* International Atomic Energy Agency.

IEA (2015). *World Energy Outlook 2015.* International Energy Agency. iea.org.

IEA (2016). *World Energy Outlook 2016.* International Energy Agency. iea.org.

IEA (2017a). *KeyWorldStatistics 2017.* International Energy Agency. iea.org.

IEA (2017b). *Future Scenarios for Energy Efficiency.* International Energy Agency. iea.org.

IEA (2017c) *Energy World Energy Outlook 2017.* International Energy Agency. iea.org.

IEA (2017d) *Energy Efficiency 2017.* International Energy Agency. iea.org.

IEA (2017e). *Energy Efficiency Statistics 2017.* International Energy Agency. iea.org.

IEA (2017f). Future Scenarios for Energy Efficiency. International Energy Agency. iea.org.

IEA (2018a) *Renewable Energy 2018.* International Energy Agency. iea.org.

IEA (2018b) CO2 emissions from fuel combustion 2018. . International Energy Agency. iea.org.

IEA (2018c). *Global Energy and CO2 status 2017.* International Energy Agency. iea.org.

IEA (2019a) *KeyworldStatistics 2019.* International Energy Agency. iea.org.

IEA (2019b) *Africa Energy Outlook 2019.* International Energy Agency. iea.org.

IEA (2019c) Renewable Energy 2019 . International Energy Agency. iea.org.

IEA (2019d) *CO2 emissions from fuel combustion 2019 .* International Energy Agency. iea.org.

IEA (2019e) *Global EV Outlook.* International Energy Agency. iea.org.

IEA (2019f) *Electricity Information.* International Energy Agency. iea.org.

IEA (2019g) *Coal Information.* International Energy Agency. iea.org.

IEA (2019h) *World Energy Outlook.* International Energy Agency. iea.org.

IEA (2019i) *Energy Efficiency 2019.* International Energy Agency. iea.org.

IEA (2019j). *Carbon capture, utilisation and storage.* International Energy Agency. iea.org.

IEA (2020a) *World Energy Outlook 2019.* International Energy Agency. iea.org.

IEA (2020b) Sustainable Recovery. International Energy Agency. iea.org.

IGU (2019). World LNG Report 2019. International Gas Union. igu.org.

IHA (2017). *Briefing: 2016 Key Trends in hydropower*. International Hydropower Association. hydropower.org.

IPCC (1992). "Climate Change 1992." The Supplementary Report of the IPCC Impacts Assessment.. ipcc.org.

IPCC (2007). "Climate Change 2007: Energy Supply: The Physical Science Basis." Contribution of Working Group 1 the Fifth Assessment Report. ipcc.org.

IPCC (2013). "Climate Change 2013: The Physical Science Basis." Contribution of Working Group 1 to the Fifth Assessment Report. ipcc.org.

IPCC (2014a). " Climate Change 2014." ipcc.org.

IPCC (2014b). " Energy Systems and Climate Change." ipcc.org.

IPCC (2018). "Summary for Policymakers: Global Warming of 1.5oC. An IPCC Special Report." ipcc.org.

IPCC-UNEP (2018b). Emissions Gap Report 2018. IPCC, UN Environment.

IPCC (2018c). *Global Warming of 1.5°C, An IPCC Special Report*

IPCC (2019). " IPCC Special Special Report on Climate Change, Desertification, Land Degradation, Sustainable Land Management, Food Security, and Greenhouse gas fluxes
in Terrestrial Ecosystems: Summary for Policymakers. ipcc.org.

IPCC (2018). Emissions Gap Report 2018. IPCC, UN Environment.

IRENA (2015). "Renewables and Electricity Storage: A Technology Roadmap for Remap 2030". *International Renewable Energy Agency*. irena.org.

IRENA (2016). "ReMAP: Roadmap for a renewable energy future, 2016 edition". *International Renewable Energy Agency*. irena.org.

IRENA (2017). "Rethinking Renewable Energy." *International Renewable Energy Agency*. irena.org.

IRENA (2018a). "Renewable capacity statistics 2018." *International Renewable Energy Agency*. irena.org.

IRENA (2018b). "GET_2018 Roadmap to 2050." *International Renewable Energy Agency*. irena.org.

IRENA (2019a). "Global energy transformation 2019." *International Renewable Energy Agency*. irena.org.

IRENA (2019b). "Renewable power generation costs in 2019." *International Renewable Energy Agency*. irena.org.

IRENA (2020). "Renewable power generation costs in 2020." *International Renewable Energy Agency*. irena.org.

ITU (2017). "The Global E-waste Monitor 2017." International telecommunications Union. Itu.int.

Myhre G. *et. al.* (2014). "Anthropogenic and natural radiative forcing. In Climate Change 2013: The physical Science Basis. ipcc.org.

IWS (2018) Internet World Stats, 2018

NOAA (2019) (The World Bank Global Gas Flaring Reduction Partnership, 2019 (NOAA, 2019) Colorado School of Mines, GGFR).

Noel-Brune, M. *et al*, (2013) "Health effects of exposure to e-waste:. www.thelancet.com/lancetgh.

Sims R. V. Gorsevski and S. Anenberg (2015). "Black Carbon Mitigation and the roe of the Global ." Environment Facility, Washington, D.C. stapgef.org.

Statistica (2020). "Number of vehicles in use worldwide 2006-2015". Stastica.com

Tester J. W. *et al.* (2005). *"Sustainable energy: choosing among options.* The MIT Press, Cambridge, MA.

United Nations (2019). "World Population Prospects 2019". un.org

UNDP (2000). *World Energy Assessment: Energy and the Challenge of Sustainability.* United Nations Development Organization. undp.org.

UNDP (2019). *Human Development Report 2019.* United Nations Development Program. undp.org.

UNEP/WMO (2011). "Integrated assessment of black carbon and tropospheric ozone." United Nations Environmental Programme and World Meteorological Organization.

UNFCCC (1992) "United Nations Framework Convention on Climate Change." unfccc.int.

UNFCCC (2015a). "Paris declaration of electro-mobility and climate change and call to action." *United Nations Framework Convention of Climate Change.* unfccc.int.

UNFAO (2018). *Transforming food and agriculture to achieve the SDGs.* fao.org.

UNFCCC (2015b). "Adoption of the Paris Agreement." *United Nations Framework Convention of Climate Change.* unfccc.int.

USEPA (2012). *Report to Congress on Black Carbon,* U/S. Environmental Protection Agency, Washington, D.C.

USGS (2000). "Coal-bed Methane: Potential and Concerns." *United States Geological Survey.* usgs.gov.

Wang, Bin *et al.* (2019). Historical change of El Nino properties sheds light on future changes of extreme El Nino." Proc. Nat. Academy of Sciences of the United States of America. Oct. 2019.

WEC (2016a). *World Energy Resources 2016.* World Energy Council, wec.org.

WEC (2016b). *World Energy Scenarios 2016.* World Energy Council, wec.org.

WEC (2016c). Energy Efficiency; a straight path towards energy sustainability" in World. wec.org.

WEC (2020) *World Energy Trilemma 2016.* World Energy Council, wec.org.

WHO (2017d). "Ambient (outdoor air quality and health." who.int.

WHO (2017a). "7 million premature deaths linked to air pollution." World Health organization. who.int. Accessed 3/28/2018.

WHO (2018). "Household air pollution and health." who.int.

Wikipedia (2019). "List of countries by carbon emissions". Wikipedia,org. Accessed 9/24/2019

Wikipedia (2020). "Carbon footprint". Wikipedia.org. Accessed 3/30/2020.

Wikipedia (2020b). "Coalbed methane. Wikipedia.org. Accessed 3/30/2020.

World Bank (2018). " Global Gas Flaring Reduction Partnership (GGFR)." worldbank.org.

World Bank (2019). "Zero routine flaring by 2030. Worldbank.org.

WRAP (2010). Waste & Resources Action Programme.wrap.co.uk

WRAP (2018). "WRAP's vision is a world in which resources are used sustainably." wrap.co.uk.

Wynes, S. And Nicholas, K. (2017). "The climate mitigation gap: education and government recommendations miss the most effective individuals". *Environ. Res. Lett.* **12** 07424.

Zehner O. (2012). "Green illusions: the dirty secrets of clean energy." books.google.com.

www.ingramcontent.com/pod-product-compliance
Lightning Source LLC
Chambersburg PA
CBHW081810200326
41597CB00023B/4209